Simplified Analytical Methods of Elastic Plates

Hideo Takabatake

Simplified Analytical Methods of Elastic Plates

Springer

Hideo Takabatake
Institute of Disaster and Environmental Science
Kanazawa Institute of Technology
Hakusan, Ishikawa, Japan

ISBN 978-981-13-0085-1 ISBN 978-981-13-0086-8 (eBook)
https://doi.org/10.1007/978-981-13-0086-8

Library of Congress Control Number: 2018957053

This Springer imprint is published by the registered company Springer Nature Singapore Pte Ltd.
The registered company address is: 152 Beach Road, #21-01/04 Gateway East, Singapore 189721, Singapore

Preface

Plates are widely used as major structural members in civil around the world. The plates support lateral external forces acting orthogonally on the plate's surface in the state of bending, in which the strains in the middle surface produced by in-plane forces are neglected.

The structural form of plates is determined from the length of the spans and the magnitude of lateral external forces. There are various plates depending on both plane (such as rectangular, circular, and irregular plans) and the variation of stiffness of bending and transverse shear (such as void slab with relatively small voids, plates stiffened with beams, plates with stepped thickness, cellular plates with relatively large voids, floating plates), in addition to normal plates with uniform thickness and uniform stiffness. The thickness of the plates is far smaller than the span of the plate. And the deflections are far smaller than the plate thickness. Thus, the thin plate dealing in this book is considered that the assumption of small deformation is always effective. So, we use the Kirchhoff-Love hypothesis and Mindlin-Reissner hypothesis for the formulation of the governing equation of plates from the point of view engineering perspective. These hypotheses can express the displacements on a general point by the displacement and rotations prescribed on the middle surface.

Except for normal plates, the stiffness of bending and transverse shear of the used plates varies in general to economically and rationally design the action of supporting loads. These variations depend on the existence of voids, attached beams, stepped thickness, etc., and appear on arbitrarily located portions. Since the treatment corresponding to each variable stiffness and mass is very complicated, the most of the previous theories are based on the equivalent plate analogy, in which the plate with the variable stiffness is replaced with a uniform thickness with equivalent stiffness.

The main key point of plate theory in two-dimensional approach which brings simply and easily development from engineering side is how to rationally express the variable stiffness and mass of plates. The author proposed an epoch-marking methodology which can express in general the irregular variation of the stiffness and mass for various plates by means of an extended Dirac function. The extended

Dirac function originally proposed by the author is defined as a function where the Dirac delta function exists continuously within the prescribed portion. This new idea extends famous Dirac delta function, which is effective on a prescribed point, to a prescribed portion. This idea has strong point which can easily deal discontinuous variation of the stiffness and mass with continuous function as for a distribution. This merit is to bring a closed-form approximate solution presented from the result of continuous treatment for static and dynamic problems of various plates.

The use of FEM in design of various plates achieves popularity due to a remarkable development of computer soft. However, for practical use the closed-form approximate solution is desired to apply to the preliminary stage of design. Since plates in civil are closely connected with human, the severe performance is demanded to remove various unpleasant phenomena due to deflection, oscillation, and propagation of noisy sound produced by live loads acting on the plate. In order to clear this obstacle and durability, plates made of reinforced concrete material are used, in general, in civil. The distinguishing trait of this structural type is that plates must always support extremely heavyweight dead loads in addition to static and dynamic live loads as compared with plates used in other fields of industry.

Therefore, it is necessary to grasp sufficiently the dynamic behavior of the designed plate in addition to the static behavior in the preliminary stage of design. The closed-form solutions proposed here are applicable to make the estimate and removal of obstacle in building slabs.

Recently, human response to vibration in building slabs has become significant in design from the performance of structures used. The closed-form solutions proposed here are applicable to make the estimate and removal of obstacle in building slabs.

The continuous treatment based on the extended Dirac function is convenient for the use of the Galerkin method for static and dynamic problems of all plates. Because the simple but exact closed-form approximate solutions are obtained from the exact or approximate orthogonal relations of the shape functions used. The Galerkin method proposed here is applicable to all static and dynamic problems of various plates by means of the same methodology.

This book presents the simplified analytical methodologies for static and dynamic problems of various elastic thin plates used in civil, in which the plate supports lateral loads in the bending state. The book is composed of six parts and 14 chapters.

Part I is composed of Chaps. 1–3 and refers to static and dynamic analyses of normal plates in which the stiffness and mass are uniform in all. Chapter 1 discusses static and dynamic analyses of rectangular normal plates. Chapter 2 presents the same problem to circular normal plates. Chapter 3 states the same problem to rectangular plates with edge beams.

Part II consists of Chaps. 4–8 and explains static and dynamic analyses of various plates which include the discontinuous variation of flexural rigidity and transverse shear stiffness, such as voided plate, cellular plate, and plates with stepped thickness.

Part III is made up of Chaps. 9 and 10 and argues about static and dynamic analyses of special plates subject to moving loads, such as plate with stepped thickness and mega floating plates.

Part IV consists of Chap. 11 and presents about the effect of dead loads on static and dynamic analyses of rectangular plates.

Part V is composed of Chaps. 12 and 13 and discusses the same problems on static and dynamic analyses of beams.

Part VI presents recent topics of plate analysis based on refined plate theory about uniform rectangular plates.

New methodologies explained in this book are based on many review papers in the literature reported by the author and coworkers and many researchers in the world.

The author hopes that this book will be contributed greatly to propose new methodologies by young researchers and students. Appreciation is hereby expressed to Mrs. Chiaki Hori and Yukiko Watanabe, who typed the text and its many alterations and corrections.

Hakusan, Japan Hideo Takabatake

Contents

Notations

A	Cross section of beam
$A_{mn\bar{m}\bar{n}}, B_{\bar{m}\bar{n}}, B_{\bar{m}\bar{n}mn} \ldots, C_{3mn\bar{m}\bar{n}}$	Coefficients
$b_{xi}, b_{yj}, h_{xi}, h_{yj}$	i and j-th stepped widths and heights, respectively
$b_{xi,j}, b_{yi,j}$	Widths in the x- and y-directions of the i, j-th void, respectively
$b_{x1}, b_{x2}, h_{x1}, h_{x2}$	Widths and heights of edge beam at $x = 0$ and $x = lx$, respectively
$b_{y1}, b_{y2}, h_{y1}, h_{y2}$	Widths and heights of edge beam at $y = 0$ and $y = ly$, respectively
c, c_{mn}	Damping coefficients
$D(x - x_i), D(y - y_i)$	Extended Dirac functions in Cartesian coordinate
$D(r - r_i), D(\phi - \phi_j)$	Extended Dirac functions in polar coordinate
D_0	Flexural rigidity for solid plate
$d(x, y)$, $d(r, \phi)$	Coefficient of flexural rigidity
E	Young's modulus
f_{mn}, g_{rmn}, $g_{\phi mn}$, f_{xm}, f_{yn}	Shape functions satisfying boundary conditions
$F_{m\bar{m}}(k_1, k_2)$, $F_{m\bar{m}}(k_1, k_2; i)$	The integrals
$F_{xss}(m, \bar{m}; i)$, $F_{xcc}(m, \bar{m}; i)$	Integrals with respect to x
$F_{yss}(n, \bar{n}; j)$, $F_{ycc}(n, \bar{n}; j)$	Integrals with respect to y
G	Share modulus of an isotropic material
h_0	Height of reference plate
h^*	Damping constant
$\tilde{h}$	Damping constant excluding the effect of voids
$h_{i,j}$	Height of the i, j-th void
h_{mn}	Damping constants of the m and n mode
h_{top}, h_{bottom}	Slab's thickness of top and bottom platelets
I	Inertia moment of beam
I_p	Rotatory inertia
$K_{\bar{m}\bar{n}mn}$, $K^{(1)}_{\bar{m}\bar{n}mn}$, $K^{(2)}_{\bar{m}\bar{n}mn}$	Coefficients
l_x, l_y	Span lengths in the x- and y-directions

m_0, ρ	Mass per unit area and mass density of solid plate
$\bar{m}$	Additional mass due to moving load
M_r, M_ϕ, $M_{r\phi}$, Q_r, Q_ϕ	Bending and twisting moments and transverse shearing forces per unit length
$M_{r\text{top}}, M_{r\text{bottom}}$	Flexural moments of top and bottom platelets
M_x, M_y, M_{xy}, Q_x, Q_y	Bending and twisting moments and transverse shearing forces
$M_{x\text{top}}, M_{x\text{bottom}}$	Bending moments of top and bottom platelets
$N_{x\text{top}}, N_{x\text{bottom}}$	Axial forces of top and bottom platelets
Q_p	Integral values of the external loads
$Q_{r\text{frame}}$, $Q_{r\text{partition}}$	Transverse shearing forces of frame and partition located parallel to radial direction
$Q_{r\text{top}}, Q_{r\text{bottom}}$	Transverse shearing forces of top and bottom platelets
$Q_{x\text{top}}, Q_{y\text{top}}$	Transverse shearing forces at top and bottom platelets
r, ϕ, z	Polar coordinate
$(r_i,\ \phi_j)$, $b_{ri,j}$, $b_{\phi i,j}$, $h_{i,j}$	Position, widths, and height of *i, j*-th void in polar coordinate
r_{xy}	Shear strain
r_0	Radius of circular plate
t	Time
T	Kinetic energy
U	Strain energy
$\overline{U}, \overline{V}, \overline{W}$	Displacement components on general point
u, v, w	Displacement components in *x*-, *y*-, and *z*-directions, respectively, on middle surface
V	Potential energy produced by external loads
V_x, V_y	Vertical edge forces
v_r^*, m_r^*	External force and external moment
v_x^*, v_y^*, m_x, m_y, m_{xy}, m_{yx}	External forces, external moments, and twisting moments, respectively
w	Lateral deflection on the middle surface
w_{mn}, θ_{mn}, $\theta_{\phi\ mn}$	Unknown displacement coefficients
x, y, z	Cartesian coordinate
$\hat{x}_i, \hat{y}_j$	Fields prescribed by the *i, j*-th void
α	Aspect ratio of span lengths l_y to l_x
α_h	Coefficient of mass
$\alpha_{Gri,j}$, $\alpha_{G\phi i,j}$	Transverse shear coefficients in the *i, j*-th region
δ	Variational operator
δ_{mn}	Kronecker delta
$\delta(x - x_i), \delta(y - y_j)$	Dirac functions in Cartesian coordinate
$\delta(r - r_i), \delta(\phi - \phi_j)$	Dirac functions in polar coordinate
Δt	Incremental time

ε_x, ε_y	Normal strain components
θ_x, θ_y; θ_r, θ_ϕ	Rotational angles in Cartesian and polar coordinate, respectively
κ	Coefficient of distribution of the shearing stress
$\kappa_r G_r h_0$; $\kappa_\phi G_\phi h_0$	Transverse shear stiffnesses
κ_x, κ_y, κ_{xy}; κ_r, κ_ϕ, $\kappa_{r\phi}$	Change of curvatures and twisting of the middle surface
λ	Constant corresponding to natural frequency
ν	Poisson's ratios
ξ, η	Supplementary variables of x and y, respectively
σ_x, σ_y	Normal stress components
τ	Shearing stress
ϕ_B, ϕ_J	Coefficients indicating the increase of rigidity by slab
Φ_{mn}, Φ_n	Functions of time
ω_{mn}, ω_{Dmn}	Natural frequencies of undamped and damped plate, respectively
ω_p	Frequency of harmonic external load

Part I
Static and Dynamic Analyses of Normal Plates

Chapter 1
Static and Dynamic Analyses of Rectangular Normal Plates

1.1 Introduction

Plates are widely used as structural members of building and support lateral external forces, which act orthogonal to the plate's surface, by bending of the plate. There are many books and theses, explained from the fundamental theory to practical applications concerning plates, for example, Refs. [1–4]. This chapter presents static and dynamic analyses of rectangular normal plates with uniform stiffness and mass in elastic problems.

We consider the differential equation for bending problem of thin plates with small deflections in Cartesian coordinate system. The shape of a plate is adequately defined by the geometry of the middle surface, in which the middle surface is prescribed as a surface which bisects the plate thickness h at each point and is initially flat. The small deflection plate theory is based on the following Kirchhoff-Love hypothesis, which is equivalent counterparts in elementary beam theory.

(1) The plate consists of elastic, homogeneous, and isotropic materials.
(2) The plate is initially flat.
(3) The thickness of the plate is far smaller than the spans of the plate.
(4) The deflections are far smaller than the plate thickness.
(5) The slopes of the deflected middle surface are smaller than unity.
(6) The planes of the plate lying initially on a normal to the middle plane remain on the normal to the middle surface of the plate after bending. It is equivalent to the disregard of the effect of transverse shearing forces on the deflection of plates.
(7) The normal stress in the transverse direction to the plate is of negligible order of the magnitude.

Thus, the small deflection theory based on the above-mentioned assumptions is inextensional plate theory which the strains in the middle surface produced by in-plane forces are neglected.

H. Takabatake, *Simplified Analytical Methods of Elastic Plates*,
https://doi.org/10.1007/978-981-13-0086-8_1

In general, the size of the thickness of the plate is at least smaller than one-tenth of the smallest span length of the plate. Also, the maximum deflection is assumed to be less than one-fiftieth of the smaller span length.

We consider rectangular plates used Cartesian coordinate system. The coordinate axes x and y indicate the longitudinal directions of the plate. The coordinate axis z is set in the vertical direction. The positive of the external and internal forces and deflection components is prescribed when they point toward the positive direction of the coordinate axes x, y, and z. However, the moments are defined as positive when the fibers located at the bottom part of the plate produce tension, as shown in Fig. 1.1.

In order to indicate simply the partial differentiation with respect to the coordinate axes x and y, we use the following simplified notations in the whole of this book:

$$\frac{\partial}{\partial x} =_{,x} \quad \frac{\partial}{\partial y} =_{,y} \quad \frac{\partial^2}{\partial x^2} =_{,xx}$$

For example, the first and second partial differentiations of a quantity Q_x with respect to x and y become $Q_{x,x}$, $Q_{x,y}$, $Q_{x,xx}$, and $Q_{x,yy}$, respectively. The suffixed x and y after the commas indicate partial differentiations with respect to x and y, respectively.

We use Cartesian coordinate system x, y, z in the thin plate, as shown in Fig. 1.1. The x- and y-axes locate on the middle surface. The horizontal displacements u and v, and vertical displacement w of the middle surface are functions of x and y. The displacements U, V, and W of a general point (x, y, z) may be written as

$$\left.\begin{aligned} U &= u(x, y) + z\alpha \\ V &= v(x, y) + z\beta \\ W &= w(x, y) + z\gamma \end{aligned}\right\} \tag{1.1}$$

in which $u(x, y)$, $v(x, y)$, and $w(x, y)$ are the displacements of the x, y, and z directions, respectively, on the middle surface; α, β, and γ are rotational angles at the middle surface. Linear strain–displacement relationships for thin plates are given as

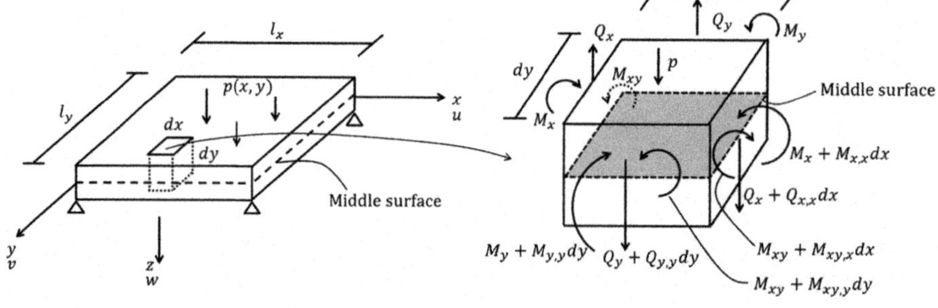

Fig. 1.1 Laterally loaded rectangular plate

$$\left.\begin{aligned} \varepsilon_x = U_{,x};\ \varepsilon_y = V_{,y};\ \varepsilon_z = W_{,z} \\ \gamma_{xy} = U_{,y} + V_{,x};\ \gamma_{yz} = V_{,z} + W_{,y} \\ \gamma_{zx} = W_{,x} + U_{,z} \end{aligned}\right\} \tag{1.2}$$

Substituting Eq. (1.1) into Eq. (1.2), we have

$$\left.\begin{aligned} \varepsilon_z &= u_{,x} + z\alpha_{,x} \\ \varepsilon_y &= v_{,y} + z\beta_{,y} \\ \varepsilon_z &= \gamma \\ \gamma_{zy} &= v_{,x} + u_{,y} + z(\beta_{,x} + \alpha_{,y}) \\ \gamma_{yz} &= w_{,y} + z\gamma_{,y} + \beta \\ \gamma_{xz} &= w_{,x} + z\gamma_{,x} + \alpha \end{aligned}\right\} \tag{1.3}$$

From the Kirchhoff-Love hypothesis, the transverse shear deformations γ_{zx} and γ_{zy} are neglected ($\gamma_{zx} = \gamma_{zy} = 0$). The thickness-wise strain ε_z is also negligible ($\varepsilon_z = 0$). Hence, the rotational angles α, β, and γ become

$$\alpha = -w_{,x};\ \beta = -w_{,y};\ \gamma = 0 \tag{1.4}$$

Thus, the displacements on a general point reduce to

$$\left.\begin{aligned} U &= u - zw_{,x} \\ V &= v - zw_{,y} \\ W &= w \end{aligned}\right\} \tag{1.5}$$

Substituting Eq. (1.5) into Eq. (1.3), nonzero strains may be written as

$$\left.\begin{aligned} \varepsilon_x &= \varepsilon_{x0} + z\kappa_x \\ \varepsilon_y &= \varepsilon_{y0} + z\kappa_y \\ \gamma_{xy} &= \gamma_{xy0} + 2z\kappa_{xy} \end{aligned}\right\} \tag{1.6}$$

in which ε_{x0}, ε_{y0}, and γ_{xy0} are in-plane normal strains and shear strain of the middle surface, respectively; and κ_x, κ_y, and κ_{xy} represent the curvature changes and warping of the deflected middle surface, respectively. They are defined as

$$\varepsilon_{x0} = u_{,x};\ \varepsilon_{y0} = v_{,y};\ \gamma_{xy0} = v_{,x} + u_{,y} \tag{1.7}$$

$$\kappa_x = -w_{,xx};\ \kappa_y = -w_{,yy};\ \kappa_{xy} = -w_{,xy} \tag{1.8}$$

In the bending problem of thin plates, the in-plane strains of the middle surface are counted out. Hence, the bending problem of thin plates based on the Kirchhoff-Love hypothesis becomes to be governed by only the lateral deflection w of the middle surface.

Figure 1.1 shows an elemental parallel piped cut out of the plate in equilibrium state subject to lateral external force p per unit area, in which the internal forces and moments to the near faces indicate positive. Negative internal forces and moments produce on its far sides in order to satisfy the equilibrium of the element.

1.2 Equilibrium Equations of the Plate Element

Since there are no forces in the x- and y-directions and no moments with respect to the axis z, the equilibrium equations of the plate element subjected to lateral forces only, as shown in Fig. 1.1, are completely defined by the following three equations. The first is the equilibrium for all the vertical forces acting on the element onto the axis z.

$$Q_{x,x}\mathrm{d}x\mathrm{d}y + Q_{y,y}\mathrm{d}y\mathrm{d}x + p\mathrm{d}x\mathrm{d}y = 0$$

Hence,

$$Q_{x,x} + Q_{y,y} + p = 0 \tag{1.9}$$

where Q_x and Q_y are the transverse shearing forces per unit length in x and y sides, respectively. The second is the equilibrium for moments of all the forces on the element with respect to the axis x.

$$M_{xy,x}\mathrm{d}x\mathrm{d}y + M_{y,y}\mathrm{d}y\mathrm{d}x - Q_y\mathrm{d}x\mathrm{d}y = 0$$

Hence,

$$M_{xy,x} + M_{y,y} - Q_y = 0 \tag{1.10}$$

where M_x, M_y, M_{xy}, and M_{yx} are the bending moments and twisting moments per unit length in x and y sides, respectively. The third is the equilibrium for moments of all the forces on the element with respect to the axis y. We obtain in the same manner

$$M_{xy,y} + M_{x,x} - Q_x = 0 \tag{1.11}$$

Substituting Eqs. (1.10) and (1.11) into Eq. (1.9) in order to eliminate the transverse shearing forces Q_x and Q_y and noticing the relation $M_{yx} = M_{xy}$, by virtue of shear stresses $\tau_{xy} = \tau_{yx}$, the Kirchhoff-Love plate theory is given as

$$M_{x,xx} + 2M_{xy,xy} + M_{y,yy} = -p \tag{1.12}$$

1.3 Relationships Among Stress, Strain, and Displacements

The two-dimensional Hook's law for elastic plates may be written as

$$\sigma_x = \frac{E}{1-\nu^2}(\varepsilon_x + \nu\varepsilon_y) \tag{1.13}$$

$$\sigma_y = \frac{E}{1-\nu^2}(\varepsilon_y + \nu\varepsilon_x). \tag{1.14}$$

in which E is Young's modulus of elasticity; ν Poisson's ratio; σ_x, σ_y, ε_x, and ε_y are components of normal stresses and normal strains in the x- and y-directions, respectively.

Indicating the deflection of the middle surface by w, the relationships between the stress and displacement are obtained by

$$\varepsilon_x = -zw_{,xx} \tag{1.15}$$

$$\varepsilon_y = -zw_{,yy} \tag{1.16}$$

The middle surface deforms from the initial flat position to the vertical displacement w due to bending. The point located at a constant distance z from the middle surface rotates due to the deformation of the middle surface, as shown in Fig. 1.2. Thus, the longitudinal displacements u and v in the x- and y-directions, respectively, are expressed by the rotational changes due to the deflection w of the middle surface as

$$u = -zw_{,x}; \quad v = -zw_{,y} \tag{1.17}$$

The shear strain γ_{xy} in the two-dimensional field is defined as

$$\gamma_{xy} = v_{,x} + u_{,y} \tag{1.18}$$

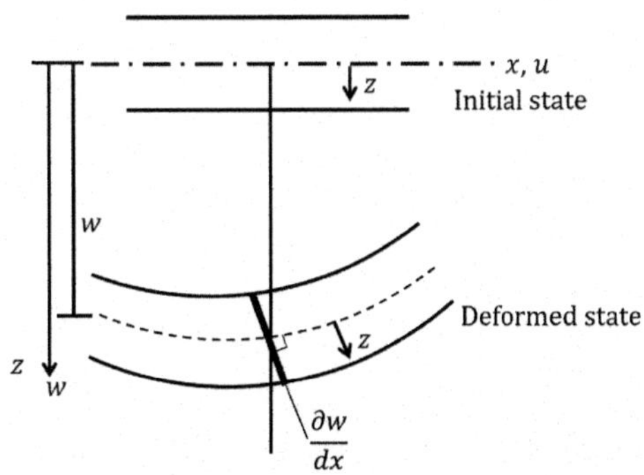

Fig. 1.2 Deformation of plate

Substituting Eq. (1.17) into the above equation, we have

$$\gamma_{xy} = -2zw_{,xy} \tag{1.19}$$

The curvature changes of the middle surface are given by

$$\kappa_x = -w_{,xx};\ \kappa_y = -w_{,yy};\ \kappa_{xy} = -w_{,xy} \tag{1.20}$$

in which κ_x, κ_y, and κ_{xy} are change of curvatures of the deflected middle surface.

Thus, the stresses σ_x and σ_y are expressed in term of w from Eqs. (1.13), (1.15), and (1.16) as

$$\sigma_x = -\frac{Ez}{1-\nu^2}(w_{,xx} + \nu w_{,yy}) \tag{1.21}$$

$$\sigma_y = -\frac{Ez}{1-\nu^2}(w_{,yy} + \nu w_{,xx}) \tag{1.22}$$

The shear stress τ $(\tau = \tau_{xy} = \tau_{yx})$ is given by

$$\tau = G\gamma_{xy} = -2Gzw_{,xy} \tag{1.23}$$

in which G is the shear modulus of the plate.

1.4 Stress Resultants and Stress Couples Expressed in Terms of *w*

The stress couples M_x, M_y and M_{xy} per unit length of the middle surface are defined by

$$M_x = \int_{-h_0/2}^{h_0/2} \sigma_x z\mathrm{d}z \tag{1.24}$$

$$M_y = \int_{-h_0/2}^{h_0/2} \sigma_y z\mathrm{d}z \tag{1.25}$$

$$M_{xy} = M_{yx} = \int_{-h_0/2}^{h_0/2} \tau z\mathrm{d}z \tag{1.26}$$

in which h_0 is the thickness of the plate.

Substituting Eqs. (1.21), (1.22), and (1.23) into the above equations, we have

$$M_x = -D_0(w_{,xx} + \nu w_{,yy}) \tag{1.27}$$

$$M_y = -D_0(w_{,yy} + \nu w_{,xx}) \tag{1.28}$$

$$M_{xy} = M_{yx} = -D_0(1 - \nu)w_{,xy} \tag{1.29}$$

in which the flexural rigidity of the plate D_0 is defined by

$$D_0 = \frac{Eh^3}{12(1 - \nu^2)} \tag{1.30}$$

Substituting Eqs. (1.27) to (1.29) into Eq. (1.12), we obtain

$$w_{,xxxx} + 2w_{,xxyy} + w_{,yyyy} = \frac{p}{D_0} \tag{1.31}$$

The above equation is equilibrium equation expressed in terms of w for static problems of rectangular plates. The above equation can be also written in the symbolic form

$$\nabla^2\nabla^2 w = \frac{p}{D_0} \tag{1.32}$$

where ∇^2 is Laplace operator of two-dimensional coordinate as

$$\nabla^2 = \frac{\partial^2}{\partial x^2} + \frac{\partial^2}{\partial y^2} \tag{1.33}$$

Equations (1.31) and (1.32) are called a nonhomogeneous biharmonic equation.

Similarly, we can express the transverse shearing forces Q_x and Q_y in term of w. From Eqs. (1.10) and (1.11) and Eqs. (1.27)–(1.29), the transverse shearing forces Q_x and Q_y are determined as

$$Q_x = M_{x,x} + M_{yx,y} = -D_0(w_{,xx} + w_{,yy})_{,x} \tag{1.34}$$

$$Q_y = M_{y,y} + M_{xy,x} = -D_0(w_{,xx} + w_{,yy})_{,y} \tag{1.35}$$

Each maximum value of the shearing stresses τ_{xz} and τ_{yz} can be given as

$$(\tau_{xz})\text{max} = \kappa\frac{Q_x}{h_0};\ (\tau_{yz})\text{max} = \kappa\frac{Q_y}{h_0} \tag{1.36}$$

in which κ is the coefficient of distribution of the shearing stresses and takes $\kappa = 1.5$ according to the parabolic distribution.

1.5 Boundary Conditions of the Bending Theory

Since Eq. (1.31) expressed in term of w is a fourth-order differential equation, two boundary conditions at each boundary are necessary. There are two kinds of the boundary condition: the geometrical boundary condition which prescribes the displacement and slope of deflection; and the mechanical boundary condition which provides the force and moment. The boundary conditions at each boundary can prescribe both whether the displacement or force and whether the slope of deflection or moment. The representative boundary conditions of plates in bending are stated as follows:

(1) Clamped support

$$w = 0;\ w_{,x} = 0 \quad \text{at } x = 0 \text{ and } x = l_x \tag{1.37}$$

$$w = 0;\ w_{,y} = 0 \quad \text{at } y = 0 \text{ and } y = l_y \tag{1.38}$$

In this case, two boundary conditions at each boundary belong to the geometrical boundary conditions.

(2) Free support

At free support edge, all vertical force, bending moment, and torsional moment must be zero.

$$Q_x = 0;\ M_x = 0;\ M_{xy} = 0 \quad \text{at } x = 0 \text{ and } x = l_x \tag{1.39}$$

$$Q_y = 0;\ M_y = 0;\ M_{yx} = 0 \quad \text{at } y = 0 \text{ and } y = l_y \tag{1.40}$$

Since the governing equation for current thin plates is the fourth-order partial differential equation, as given in Eq. (1.31), the necessary number of the boundary conditions at each edge for each one direction is enough two. Hence, Eqs. (1.39) and (1.40) are overabundant for the demanded boundary condition. In order to solve this contradiction, Kirchhoff considered to reduce the number of the internal forces from three to two by introducing Kirchhoff's supplementary forces which replace the torsional moments with statically equivalent coupled shearing forces, as shown in Fig. 1.3. The torsional moment M_{xy} acting on the free edge of x-face is considered to change with the incremental value of $M_{xy,y}$ per unit length. Replacing this torsional moment M_{xy} in infinitesimal element dy by statically equivalent coupled shearing force M_{xy} dy/dy, as shown in Fig. 1.3, the value of the coupled shearing forces is M_{xy}. These coupled shearing forces cancel out at the adjoining elements, except for their incremental parts; $M_{xy,y}\mathrm{d}y$ and $M_{yx,x}\mathrm{d}x$. Kirchhoff defined the additional shearing forces as Kirchhoff's supplemental forces:

$$Q_x^* = M_{xy,y};\ Q_y^* = M_{yx,y} \tag{1.41}$$

Hence, the vertical forces at free edges reduce to

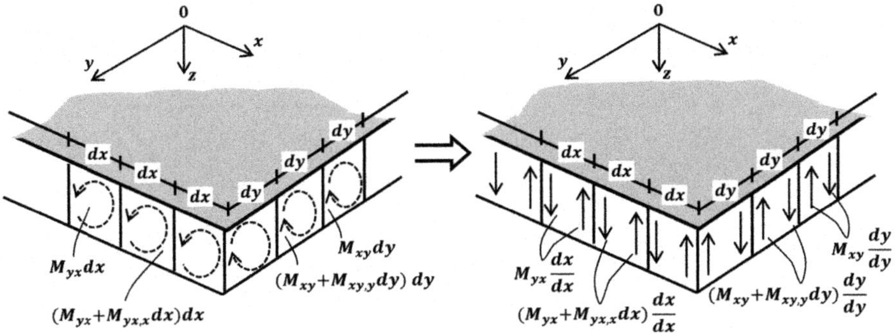

Fig. 1.3 Kirchhoff's supplementary forces

$$V_x = Q_x + Q_x^* = -D_0[w_{,xxx} + (2 - \nu)w_{,xyy}] \tag{1.42}$$

$$V_y = Q_y + Q_y^* = -D_0[w_{,yyy} + (2 - \nu)w_{,xxy}] \tag{1.43}$$

by using Eqs. (1.34), (1.35), and (1.41).

Hence, the boundary conditions at free edges become

$$M_x = 0;\ V_x = 0 \quad \text{at } x = 0 \text{ and } x = l_x \tag{1.44}$$

$$M_y = 0;\ V_y = 0 \quad \text{at } y = 0 \text{ and } y = l_y \tag{1.45}$$

Thus, the boundary conditions expressed in term of w for free support at edge can be obtained by applying Eqs. (1.27), (1.28), (1.42), and (1.43) to Eqs. (1.44), and (1.45)

$$w_{,xx} + \nu w_{,yy} = 0; \quad w_{,xxx} + (2 - \nu)w_{,xyy} = 0 \quad \text{at } x = 0 \text{ and } x = l_x \tag{1.46}$$

$$w_{,yy} + \nu w_{,xx} = 0; \quad w_{,yyy} + (2 - \nu)w_{,xxy} = 0 \quad \text{at } y = 0 \text{ and } y = l_y \tag{1.47}$$

These boundary conditions for free support belong to the mechanical boundary conditions.

The Kirchhoff's supplementary forces are infinitesimal in the case which the adjoining elements exist because of canceling out at the adjoining elements. However, it must be noticed to produce the large shearing forces at the corner of a rectangular plate with all simply supported edges. The corner lifts up due to the additional corner force R_0, as shown in Fig. 1.4. The additional corner force R_0 is given in Eq. (1.50).

(3) Simply support

$$w = 0;\ M_x = -D_0(w_{,xx} + \nu w_{,yy}) = 0 \quad \text{at } x = 0 \text{ and } x = l_x \tag{1.48}$$

$$w = 0;\ M_y = -D_0(w_{,yy} + \nu w_{,xx}) = 0 \quad \text{at } y = 0 \text{ and } y = l_y \tag{1.49}$$

These boundary conditions belong to both the geometrical boundary condition for deflection and the mechanical one for moment.

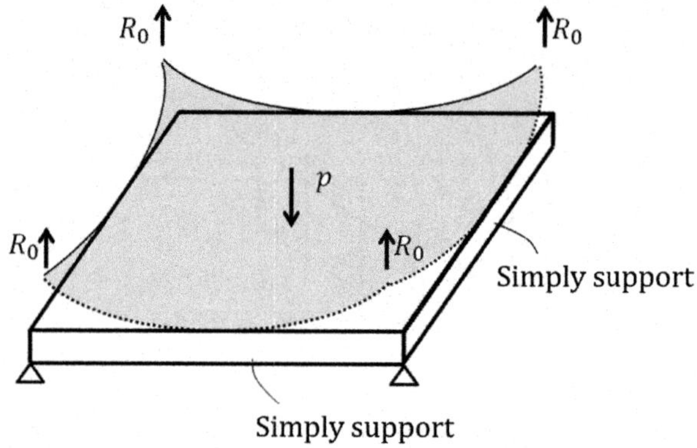

Fig. 1.4 Additional corner force at simply supported plates

When all edges of the rectangular plate are simply support, the additional corner force R_0 at the corners of rectangular plate is produced by the action of torsional moments add up instead of canceling as shown in Fig. 1.4.

$$R_0 = 2M_{xy} = -2D_0(1-\nu)w_{,xy} \tag{1.50}$$

Since this additional corner force is produced by the existence of the torsional moments at edge, such as simply support edges, the corners of the rectangular plate must be anchored to prevent lifting up.

(4) Elastic support and restraint

When the vertical restrain of the edge along $x = 0$ or $x = l_x$ is elastic support with the spring constant k_x, such as edge beam, we can express it as elastic support as follows:

$$w = \frac{V_x}{k_x(y)} = \mp\frac{D_0}{k_x(y)}[w_{,xxx} + (2-\nu)\, w_{,xyy}] \quad \text{at } x = 0 \text{ and } x = l_x \tag{1.51}$$

in which $\mp$ takes minus $(-)$ at $x = 0$ and plus at $x = l_x$.

If the edge is supported by an edge beam with the flexural stiffness EI_b, the edge beam subjects to vertical distributed load produced by the vertical edge force V_x in Eq. (1.51). It must be noticed that the distributed force transmitted from the plate to the edge beam is $-V_x$ where I_b is the moment of inertia of the edge beam. Hence, Eq. (1.51) reduces to

$$EI_b w_{,yyyy} = \mp D_0[w_{,xxx} + (2-\nu)w_{,xyy}] \quad \text{at } x = 0 \text{ and } x = l_x \tag{1.52}$$

Next, we consider the second boundary condition for moment or rotation at the edge. When the rotation along $x=0$ and $x = l_x$ at the edge does not restrain, such as edge beam without torsional rigidity, we have for rotation

$$M_x = D_0\,[w_{,xx} + \nu w_{,yy}] = 0 \quad \text{at } x = 0 \text{ and } x = l_x \tag{1.53}$$

If the rotation at the edge is restrained by an edge beam with torsional rigidity $k_{Rx}(y)$, Eq. (1.53) must be replaced with

$$w_{,x} = \mp\frac{D_0}{k_{Rx}(y)}\,[w_{,xx} + \nu w_{,yy}] \quad \text{at } x = 0 \text{ and } x = l_x \tag{1.54}$$

The above equations have presented the boundary conditions at $x=0$ and $x = l_x$. Similar expressions can be written for edges at $y=0$ and $y = l_y$.

$$w = \frac{V_y}{k_y(x)} = \mp\frac{D_0}{k_y(x)}\,[w_{,yyy} + (2-\nu)w_{,xxy}] \quad \text{at } y = 0 \text{ and } y = l_y \tag{1.55}$$

$$w_{,y} = \mp\frac{D_0}{k_{Ry}(x)}[w_{,yy} + \nu w_{,xx}] \quad \text{at } y = 0 \text{ and } y = l_y \tag{1.56}$$

In Eqs. (1.51), (1.54), (1.55), and (1.56), changing the values of spring constants $k_x(y)$ and $k_y(x)$ and restraints k_{Rx} and k_{Ry}, we have obtained the analytical formulation of the various boundary conditions.

1.6 Analytical Method of Static Rectangular Plates Used the Galerkin Method

There are many analytical methods to solve the static problem of rectangular elastic plates. The solution is given by solving the equilibrium equation (1.31) under the prescribed boundary conditions. The governing equation (1.31) is a linear partial differential equation of the fourth order with constant coefficients. For various load forms and various boundary conditions, it is difficult to find out the closed-form solution. So, we use the analytical method based on the Galerkin method. The Galerkin method regards as the general methodology which is applicable to rectangular and circular plates with various boundary conditions, if the appropriate shape functions satisfying the specified boundary conditions are given beforehand.

The deflections $w(x, y)$ can be expressed by a series expansion as follows:

$$w(x, y) = \sum_{m=1}\sum_{n=1} w_{mn}\, f_{mn}(x, y) \tag{1.57}$$

in which $f_{mn}(x, y)$ are shape functions satisfying the specified boundary conditions and w_{mn} indicate unknown displacement coefficients. The variation $\delta\, w$ of the deflection $w\,(x, y)$ is written as

$$\delta\, w(x, y) = \sum_{m=1}\sum_{n=1} \delta w_{mn}\, f_{mn}(x, y) \tag{1.58}$$

Representing the governing equation (1.31) of the plate with the notation Q, the Galerkin equation may be expressed as

$$\int_0^{l_x}\int_0^{l_y} \delta w \, Q \mathrm{d}x\mathrm{d}y = 0 \tag{1.59}$$

The substitution of Eqs. (1.57) and (1.58) into Eq. (1.59) reduces to

$$\int_0^{l_x}\int_0^{l_y} \sum_{\bar{m}=1}\sum_{\bar{n}=1} \delta w_{\bar{m}\bar{n}} f_{\bar{m}\bar{n}}(x, y) \times \left\{ \left[\sum_{m=1}\sum_{n=1} w_{mn}(f_{mn,xxxx} + 2f_{mn,xxyy} + f_{mn,yyyy}) \right] - \frac{p}{D_0} \right\} \mathrm{d}x\mathrm{d}y = 0 \tag{1.60}$$

The condition which the above equation always becomes effective for arbitral variation $\delta w_{\bar{m}\bar{n}}$ must be satisfied the following equation:

$$\delta w_{\bar{m}\bar{n}} : \sum_{m=1}\sum_{n=1} w_{mn} \int_0^{l_x}\int_0^{l_y} (f_{mn,xxxx} + 2f_{mn,xxyy} + f_{mn,yyyy}) f_{\bar{m}\bar{n}} \mathrm{d}x\mathrm{d}y = \int_0^{l_x}\int_0^{l_y} \frac{p}{D_0} f_{\bar{m}\bar{n}} \mathrm{d}x\mathrm{d}y \tag{1.61}$$

For simplicity, we use the following notations for the integral term in the above equation:

$$A_{\bar{m}\bar{n}mn} = \int_0^{l_x}\int_0^{l_y} (f_{mn,xxxx} + 2f_{mn,xxyy} + f_{mn,yyyy}) f_{\bar{m}\bar{n}} \mathrm{d}x\mathrm{d}y \tag{1.62}$$

$$P_{\bar{m}\bar{n}} = \int_0^{l_x}\int_0^{l_y} \frac{p}{D_0} f_{\bar{m}\bar{n}} \mathrm{d}x\mathrm{d}y \tag{1.63}$$

$A_{\bar{m}\bar{n}mn}$ and $P_{\bar{m}\bar{n}}$ have constant values for the prescribed indexes $m, n, \bar{m}$, and $\bar{n}$. Hence, Eq. (1.61) may be expressed in the simplified form as

$$\delta w_{\bar{m}\bar{n}} : \sum_{m=1}\sum_{n=1} w_{mn} A_{\bar{m}\bar{n}mn} = P_{\bar{m}\bar{n}} \tag{1.64}$$

The unknown displacement coefficients w_{mn} are obtained from solving a set of algebraic linear equations given by Eq. (1.64), in which $\bar{m}$ and $\bar{n}$ indicate the row of the set of algebraic linear equations and on the other hand m and n give the column in the square matrix $A_{\bar{m}\bar{n}mn}$ and the column vectors w_{mn} and $P_{\bar{m}\bar{n}}$. The unknown displacement coefficients w_{mn} are easily obtained from solving this set of simultaneous equations with multiple unknowns. This calculation can be carried out by the use of computer if suitable shape functions which satisfy the prescribed boundary conditions of the plate are selected previously.

In general, the square matrix $A_{\bar{m}\bar{n}mn}$ has nonzero value in the diagonal and nondiagonal terms. However, when the suitable shape functions which satisfy the following orthogonal relations are selected, the square matrix $A_{\bar{m}\bar{n}mn}$ has the value in only the diagonal terms

$$\int_0^{l_x} f_m(x) f_{\bar{m}}(x) \mathrm{d}x = \delta_{m\bar{m}} a_m \tag{1.65}$$

$$\int_0^{l_y} f_n(y) f_{\bar{n}}(y) \mathrm{d}y = \delta_{n\bar{n}} a_n \tag{1.66}$$

in which $\delta_{m\bar{m}}$ and $\delta_{n\bar{n}}$ are the Kronecker deltas; and a_m and a_n are constant integral values. The Kronecker delta $\delta_{m\bar{m}}$ takes the value of 1 for $m = \bar{m}$ and the value of 0 for $m \neq \bar{m}$. Thus, Eq. (1.64) reduces to uncoupled form and the displacement coefficients w_{mn} are given in closed form as

$$w_{mn} = \frac{1}{A_{mnmn}} P_{mn} \tag{1.67}$$

The above solutions are also considered to be the approximate solutions in the case of the selection of shape functions which do not satisfy the orthogonal relations given in Eqs. (1.65) and (1.66).

1.7 Selection of Shape Functions for Static Problems

The shape functions $f_{mn}(x, y)$ consist of shape function $f_m(x)$ in the x-direction and $f_n(y)$ in the y-direction and have the relation $f_{mn}(x, y) = f_m(x) f_n(y)$. The shape functions $f_{mn}(x, y)$ represented in terms of the series expansion must satisfy the geometrical boundary at the least. We consider the shape functions which satisfy the boundary conditions at $x = 0$ and $x = l_x$. For the boundary conditions at $y = 0$ and $y = l_y$, the similar results are obtained.

(1) Shape function for simply supported boundary condition at $x = 0$ and l_x:

$$f_m(x) = \sin \frac{m\pi x}{l_x} \quad (m = 1, 2, 3, \ldots) \tag{1.68}$$

(2) Shape function for clamped boundary condition at $x=0$ and l_x:

$$f_m(x) = \frac{1}{2}\left(1 - \cos\frac{2m\pi x}{l_x}\right) \quad (m = 1, 3, 5, \ldots) \tag{1.69}$$

$$f_m(x) = \sin\frac{\pi x}{l_x} \sin\frac{m\pi x}{l_x} \tag{1.70}$$

Szilard [3] presented shape functions for approximate analysis of rectangular plates with various boundary conditions. Once the displacement coefficients w_{mn} are determined from Eq. (1.64) or (1.67), the deflection w is obtained from Eq. (1.57). Also, the stress couples M_x, M_y, and $M_{xy} = M_{yx}$, par unit length of the middle surface are derived from Eqs. (1.27) to (1.29), respectively. The transverse shearing forces Q_x and Q_y are given from Eqs. (1.34) and (1.35), respectively.

1.8 Free Transverse Vibrations of Plates Without Damping

The equation of free transverse vibrations of a rectangular plate without damping is expressed by adding inertia term to Eq. (1.31) for the equilibrium equation of the static problem, in which the term of external loads p is removed, as

$$m_0\ddot{w} + D_0(w_{,xxxx} + 2w_{,xxyy} + w_{,yyyy}) = 0 \tag{1.71}$$

in which m_0 is mass per unit area ($m_0 = \rho h_0$, ρ is the density, and h_0 is the thickness of uniform plates). We also use the symbolic notation to indicate simply the differentiation with respect to time t by superscript dot as follows:

$$\dot{w} = \frac{\partial w}{\partial t} \quad \ddot{w} = \frac{\partial^2 w}{\partial t^2}$$

The deflection w is function of x, y, and time t. The method of separation of variables is employed, assuming that

$$w(x, y, t) = W(x, y)\Phi(t) \tag{1.72}$$

in which $W(x, y)$ is a function of x and y, and $\Phi(t)$ is a function of time t. Substituting Eq. (1.72) into Eq. (1.71) for the transverse vibration and then separating the variables, we have obtain

$$\frac{\ddot{\Phi}(t)}{\Phi(t)} = -\frac{D_0}{m_0 W}(W_{,xxxx} + 2W_{,xxyy} + W_{,yyyy}) = -\omega^2$$

It must have the relation of a constant $-\omega^2$ that the above two equations concerning two kinds of the different variables, time t and coordinates x and y, are equal. Hence, the two equations satisfied by $W(x, y)$ and $\Phi(t)$ become as follows:

$$\ddot{\Phi}(t) + \omega^2 \Phi(t) = 0 \tag{1.73}$$

$$W_{,xxxx} + 2W_{,xxyy} + W_{,yyyy} - \frac{\omega^2 m_0 W}{D_0} = 0 \tag{1.74}$$

in which ω is a constant.

The natural frequencies of a normal plate are also presented by means of the Galerkin method. $W(x, y)$ is expressed by a power series expansion as follows:

$$W(x, y) = \sum_{m=1} \sum_{n=1} W_{mn} f_{mn}(x, y) \tag{1.75}$$

in which W_{mn} are the unknown coefficients, and $f_{mn}(x, y)$ are functions satisfying the specified boundary conditions of the plates. The variation of Eq. (1.75) is expressed as

$$\delta W(x, y) = \sum_{m=1} \sum_{n=1} \delta W_{mn} f_{mn}(x, y) \tag{1.76}$$

For simplicity, we indicate Eq. (1.74) with the notational symbol Q as

$$Q = W_{,xxxx} + 2W_{,xxyy} + W_{,yyyy} - \frac{\omega^2 m_0 W}{D_0} = 0 \tag{1.77}$$

Applying the Galerkin method in the above equation is

$$\int_0^{l_x} \int_0^{l_y} \delta W \cdot Q \mathrm{d}x\mathrm{d}y = 0 \tag{1.78}$$

The substitution of Eqs. (1.76) and (1.77) in the above equation becomes

$$\sum_{\bar{m}=1} \sum_{\bar{n}=1} \delta W_{\bar{m}\bar{n}} \sum_{m=1} \sum_{n=1} W_{mn} \left\{ \int_0^{l_x} \int_0^{l_y} (f_{mn,xxxx} + 2f_{mn,xxyy} + f_{mn,yyyy}) f_{\bar{m}\bar{n}} \mathrm{d}x\mathrm{d}y - \omega^2 \int_0^{l_x} \int_0^{l_y} \frac{m_0}{D_0} f_{mn} f_{\bar{m}\bar{n}} \mathrm{d}x\mathrm{d}y \right\} = 0 \tag{1.79}$$

For simplicity, we use the following notations

$$A_{\bar{m}\bar{n}mn} = \int_0^{l_x}\int_0^{l_y} (f_{mn,xxxx} + 2f_{mn,xxyy} + f_{mn,yyyy}) f_{\bar{m}\bar{n}} \mathrm{d}x\mathrm{d}y \tag{1.80}$$

$$B_{\bar{m}\bar{n}mn} = \int_0^{l_x}\int_0^{l_y} \frac{m_0}{D_0} f_{mn} f_{\bar{m}\bar{n}} \mathrm{d}x\mathrm{d}y \tag{1.81}$$

$A_{\bar{m}\bar{n}mn}$ and $B_{\bar{m}\bar{n}mn}$ have constant values for the prescribed indexes $m, n, \bar{m}$, and $\bar{n}$, similar to the static problem. Thus, Eq. (1.79) reduces to a concise style

$$\sum_{\bar{m}=1}\sum_{\bar{n}=1} \delta W_{\bar{m}\bar{n}} \sum_{m=1}\sum_{n=1} [W_{mn}(A_{\bar{m}\bar{n}mn} - \omega^2 B_{\bar{m}\bar{n}mn})] = 0 \tag{1.82}$$

In order to that the above equation applies for arbitral variation $\delta W_{\bar{m}\bar{n}}$, the Galerkin equation must satisfy the following equations for the indexes $\bar{m}$ and $\bar{n}$.:

$$\sum_{m=1}\sum_{n=1} [W_{mn}(A_{\bar{m}\bar{n}mn} - \omega^2 B_{\bar{m}\bar{n}mn})] = 0 \text{ for } \bar{m} \text{ and } \bar{n} \tag{1.83}$$

The above equations are a system of linear, homogeneous, simultaneous algebraic equations with respect to the unknown displacement coefficients W_{mn}. The values of the coefficients $A_{\bar{m}\bar{n}mn}$ and $B_{\bar{m}\bar{n}mn}$, in general, depend on the selected shape functions $f_{mn}(x, y)$ and appear in diagonal and nondiagonal terms. Equation (1.83) corresponds to well-known eigenvalue problem. Solving Eq. (1.83) as eigenvalue problem, the i-th natural frequency ω_i of the plate is determined numerically.

The natural frequencies of a rectangular normal plate are now obtained from numerical computations. Then, we consider approximate expressions to directly obtain the natural frequencies. If the selected shape functions $f_{xm}(x)$ and $f_{yn}(y)$, which are composed of the shape function $f_{mn}(x, y)$, satisfy the orthogonal relations, the square matrices $A_{\bar{m}\bar{n}mn}$ and $B_{\bar{m}\bar{n}mn}$ take the values in only diagonal terms, and the nondiagonal terms are absent. Even if the selected shape functions do not perfectly satisfy the orthogonal relations, the main behavior of the plate is now dominated by the diagonal terms of the square matrices $A_{\bar{m}\bar{n}mn}$ and $B_{\bar{m}\bar{n}mn}$. Hence, taking only the diagonal terms in $A_{\bar{m}\bar{n}mn}$ and $B_{\bar{m}\bar{n}mn}$ into consideration, Eq. (1.83) becomes of an uncoupled-form. Thus, the approximate values of ω_i are obtained as

$$\omega_i \approx \sqrt{\frac{B_{mnmn}}{A_{mnmn}}} \tag{1.84}$$

in which ω_i is the ith natural frequency arranged in large value (rad/s).

The shape functions for notable boundary conditions are recommended as follows:

For simply supported rectangular plates [2],

$$f_{xm}(x) = \sin\frac{m\pi x}{l_x} \tag{1.85}$$

For clamped rectangular plates [2],

$$f_{xm}(x) = \cosh\left(\frac{\lambda_m x}{l_x}\right) - \cos\left(\frac{\lambda_m x}{l_x}\right) - \frac{\cosh(\lambda_m) - \cos(\lambda_m)}{\sinh(\lambda_m) - \sin(\lambda_m)} \times \left[\sinh\left(\frac{\lambda_m x}{l_x}\right) - \sin\left(\frac{\lambda_m x}{l_x}\right)\right] \tag{1.86}$$

in which λ_m is well-known constant as given by

$$\lambda_1 = 4.7300;\ \lambda_2 = 7.8532;\ \lambda_3 = 10.9956;\ \lambda_4 = 14.1372$$

For m large $\lambda_m \approx (2m + 1)\pi/2$.

1.9 Forced Vibrations of Rectangular Plates

We consider forced transverse vibrations of a normal rectangular plate with damping. The equation of motion for forced vibrations of the plate with damping may be extended from Eq. (1.31) to

$$m_0\ddot{w} + c\dot{w} + D_0(w_{,xxxx} + 2w_{,xxyy} + w_{,yyyy}) - p(x, y, t) = 0 \tag{1.87}$$

in which c is damping coefficient.

The general solution of Eq. (1.87) is assumed to be of the form

$$w(x, y, t) = \sum_{m=1}\sum_{n=1} f_{mn}(x, y)\Phi_{mn}(t) \tag{1.88}$$

in which $\Phi_{mn}(t)$ are unknown functions of time t; and $f_{mn}(x, y)$ indicate the natural functions satisfying both Eq. (1.74) and the specified boundary conditions of current plate. The damping coefficient c is assumed to be also series expansion c_{mn} corresponding to the shape function f_{mn}. Substituting Eq. (1.88) into Eq. (1.87), we have

$$\sum_{m=1}\sum_{n=1} [m_0 f_{mn}\ddot{\Phi}_{mn} + c_{mn} f_{mn}\dot{\Phi}_{mn} + D_0(f_{mn,xxxx} + 2f_{mn,xxyy} + f_{mn,yyyy})\Phi_{mn}] - p(x, y, t) = 0 \tag{1.89}$$

The shape functions f_{mn} must be satisfied Eq. (1.74) as

$$f_{mn,xxxx} + 2f_{mn,xxyy} + f_{mn,yyyy} - \frac{\omega_{mn}^2 m_0 f_{mn}}{D_0} = 0 \tag{1.90}$$

The substitution of Eq. (1.90) into Eq. (1.89) becomes

$$\sum_{m=1}\sum_{n=1} f_{mn}\left[m_0\ddot{\Phi}_{mn} + c_{mn}\dot{\Phi}_{mn} + \omega_{mn}^2 m_0\Phi_{mn}\right] - p(x, y, t) = 0 \tag{1.91}$$

The relationships between the damping coefficient c_{mn} and the damping constant h_{mn} for the m- and n-th mode shape of the plate, corresponding the shape function f_{mn}, may be written as

$$\frac{c_{mn}}{m_0} = 2h_{mn}\omega_{mn} \tag{1.92}$$

Applying Eq. (1.92) into Eq. (1.91), we obtain

$$\sum_{m=1}\sum_{n=1} m_0 f_{mn}[\ddot{\Phi}_{mn} + 2h_{mn}\omega_{mn}\dot{\Phi}_{mn} + \omega_{mn}^2\Phi_{mn}] - p(x, y, t) = 0 \tag{1.93}$$

Representing Eq. (1.93) with the notation Q, the Galerkin equation of the above equation may be expressed in

$$\int_0^{l_x}\int_0^{l_y} \delta w \cdot Q\mathrm{d}x\mathrm{d}y = 0 \tag{1.94}$$

The variation of w is written from Eq. (1.88) as

$$\delta w = \sum_{\bar{m}=1}\sum_{\bar{n}=1} f_{\bar{m}\bar{n}}(x, y)\delta\Phi_{\bar{m}\bar{n}}(t) \tag{1.95}$$

Hence, the Galerkin equation of the above equation becomes

$$\sum_{\bar{m}=1}\sum_{\bar{n}=1}\delta\Phi_{\bar{m}\bar{n}}\left\{\sum_{m=1}\sum_{n=1} m_0\left[\ddot{\Phi}_{mn} + 2h_{mn}\omega_{mn}\dot{\Phi}_{mn} + \omega_{mn}^2\Phi_{mn}\right]\right.$$
$$\left.\times\int_0^{l_x}\int_0^{l_y} f_{\bar{m}\bar{n}} f_{mn}\mathrm{d}x\mathrm{d}y - \int_0^{l_x}\int_0^{l_y} p(x, y, t) f_{\bar{m}\bar{n}}\mathrm{d}x\mathrm{d}y\right\} = 0 \tag{1.96}$$

From the condition which the above equation is always applicable for arbitral variation $\delta\Phi_{\bar{m}\bar{n}}$, we have the following expressions for index $\bar{m}$ and $\bar{n}$.

$$\delta\Phi_{\bar{m}\bar{n}} : \sum_{m=1}\sum_{n=1} m_0[\ddot{\Phi}_{mn} + 2h_{mn}\omega_{mn}\dot{\Phi}_{mn} + \omega_{mn}^2\Phi_{mn}]\int_0^{l_x}\int_0^{l_y} f_{\bar{m}\bar{n}} f_{mn}\mathrm{d}x\mathrm{d}y$$
$$-\int_0^{l_x}\int_0^{l_y} p(x, y, t) f_{\bar{m}\bar{n}}\mathrm{d}x\mathrm{d}y = 0 \tag{1.97}$$

The above equations are a set of differential equation with coupled form for indexes $m, n, \bar{m}$, and $\bar{n}$. Its treatment is very troublesome. So, we consider a simplified equation of motion by approximately replacing from the coupled form of Eq. (1.97) to uncoupled form. Using the normalized shape function $f_{mn}(x, y)$ and the following orthogonal relations for the normalized shape functions,

$$\int_0^{l_x} f_m(x) f_{\bar{m}}(x) \mathrm{d}x = \delta_{m\bar{m}}; \quad \int_0^{l_y} f_n(y) f_{\bar{n}}(y) \mathrm{d}y = \delta_{n\bar{n}} \tag{1.98}$$

we have the simplified equation of motion in uncoupled form

$$\delta \Phi_{mn}; \ddot{\Phi}_{mn} + 2h_{mn}\omega_{mn}\dot{\Phi}_{mn} + \omega_{mn}^2 \Phi_{mn} = Q_{mn} \tag{1.99}$$

in which the term of external load Q_{mn} is defined as

$$Q_{mn}(t) = \frac{1}{a_{mn} m_0} \int_0^{l_x} \int_0^{l_y} p(x, y, t) f_{mn}(x, y) \mathrm{d}x\mathrm{d}y \tag{1.100}$$

and is a function of time t. When the normalized shape functions are used, the orthogonal relations take the value one, as given in Eq. (1.98). Hence, a_{mn} in Eq. (1.100) takes $a_{mn} = 1$ for the normalized shape function. If the normalized shape functions are not used, the orthogonal relations have the value a_{mn}, as follows:

$$\int_0^{l_x} \int_0^{l_y} f_{mn} f_{\bar{m}\bar{n}} \mathrm{d}x\mathrm{d}y = \delta_{m\bar{m}} \delta_{n\bar{n}} a_{mn} \tag{1.101}$$

in which a_{mn} is constant which the relation $a_{mn} = a_m a_n$ has. In the above-mentioned process from Eqs. (1.96) to (1.99), we used the Galerkin method. However, this logical development is also accomplished by the use of the orthogonal relationships of the shape functions.

The general solution of Eq. (1.99) is

$$\begin{aligned} \Phi_{mn}(t) &= \exp(-h_{mn}\omega_{mn}t)[C_1 \sin \omega_{Dmn}t + C_2 \cos \omega_{Dmn}t] \\ &+ \frac{1}{a_{mn} m_0 \omega_{Dmn}} \int_0^t \exp[-h\omega_{mn}(t-\tau)] \sin[\omega_{Dmn}(t-\tau)] Q_{mn}(\tau) \mathrm{d}\tau \end{aligned} \tag{1.102}$$

in which ω_{mn} are the natural frequencies of the undamped rectangular plates; and ω_{Dm} are the natural frequencies of the damped plates. The relationship between ω_{mn} and ω_{Dmn} is

$$\omega_{Dmn} = \omega_{mn}\sqrt{1 - h_{mn}^2} \tag{1.103}$$

Equation (1.102) is the same as the Duhamel integral in beams. The Duhamel integral in Eq. (1.102) may be calculated approximately by the method, as given in Eqs. (3.68)–(3.70). On the other hand, the simplified Eq. (1.99) is also calculated by the use of appropriate dynamic calculation method, such as step-by-step integration methods, based on linear acceleration method, Wilson-θ method, constant acceleration method, Newmark-β method. [5, 6]. Once $\Phi_{mn}(t)$ are obtained, the dynamic deflections are determined from Eq. (1.88).

The damping constants h_{mn} take the value corresponding to the m- and n-th mode shape of the plate. Usually the damping constant of a rectangular plate, such as reinforced concrete slabs, is approximately considered to be independent of the natural modes. Hence, all damping constants h_{mn} equal to the damping constant of the first mode with $m = n = 1$.

1.10 Dynamic Response of Sinusoidal Dynamic Loads

We consider the dynamic solution of a rectangular normal plate, subjected to the following harmonic external load:

$$p(x, y, t) = p_{xy}(x, y) \sin \omega_p t \tag{1.104}$$

in which $p_{xy}(x, y)$ is a function of the external loads with respect to x and y; and ω_p is the frequency of the sinusoidal external loads. Then the notation $Q_{mn}(t)$ can be written as

$$Q_{mn}(t) = \sin \omega_p t \; Q_p(m, n) \tag{1.105}$$

in which, for simplicity, the notation $Q_p(m, n)$ is defined as

$$Q_p(m, n) = \int_0^{l_x}\int_0^{l_y} p_{xy}(x, y) f_{mn}(x, y) \mathrm{d}x\mathrm{d}y \tag{1.106}$$

If at $t=0$, the following initial conditions are

$$w(x, y, 0) = 0; \;\; \dot{w}(x, y, 0) = 0 \tag{1.107}$$

then $\Phi_{mn}(t)$ with the indexes m and n becomes from Eq. (1.102)

$$\Phi_{mn}(t) = \frac{Q_p(m,n)}{a_{mn}m_0\omega_{Dmn}}\frac{1}{2}\left\{\frac{h_{mn}\omega_{mn}\cos(\omega_p t) + (\omega_p + \omega_{Dmn})\sin(\omega_p t)}{(h_{mn}\omega_{mn})^2 + (\omega_p + \omega_{Dmn})^2}\right.$$
$$- \frac{h_{mn}\omega_{mn}\cos(\omega_p t) + (\omega_p - \omega_{Dmn})\sin(\omega_p t)}{(h_{mn}\omega_{mn})^2 + (\omega_p - \omega_{Dmn})^2}$$
$$+ \exp(-h_{mn}\omega_{mn}t)\left[\frac{(\omega_p + \omega_{Dmn})\sin(\omega_{Dmn}t) - h_{mn}\omega_{mn}\cos(\omega_{Dmn}t)}{(h_{mn}\omega_{mn})^2 + (\omega_p + \omega_{Dmn})^2}\right.$$
$$\left.\left.\frac{(\omega_p - \omega_{Dmn})\sin(\omega_{Dmn}t) - h_{mn}\omega_{mn}\cos(\omega_{Dmn}t)}{(h_{mn}\omega_{mn})^2 + (\omega_p - \omega_{Dmn})^2}\right]\right\} \tag{1.108}$$

Thus, the closed-form approximate solutions for a damped uniform plate subject to a harmonic external load have been presented. The closed-form solution for undamped plates is given by setting, as $h_{mn} \rightarrow 0$, $\omega_{Dmn} \rightarrow \omega_{mn}$ and $\exp(-h_{mn}\omega_{mn}t) \rightarrow 1$ in Eq. (1.108) as follows:

$$\Phi_{mn}(t) = \frac{Q_p(m,n)}{m_0\omega_{mn}}\frac{\omega_p \sin\omega_{mn}t - \omega_{mn}\sin\omega_p t}{\omega_p^2 - \omega_{mn}^2} \tag{1.109}$$

When the frequency of the external loads, ω_p, is near the frequency of the plate, the resonance arises because the denominator of Eq. (1.109) approaches to zero. Inversely, we can find out experimentally the frequencies of the plate by means of this phenomenon, when observing the vibration state of the plate by gradually changing the frequency of the external force.

1.11 Conclusions

In this chapter, we have presented the static and dynamic analyses under the Kirchhoff-Love hypothesis for rectangular normal plates subjected to general vertical loads, by means of the Galerkin methods. The approximate solutions for the static, free vibration, and forced vibration have also been proposed by the use of orthogonal relations on shape functions used. The methodology and solutions demonstrated here are confirmed from numerical computations to exactness of analytical methodology for various plates as after mentioned.

References

1. Timoshenko, S., Woinowsky-Krieger, S.: Theory of Plates and Shells. McGraw-Hill Book Company, New York, NY (1959)
2. Volterra, E., Zachmanoglou, E.C.: Dynamics of Vibrations. Charles, E., Merrill Books, Inc., Columbus, OH (1965)
3. Szilard, R.: Theory and Analysis of Plates-Classical and Numerical Methods. Prentice-Hall, Englewood Cliffs, NJ (1974)
4. Gorman, D.J.: Free Vibration Analysis of Rectangular Plates. Elsevier-North, Holland, Amsterdam (1982)
5. Mario, P.: Structural Dynamics: Theory and Computation, 3rd edn. Van Nostrand Reinhold, New York, NY (1991)
6. Buchholdt, H.A.: Structural Dynamics for Engineers. Thomas Telford, London (1997)

Chapter 2
Static and Dynamic Analyses of Circular Normal Plates

2.1 Introduction

In the preceding chapter, the static and dynamic problems of normal (uniform) rectangular thin plates have been presented. This chapter discusses the same problems for circular normal plates under Kirchhoff-Love hypothesis by transforming from the Cartesian coordinate system x, y, z to the polar coordinate one r, ϕ, z.

For circular plates, it is convenient to use a polar coordinate system instead of the Cartesian coordinate one. So, we consider the geometrical relationships between the Cartesian and polar coordinates. Arbitrary point B, which locates in a circular plate, as shown in Fig. 2.1, is expressed by $B(r, \phi, z)$ for the polar coordinate system and $B(x, y, z)$ for the Cartesian coordinate one.

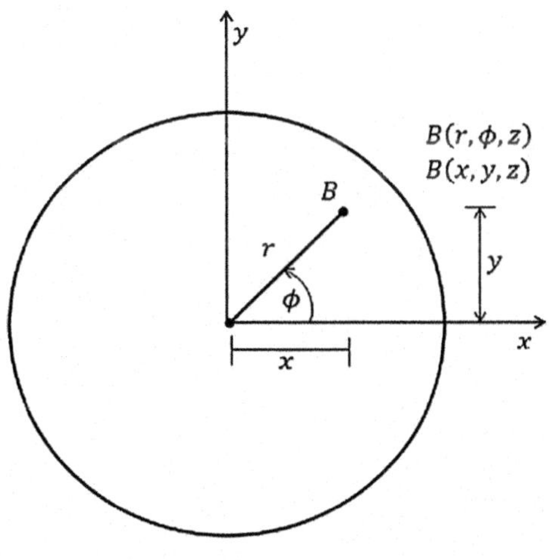

Fig. 2.1 Relationship between Cartesian and polar coordinates

H. Takabatake, *Simplified Analytical Methods of Elastic Plates*,
https://doi.org/10.1007/978-981-13-0086-8_2

The polar coordinate system is expressed by the radius r, angle ϕ, and vertical axis z. We hold the following expressions:

$$x = r\cos\phi \quad y = r\sin\phi \tag{2.1}$$

$$r^2 = x^2 + y^2 \tag{2.2}$$

in which ϕ is the angle between x-axis and radius r, as shown in Fig. 2.1. From Eq. (2.1), the axis x is a function of r and ϕ. The derivative of $w\,(r, \phi)$ with respect to x can be transformed into derivatives with respect to r and ϕ as follows:

$$\frac{\partial w}{\partial x} = \frac{\partial w}{\partial r}\frac{\partial r}{\partial x} + \frac{\partial w}{\partial \phi}\frac{\partial \phi}{\partial x} \tag{2.3}$$

From the differentiations of Eq. (2.2) with respect to x and y, we have

$$\frac{\partial r}{\partial x} = \frac{x}{r} = \cos\phi; \ \frac{\partial r}{\partial y} = \frac{y}{r} = \sin\phi \tag{2.4}$$

Since there is the relation $\tan\phi = y/x$, the differentiations with respect to x and y become as given by [1, 2, 3],

$$\sec^2\phi\frac{\partial \phi}{\partial x} = -\frac{y}{x^2}; \ \sec^2\phi\frac{\partial \phi}{\partial y} = -\frac{1}{x} \tag{2.5}$$

Hence,

$$\begin{aligned} \frac{\partial \phi}{\partial x} &= -\frac{y}{x^2}\cos^2\phi = -\frac{r\sin\phi}{(r\cos\phi)^2}\cos^2\phi = -\frac{\sin\phi}{r} \\ \frac{\partial \phi}{\partial y} &= \frac{1}{x}\cos^2\phi = \frac{1}{r\cos\phi}\cos^2\phi = -\frac{\cos\phi}{r} \end{aligned} \tag{2.6}$$

Using Eqs. (2.4) and (2.6), the relationships of the first-order differentiations between the Cartesian and polar coordinate systems are

$$\begin{aligned} \frac{\partial}{\partial x} &= \frac{\partial}{\partial r}\frac{\partial r}{\partial x} + \frac{\partial}{\partial \phi}\frac{\partial \phi}{\partial x} = \cos\phi\frac{\partial}{\partial r} - \frac{\sin\phi}{r}\frac{\partial}{\partial \phi} \\ \frac{\partial}{\partial y} &= \frac{\partial}{\partial r}\frac{\partial r}{\partial y} + \frac{\partial}{\partial \phi}\frac{\partial \phi}{\partial y} = \sin\phi\frac{\partial}{\partial r} + \frac{\cos\phi}{r}\frac{\partial}{\partial \phi} \end{aligned} \tag{2.7}$$

Furthermore, the second-order differentiations are obtained from the above equations as

$$
\begin{aligned}
\frac{\partial^2}{\partial x^2} &= \cos\phi \frac{\partial}{\partial r}\left(\frac{\partial}{\partial x}\right) - \frac{\sin\phi}{r}\frac{\partial}{\partial \phi}\left(\frac{\partial}{\partial x}\right) \\
&= \cos^2\phi \frac{\partial^2}{\partial r^2} - \frac{\sin 2\phi}{r}\frac{\partial^2}{\partial r \partial \phi} + \frac{\sin^2\phi}{r}\frac{\partial}{\partial r} + \frac{\sin 2\phi}{r^2}\frac{\partial}{\partial \phi} + \frac{\sin^2\phi}{r^2}\frac{\partial^2}{\partial \phi^2} \\
\frac{\partial^2}{\partial y^2} &= \sin\phi \frac{\partial}{\partial r}\left(\frac{\partial}{\partial y}\right) + \frac{\cos\phi}{r}\frac{\partial}{\partial \phi}\left(\frac{\partial}{\partial y}\right) \\
&= \sin^2\phi \frac{\partial^2}{\partial r^2} + \frac{\sin 2\phi}{r}\frac{\partial^2}{\partial r \partial \phi} + \frac{\cos^2\phi}{r}\frac{\partial}{\partial r} - \frac{\sin 2\phi}{r^2}\frac{\partial}{\partial \phi} + \frac{\cos^2\phi}{r^2}\frac{\partial^2}{\partial \phi^2} \\
\frac{\partial^2}{\partial x \partial y} &= \cos\phi \frac{\partial}{\partial r}\left(\frac{\partial}{\partial y}\right) - \frac{\sin\phi}{r}\frac{\partial}{\partial \phi}\left(\frac{\partial}{\partial y}\right) \\
&= \frac{1}{2}\sin 2\phi \frac{\partial^2}{\partial r^2} - \frac{1}{2}\frac{\sin 2\phi}{r^2}\frac{\partial^2}{\partial \phi^2} + \frac{\cos 2\phi}{r}\frac{\partial^2}{\partial r \partial \phi} - \frac{\cos 2\phi}{r^2}\frac{\partial}{\partial \phi} - \frac{\sin 2\phi}{2r}\frac{\partial}{\partial r}
\end{aligned}
\tag{2.8}
$$

The Laplace operator becomes

$$
\nabla^2 = \frac{\partial^2}{\partial x^2} + \frac{\partial^2}{\partial y^2} = \frac{\partial^2}{\partial r^2} + \frac{1}{r}\frac{\partial}{\partial r} + \frac{1}{r^2}\frac{\partial^2}{\partial \phi^2} \tag{2.9}
$$

The expression of the biharmonic equation is

$$
\nabla^2\nabla^2 w = \left(\frac{\partial^2}{\partial r^2} + \frac{1}{r}\frac{\partial}{\partial r} + \frac{1}{r^2}\frac{\partial^2}{\partial \phi^2}\right)\left(\frac{\partial^2 w}{\partial x^2} + \frac{1}{r}\frac{\partial w}{\partial r} + \frac{1}{r^2}\frac{\partial^2 w}{\partial \phi^2}\right) \tag{2.10}
$$

For simplicity, the partial differentiations with respect to the polar coordinates r and ϕ are indicated by the following simplified notations.

$$
\frac{\partial}{\partial r} =_{,r}\,; \quad \frac{\partial}{\partial \phi} =_{,\phi}
$$

From the above-mentioned relationships between the Cartesian and polar coordinate systems, we have the following expressions:

$$
w_{,x} = \cos\phi\, w_{,r} - \frac{1}{r}\sin\phi\, w_{,\phi} \tag{2.11}
$$

$$
w_{,y} = \sin\phi\, w_{,r} + \frac{1}{r}\cos\phi\, w_{,\phi} \tag{2.12}
$$

$$
\begin{aligned}
w_{,xx} =& \cos^2\phi\, w_{,rr} + \frac{1}{r^2}\sin^2\phi\, w_{,\phi\phi} + \frac{1}{r}\sin^2\phi\, w_{,r} \\
& - \frac{1}{r}\sin 2\phi\, w_{,r\phi} + \frac{1}{r^2}\sin 2\phi\, w_{,\phi}
\end{aligned}
\tag{2.13}
$$

$$w_{,yy} = \sin^2\phi\, w_{,rr} + \frac{1}{r^2}\cos^2\phi\, w_{,\phi\phi} + \frac{1}{r}\cos^2\phi\, w_{,r}$$
$$+ \frac{1}{r}\sin 2\phi\, w_{,r\phi} - \frac{1}{r^2}\sin 2\phi w_{,\phi} \tag{2.14}$$

$$w_{,xy} = \frac{1}{2}\sin 2\phi\, w_{,rr} - \frac{1}{r^2}\cos 2\phi\, w_{,\phi} - \frac{1}{2r^2}\sin 2\phi\, w_{,\phi\phi}$$
$$- \frac{1}{2r}\sin 2\phi\, w_{,r} + \frac{1}{r}\cos 2\phi w_{,r\phi} \tag{2.15}$$

2.2 Governing Equations of Uniform Circular Plates

The governing equation expressed in terms of polar coordinate can be obtained from transforming Eq. (1.32) for the previous rectangular plates as

$$\nabla^4 w = \left(\frac{\partial^2}{\partial r^2} + \frac{1}{r}\frac{\partial}{\partial r} + \frac{1}{r^2}\frac{\partial^2}{\partial \phi^2}\right)\left(w_{,rr} + \frac{1}{r}w_{,r} + \frac{1}{r^2}w_{,\phi\phi}\right) = \frac{p(r,\phi)}{D_0} \tag{2.16}$$

Explicit expression of Eq. (2.16) is also given by modifying Eq. (5.21) being the general governing equation of circular plates with the variation of mass and rigidity, proposed in Chap. 5, in which the coefficients α_h and d are replaced with 1 and the dynamic terms are neglected.

The expressions for internal moments and shearing forces can be transformed into polar coordinates as

$$M_r = -D_0\left[w_{,rr} + \nu\left(\frac{1}{r^2}w_{,\phi\phi} + \frac{1}{r}w_{,r}\right)\right] \tag{2.17}$$

$$M_\phi = -D_0\left(\frac{1}{r}w_{,r} + \frac{1}{r^2}w_{,\phi\phi} + \nu\, w_{,rr}\right) \tag{2.18}$$

$$M_{r\phi} = M_{\phi r} = -(1-\nu)D_0\left(\frac{1}{r}w_{,r\phi} - \frac{1}{r^2}w_{,\phi}\right) \tag{2.19}$$

$$Q_r = -D_0\frac{\partial}{\partial r}\left(w_{,rr} + \frac{1}{r^2}w_{,\phi\phi} + \frac{1}{r}w_{,r}\right) \tag{2.20}$$

$$Q_\phi = -D_0\frac{1}{r}\frac{\partial}{r\phi}\left(w_{,rr} + \frac{1}{r^2}w_{,\phi\phi} + \frac{1}{r}w_{,r}\right) \tag{2.21}$$

The vertical edge forces are obtained from the transformation of Eqs. (1.42) and (1.43) as

$$V_r = Q_r + \frac{1}{r}M_{r\phi,\phi} - D_0\left[\frac{\partial}{\partial r}\nabla^2 w + \frac{1-\nu}{r}\frac{\partial}{\partial \phi}\left(\frac{1}{r}w_{,r\phi} - \frac{1}{r^2}w_{,\phi}\right)\right] \tag{2.22}$$

$$V_\phi = Q_\phi + M_{\phi r,r} = -D_0\left[\frac{1}{r}\frac{\partial}{\partial \phi}\nabla^2 w + (1-\nu)\frac{\partial}{\partial r}\left(\frac{1}{r}w_{,r\phi} - \frac{1}{r^2}w_{,\phi}\right)\right] \tag{2.23}$$

2.3 Governing Equations of Circular Plates Subjected to Rotationally Symmetric Loading

Since the governing equation of circular normal plates subject to a general external load is complicated, as given in Eq. (2.16), we consider a simple problem which can be analytically solved by the closed-form solutions. When the lateral loads act radially symmetric with respect to the origin of the polar coordinate system on circular plates with the uniform boundary conditions, the deflection is also rotationally symmetric and is independent from ϕ. Hence, Laplacian operator (2.9) is simplified

$$\nabla_r^2 = \frac{\mathrm{d}^2}{\mathrm{d}r^2} + \frac{1}{r}\frac{\mathrm{d}}{\mathrm{d}r} \tag{2.24}$$

and the governing equation becomes

$$\nabla_r^2\nabla_r^2 w = w_{,rrrr} + \frac{2}{r}w_{,rrr} - \frac{1}{r^2}w_{,rr} + \frac{1}{r^3}w_{,r} = \frac{p(r)}{D_0} \tag{2.25}$$

in which ∇_r^2 represents Laplacian operator to rotationally symmetric circular plates.

The internal moments and shearing forces become from Eqs. (2.17) to (2.21) by neglecting the differentiation with respect to ϕ as follows:

$$M_r = -D_0\left(w_{,rr} + \frac{\nu}{r}\,w_{,r}\right) \tag{2.26}$$

$$M_\phi = -D_0\left(\nu w_{,rr} + \frac{1}{r}\,w_{,r}\right) \tag{2.27}$$

$$M_{r\phi} = M_{\phi r} = 0 \tag{2.28}$$

$$Q_r = V_r = -D_0\left(w_{,rrr} + \frac{1}{r}w_{,rr} - \frac{1}{r^2}w_{,r}\right);\ Q_\phi = 0 \tag{2.29}$$

The general solution of Eq. (2.25) is the sum of complemental solution w_c and particular solution w_p, namely

$$w = w_\mathrm{c} + w_\mathrm{p} \tag{2.30}$$

The complemental solution w_c of the homogeneous differential equation $\nabla_r^2\nabla_r^2 w_\mathrm{c} = 0$ may be expressed as

$$w_c = C_1 + C_2 r^2 + C_3 l_n r + C_4 r^2 l_n r \tag{2.31}$$

where C_1, C_2, C_3, and C_4 are constants determined from boundary conditions. If the deflections and moments at the center of the plate are not infinitely large, the constants C_3 and C_4 must be zero, and Eq. (2.31) reduces to

$$w_c = C_1 + C_2 r^2 \tag{2.32}$$

On the other hand, the particular solution w_p is easily obtained through moment-sum. The moment-sum in terms of polar coordinate is defined as

$$M_{\mathrm{sum}} = \frac{M_r + M_\phi}{1 + v} \tag{2.33}$$

The substitution of Eqs. (2.26) and (2.27) into the above equation becomes

$$M_{\mathrm{sum}} = -D_0\left(w_{,rr} + \frac{1}{r}w_{,r}\right) = -D_0\nabla_r^2 w \tag{2.34}$$

The governing equation (2.25) may be transformed by the use of Eq. (2.34) as

$$\nabla_r^2 M_{\mathrm{sum}} = -p(r) \tag{2.35}$$

The above equation may be rewritten as

$$\nabla_r^2 M_{\mathrm{sum}} = M_{\mathrm{sum},rr} + \frac{1}{r}M_{\mathrm{sum},r} = \frac{1}{r}(rM_{\mathrm{sum},r})_{,r} = -p(r) \tag{2.36}$$

Twice integrations of Eq. (2.36) with respect to r are

$$M_{\mathrm{sum},r} = -\frac{1}{r}\int p(r)r\mathrm{d}r \tag{2.37}$$

$$M_{\mathrm{sum}} = -\int \frac{1}{r}\left[\int p(r)r\mathrm{d}r\right]\mathrm{d}r \tag{2.38}$$

Thus, the moment-sum is obtained from Eq. (2.38) for a prescribed rotational symmetry load $p(r)$.

Since the moment-sum is known, we can obtain the deflections from Eq. (2.34).

$$\nabla_r^2 w_p = w_{p,rr} + \frac{1}{r}w_{p,r} = \frac{1}{r}(rw_{p,r})_{,r} = -\frac{M_{\mathrm{sum}}}{D_0} \tag{2.39}$$

Similarly, carrying out twice integrations with respect to r, the particulate solution w_p is obtained

$$w_p = -\frac{1}{D_0}\int \frac{1}{r}\left[\int M_{\mathrm{sum}}\, r\mathrm{d}r\right]\mathrm{d}r \tag{2.40}$$

Example Let us consider the internal forces and the deflection of a uniform circular plate with simply supported at the edge subject to uniformly distributed load $p = p_0$, as shown in Fig. 2.2.

(1) The moment-sum is obtained from Eq. (2.38)

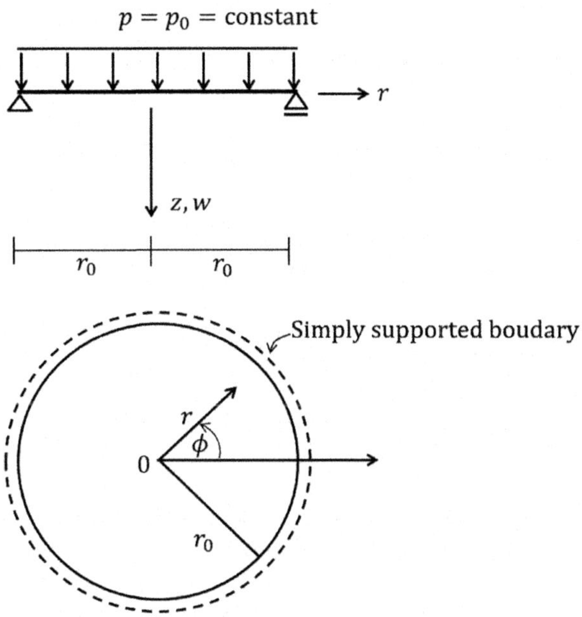

Fig. 2.2 Uniformly loaded circular plate with simply supported boundary

$$M_{\text{sum}} = -p_0 \int \frac{1}{r}\left[\int r\,\mathrm{d}r\right]\mathrm{d}r = -\frac{p_0 r^2}{4} \tag{2.41}$$

(2) The particular solution is determined from Eq. (2.40).

$$w_p = -\frac{1}{D_0}\int \frac{1}{r}\left[\int -\frac{p_0 r^2}{4} r\,\mathrm{d}r\right]\mathrm{d}r = -\frac{p_0 r^4}{64 D_0} \tag{2.42}$$

(3) The general solution is from Eq. (2.30)

$$w = C_1 + C_2 r^2 + \frac{p_0 r^4}{64 D_0} \tag{2.43}$$

(4) The boundary conditions are given as

$$w = 0 \quad \text{at} \quad r = r_0 \tag{2.44}$$

$$M_r = -D_0\left(w_{,rr} + \frac{\nu}{r} w_{,r}\right) = 0 \quad \text{at} \quad r = r_0 \tag{2.45}$$

From Eqs. (2.44) and (2.45), we have

$$C_1 + C_2 r_0^2 + \frac{p_0 r_0^4}{64 D_0} = 0$$

$$2C_2 + \frac{12p_0r_0^2}{64D_0} + \frac{\nu}{r_0}\left(2C_2r_0 + \frac{4p_0r_0^3}{64D_0}\right) = 0$$

Hence, we can determine two integral constants C_1 and C_2 from the above equations

$$C_1 = -\frac{p_0r_0^4}{64D_0}\left(1 - \frac{6+2\nu}{1+\nu}\right);\ C_2 = -\frac{p_0r_0^2}{64D_0}\left(\frac{6+2\nu}{1+\nu}\right) \tag{2.46}$$

(5) Thus, the deflection w is given from Eqs. (2.43) and (2.46)

$$w(r) = \frac{p_0r_0^4}{64D_0}\left\{-1 + \frac{6+2\nu}{1+\nu}\left[1 - \left(\frac{r}{r_0}\right)^2\right] + \left(\frac{r}{r_0}\right)^4\right\} \tag{2.47}$$

Since Poisson's ratio ν is smaller than 1, a simplified deflection which neglects the effect of Poisson's ratio reduces to

$$w(r) \approx \frac{p_0r_0^4}{64D_0}\left[5 - 6\left(\frac{r}{r_0}\right)^2 + \left(\frac{r}{r_0}\right)^4\right] \tag{2.48}$$

(6) The maximum deflection at the center of the plate is from Eqs. (2.47) and (2.48)

$$w_{\max} = \frac{p_0r_0^4}{64D_0}\left(-1 + \frac{6+2\nu}{1+\nu}\right) \approx \frac{5p_0r_0^4}{64D_0} \tag{2.49}$$

This value is equal to five times of the maximum deflection of the same plate with clamped boundary. This ratio equals to the ratio of the maximum deflections of beams with the same two boundary conditions, simply support and clamped, subject to uniform distributed vertical loads.

(7) The internal forces are obtained from Eqs. (2.26) and (2.47) as

$$M_r = \frac{p_0r_0^2}{64}\left\{\frac{2(6+2\nu)}{1+\nu}\left[1 + \nu\left(\frac{r_0}{r}\right)^2\right] - 12\left(\frac{r}{r_0}\right)^2 - 4\nu\left(\frac{r}{r_0}\right)^4\right\} \tag{2.50}$$

$$M_\phi = \frac{p_0r_0^2}{64}\left\{\frac{2(6+2\nu)}{1+\nu}\left[\nu + \left(\frac{r_0}{r}\right)^2\right] - 12\nu\left(\frac{r}{r_0}\right)^2 - 4\left(\frac{r}{r_0}\right)^4\right\} \tag{2.51}$$

$$Q_r = -\frac{p_0r}{2} \tag{2.52}$$

2.4 Conclusions

In this chapter, we have described the governing equation for a general circular uniform plate subject to general vertical loads under the Kirchhoff-Love hypothesis. As already stated, the explicit expression of Eq. (2.16) is given by replacing two coefficients α_h and d with all 1 in Eq. (5.21) of Chap. 5. On the other hand, the advanced governing equations for a general circular normal plate under the Mindlin-Reissner hypothesis will be given by Eqs. (7.89)–(7.91) in Chap. 7.

References

1. Timoshenko, S., Woinowsky-Krieger, S.: Theory of Plates and Shells. McGraw-Hill Book Company, New York, NY (1959)
2. Volterra, E., Zachmanoglou, E.C.: Dynamics of Vibrations. Charles E., Merrill Books Inc., Columbus, OH (1965)
3. Szilard, R.: Theory and Analysis of Plates-Classical and Numerical Methods. Prentice-Hall, Englewood Cliffs, NJ (1974)

Chapter 3
Static and Dynamic Analyses of Rectangular Normal Plates with Edge Beams

3.1 Introduction

The behavior of thin plates subject to lateral loads is remarkably influenced on the boundary conditions of the edges. There are many papers concerning solutions and/or analytical methods for various kinds of plates and boundary conditions, as presented by Leissa [1, 2], Takabatake et al. [3, 4, 5, 6, 7, 8], Laura and Grossi [9], Laura et al. [10], Leissa et al. [11], Zhou [12], Warburton and Edney [13]. Gorman [14] demonstrated a general solution for free transverse vibration of rectangular plates with arbitrarily distributed lateral and rotational elastic edge supported by means of the method of superposition.

Lateral and rotational elastic edge supports are proposed in order to generalize the boundary conditions. However, for building slabs supported with edge beams, the relationships between edge beams and lateral and rotational edge supports are unclear. The purpose of this chapter is to propose a simplified analysis for static and dynamic problems of a rectangular normal plate with edge beams by the use of the Galerkin method.

The accuracy of the Galerkin method depends on the shape functions used. It is difficult to find out the practical shape functions satisfying the boundary conditions of plates with edge beams. So, the translational stiffness and rotational stiffness of edges elastically restrained must satisfy the relationships between these boundary conditions. The shape functions for the original plate supported with edge beams are substituted with the shape functions of a beam supported with equivalent translational stiffness and torsional stiffness at both the ends, as shown in Fig. 3.2. The improvement of the shape functions used is achieved from convergence of the maximum deflection and of the natural frequency. Thus, approximate but accurate solutions for static and dynamic problems of a rectangular normal plate with edge beams are proposed by means of the Galerkin method. The analytical methodology proposed in this chapter is mainly based on Ref. [15].

H. Takabatake, *Simplified Analytical Methods of Elastic Plates*,
https://doi.org/10.1007/978-981-13-0086-8_3

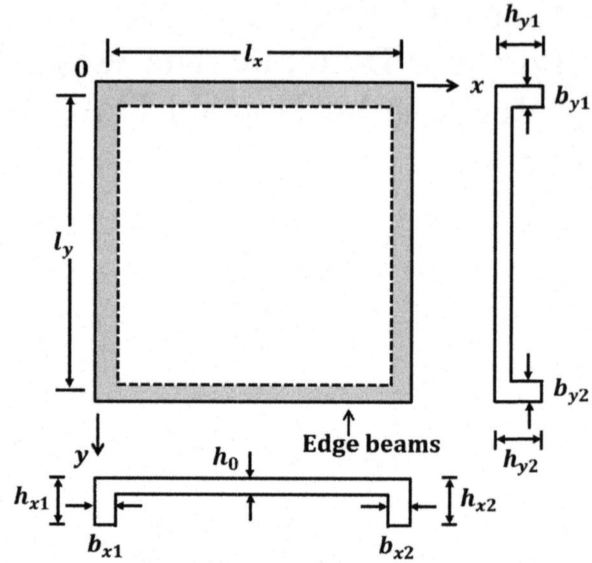

Fig. 3.1 Coordinates of a rectangular normal plate with edge beams [15]

3.2 Governing Equations of a Normal Plate with Edge Beams

We consider transverse bending problems of an isotropic rectangular uniform thin plate with edge beams, as shown in Fig. 3.1. We also assume that the Kirchhoff-Love hypothesis is valid. Cartesian coordinate systems x, y, z are employed. The widths and heights of edge beams located on edge lines at $x=0$ and $x=l_x$ are b_{x1}, b_{x2}, h_{x1}, and h_{x2}, respectively. Similarly, the sizes of edge beams located on edge lines at $y=0$ and $y=l_y$ are b_{y1}, b_{y2}, h_{y1}, and h_{y2}, respectively, in which those edge beams are assumed to be uniform.

The equation of motion of the current rectangular uniform thin plate may be given from Eq. (1.87) as

$$m_0\ddot{w} + c\dot{w} + D_0(w_{,xxxx} + 2w_{,xxyy} + w_{,yyyy}) - p = 0 \tag{3.1}$$

in which m_0 is mass per unit area; c is damping coefficient; D_0 is the flexural rigidity of the plate which is given by $D_0 = Eh_0^3/12(1-\nu)$, and p is external lateral loads per unit area. On the other hand, the boundary conditions of the plate supported with edge beams are written by Vinson [16], as

$$-D_0[w_{,xxx} + (2-\nu)w_{,xyy}] = (EI_{x1})^b w_{,yyyy} \quad \text{at}\, x = 0 \tag{3.2}$$

$$-D_0[w_{,xx} + \nu w_{,yy}] = (GJ_{x1})^b w_{,xyy} \quad \text{at}\, x = 0 \tag{3.3}$$

$$-D_0[w_{,xxx} + (2-\nu)w_{,xyy}] = -(EI_{x2})^b w_{,yyyy} \quad \text{at}\, x = l_x \tag{3.4}$$

$$-D_0[w_{,xx} + \nu w_{,yy}] = -(GJ_{x2})^b w_{,xyy} \quad \text{at } x = l_x \tag{3.5}$$

at $x=0$ and $x = l_x$,

$$-D_0[w_{,yyy} + (2-\nu)w_{,xxy}] = (EI_{y1})^b w_{,xxxx} \quad \text{at } y = 0 \tag{3.6}$$

$$-D_0[w_{,yy} + \nu w_{,xx}] = (GJ_{y1})^b w_{,xxy} \quad \text{at } y = 0 \tag{3.7}$$

$$-D_0[w_{,yyy} + (2-\nu)w_{,xxy}] = -(EI_{y2})^b w_{,xxxx} \quad \text{at } y = l_y \tag{3.8}$$

$$-D_0[w_{,yy} + \nu w_{,xx}] = -(GJ_{y2})^b w_{,xxy} \quad \text{at } y = l_y \tag{3.9}$$

at $y=0$ and $y = l_y$, and

$$w = 0 \tag{3.10}$$

at corners, where $(EI_{x1})^b, (EI_{x2})^b, \cdots$ and $(GJ_{x1})^b, (GJ_{x2})^b \cdots$ represent the bending stiffness and torsional stiffness of the edge beams, respectively. The superscript, b, indicates that it is the quantity related to the edge beams.

Since the above boundary conditions for edge beams are complicated, it is difficult to find out the suitable shape functions satisfying the boundary conditions. So, we substitute the shape functions satisfying the above boundary conditions with the shape function of beams supported by translational and rotational stiffness at both the ends, as shown in Fig. 3.2. With this approximate treatment, we can easily obtain the shape functions for the static and free vibration problems of thin plates supported by the boundary edges with translational and rotational stiffnesses.

First, the boundary conditions supported by equivalent translational and rotational flexibility coefficients, as shown in Fig. 3.2, are written as

$$-D_0[w_{,xxx} + (2-\nu)w_{,xyy}] = k_{x1} w \tag{3.11}$$

$$-D_0[w_{,xx} + \nu w_{,yy}] = -R_{x1}\, w_{,x} \tag{3.12}$$

at $x = 0$,

$$-D_0[w_{,xxx} + (2-\nu)w_{,xyy}] = -k_{x2}\, w \tag{3.13}$$

$$-D_0[w_{,xx} + \nu w_{,yy}] = R_{x2}\, w_{,x} \tag{3.14}$$

at $x = l_x$,

$$-D_0[w_{,yyy} + (2-\nu)w_{,xxy}] = k_{y1}\, w \tag{3.15}$$

$$-D_0[w_{,yy} + \nu w_{,xx}] = -R_{y1}\, w_{,y} \tag{3.16}$$

at $y=0$, and

$$-D_0[w_{,yyy} + (2-\nu)w_{,xxy}] = -k_{y2}\, w \tag{3.17}$$

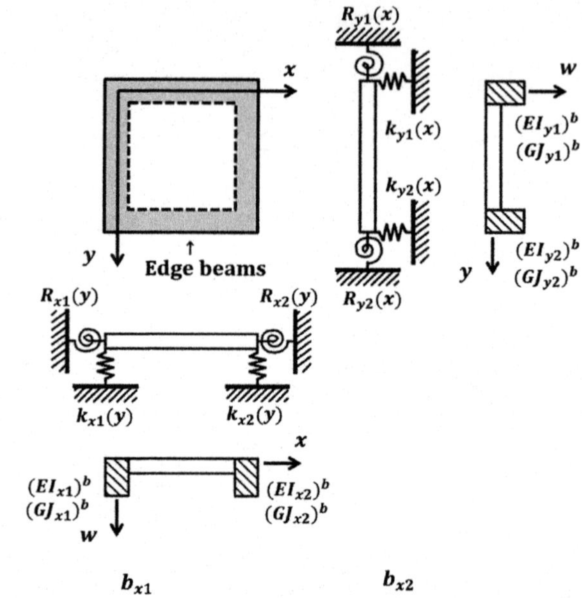

Fig. 3.2 Relationships between edge beams and translational and rotational stiffnesses in plates [15]

$$-D_0[w_{,yy} + \nu w_{,xx}] = R_{y2}\, w_y \tag{3.18}$$

at $y = l_y$, where $k_{x1}, k_{x2}, \ldots, k_{y2}$ are translational coefficients and $R_{x1}, R_{x2}, \ldots, R_{y2}$ are rotational coefficients.

When the translational and rotational flexibility coefficients are equal to the stiffness of edge beams, the following relationships must be given from the comparison of Eqs. (3.2)–(3.9) and (3.11)–(3.18):

$$(EI_{x1})^b w_{,yyyy} = k_{x1} w \tag{3.19}$$

$$(GJ_{x1})^b w_{,xyy} = -R_{x1} w_{,x} \tag{3.20}$$

at $x = 0$,

$$-(EI_{x2})^b w_{,yyyy} = -k_{x2} w \tag{3.21}$$

$$-(GJ_{x2})^b w_{,xyy} = R_{x2} w_{,x} \tag{3.22}$$

at $x = l_x$,

$$(EI_{y1})^b w_{,xxxx} = k_{y1} w \tag{3.23}$$

$$(GJ_{y1})^b w_{,xxy} = -R_{y1} w_{,y} \tag{3.24}$$

at $y = 0$, and

$$-(EI_{y2})^b w_{,xxxx} = -k_{y2} w \tag{3.25}$$

$$-(GJ_{y2})^b w_{,xxy} = R_{y2} w_{,y} \tag{3.26}$$

at $y = l_y$.

Since the translational and rotational flexibility coefficients of the plate with edge beams depend on the deflection of the plate, they are variable along each edge line. This means that the shape functions $f_m(x)$ and $f_n(y)$ of the current plates are not uniform and change along the edge beams. This makes it difficult to find out an appropriate shape function. As an approximate method to solve this problem, we assume that the shape functions $f_m(x)$ and $f_n(y)$ of the plates with edge beams do not change along the edge beams. The shape functions selected under this assumption are represented by each one in the x- and y-directions. We also assume that these selected shape functions can be substituted by the shape function of the beam with equivalent translation and rotational flexibility constants at the both edges. In order to improve the above approximation with respect to the shape functions, iterative calculation is performed to adjust the translation and rotational flexibility constants of beams at the edges until the maximum deflection of the plate with edge beams converges.

Thus, the deflection $w(x, y)$ of the plate is expanded as

$$w = \sum_{m=1} \sum_{n=1} w_{mn} f_m(x) f_n(y) \tag{3.27}$$

in which $f_m(x)$ and $f_n(y)$ are shape functions of plates supported with the boundary edges which have equivalent translational and rotational stiffnesses. Approximately, only the first term ($m=1$ and $n=1$) in Eq. (3.27) is considered. Substituting Eq. (3.27) into (3.19) and integrating between 0 and l_y after multiplying both sides of the result by $f_n(y)$, the translational flexibility k_{x1} is obtained as

$$k_{x1} \approx (EI_{x1})^b \frac{F_y(1,\ 1;4,\ 0)}{F_y(1,\ 1;0,\ 0)} \tag{3.28}$$

Similarly,

$$R_{x1} \approx -(GJ_{x1})^b \frac{F_y(1,\ 1;2,\ 0)}{F_y(1,\ 1;0,\ 0)} \tag{3.29}$$

$$k_{x2} \approx (EI_{x2})^b \frac{F_y(1,\ 1;4,\ 0)}{F_y(1,\ 1;0,\ 0)} \tag{3.30}$$

$$R_{x2} \approx -(GJ_{x2})^b \frac{F_y(1,\ 1;2,\ 0)}{F_y(1,\ 1;0,\ 0)} \tag{3.31}$$

$$k_{y1} \approx (EI_{y1})^b \frac{F_x(1,\ 1;4,\ 0)}{F_x(1,\ 1;0,\ 0)} \tag{3.32}$$

$$R_{y1} \approx -(GJ_{y1})^b \frac{F_x(1,\ 1;2,\ 0)}{F_x(1,\ 1;0,\ 0)} \tag{3.33}$$

$$k_{y2} \approx (EI_{y2})^b \frac{F_x(1,\ 1;4,\ 0)}{F_x(1,\ 1;0,\ 0)} \tag{3.34}$$

$$R_{y2} \approx -(GJ_{y2})^b \frac{F_x(1,\ 1;2,\ 0)}{F_x(1,\ 1;0,\ 0)} \tag{3.35}$$

where $F_y(n, \bar{n}; k_1, k_2)$ and $F_x(m, \bar{m}; k_1, k_2)$ are a simplified notation and are expressed in general form by

$$F_y(n, \bar{n}; k_1, k_2) \equiv \int_0^{l_y} f_n(y)^{(k_1)} f_{\bar{n}}(y)^{(k_2)} \mathrm{d}y \tag{3.36}$$

$$F_x(m, \bar{m}; k_1, k_2) \equiv \int_0^{l_x} f_m(x)^{(k_1)} f_{\bar{m}}(x)^{(k_2)} \mathrm{d}x \tag{3.37}$$

in which the superscripts enclosed within parentheses represent the differential order. Hence, the third and fourth indexes in $F_y(n, \bar{n}; k_1, k_2)$ indicate the differential order of shape functions $f_n(y)$ and $f_{\bar{n}}(y)$, respectively. Index 0 means no differential.

Although the value of the flexibility constants is determined from the convergence of iteration calculations, the initial value is necessary. Assuming that the deflection of the edge beam is expressed by a sinusoidal function, the initial values may be given as

$$k_{xi} \approx (EI_{xi})^b \left(\frac{\pi}{l_y}\right)^4 \quad (\text{for } i = 1, 2) \tag{3.38}$$

$$R_{xi} \approx (GJ_{xi})^b \left(\frac{\pi}{l_y}\right)^2 \quad (\text{for } i = 1, 2) \tag{3.39}$$

$$k_{yi} \approx (EI_{yi})^b \left(\frac{\pi}{l_x}\right)^4 \quad (\text{for } i = 1, 2) \tag{3.40}$$

$$R_{yi} \approx (GJ_{yi})^b \left(\frac{\pi}{l_x}\right)^2 \quad (\text{for } i = 1, 2) \tag{3.41}$$

Since the edge beam is connected rigidly with the plate, the bending and torsional rigidities of the edge beam increase due to the effect of plate (slab) as given as

$$(EI_{x1})^b = \phi_B (EI_{x1})_0^b \tag{3.42}$$

$$(GJ_{x1})^b = \phi_J (GJ_{x1})_0^b \tag{3.43}$$

in which the coefficients ϕ_B and ϕ_J indicate the increase of the bending and torsional rigidities due to the effect of the slab, respectively, and $(EI_{x1})_0^b$ and $(GJ_{x1})_0^b$ are the bending rigidity and torsional rigidity, respectively, of only the edge beam. From engineering assumption, as shown in the usual approximation for effective width of reinforced concrete beams, these coefficients are assumed to be the values $\phi_B = 1.5$ and $\phi_J = 1.5$ as one side of the end beam is in contact with the plate twisting, while $\phi_B = 2$ and $\phi_J = 2$, as both the sides of the edge beam are contact with the plate.

3.3 Static Analysis Used the Galerkin Method

We consider the static solution of a rectangular normal plate with edge beams by means of the Galerkin method. The methodology is the same as the presentation stated in Chap. 1. The difference is to find out a suitable shape function for a thin plate with edge beams by the iteration. The deflection w is expressed by

$$w(x, y) = \sum_{m=1} \sum_{n=1} w_{mn} f_{mn}(x, y) \tag{3.44}$$

in which w_{mn} are unknown displacement coefficients and f_{mn} are shape functions satisfying the boundary conditions.

Carrying out the same procedure stated already in Chap. 1, the Galerkin equation for the static problem from Eq. (3.1) can be written as

$$\delta w_{\bar{m}\bar{n}} : \sum_{m=1} \sum_{n=1} w_{mn} A_{\bar{m}\bar{n}mn} = B_{\bar{m}\bar{n}} \tag{3.45}$$

in which the coefficients $A_{\bar{m}\bar{n}mn}$ and $B_{\bar{m}\bar{n}}$ are defined as follows:

$$A_{\bar{m}\bar{n}mn} = \int_0^{l_x} \int_0^{l_y} (f_{mn,xxxx} + 2 f_{mn,xxyy} + f_{mn,yyyy}) f_{\bar{m}\bar{n}} \mathrm{d}x\, \mathrm{d}y \tag{3.46}$$

$$B_{\bar{m}\bar{n}} = \int_0^{l_x} \int_0^{l_y} \frac{p}{D_0} f_{\bar{m}\bar{n}} \mathrm{d}x\, \mathrm{d}y \tag{3.47}$$

The shape functions $f_{mn}(x, y)$ of thin plates with edge beams are already assumed to be expressed by shape functions of beam supported with equivalent translational stiffness and rotational stiffness at both ends, as given from Eqs. (3.80) and (3.81). So, the shape function $f_{mn}(x, y)$ of the current plate may be expressed as

$$f_{mn}(x, y) \approx f_m(x) f_n(y) \tag{3.48}$$

The shape functions used here depend on the translational stiffness and rotational stiffness at both ends of the replaced beams, but do not include directly the effect of the stiffness of the edge beam attached to the plate. So, the improvement of the shape functions is accomplished by the following processes:

Step 1: Determine initial values of edge stiffness of beam from Eqs. (3.38) to (3.41);
Step 2: Determine the shape function of the beam with the above edge stiffness from Eqs. (3.80) or (3.81). The shape function of the plate with edge beams is substituted by the calculated shape function of the beam corresponding to the above stiffness;
Step 3: Calculate the maximum value of deflections from the Galerkin Eq. (3.45);
Step 4: Determine new edge stiffness of the beam by substituting the shape functions found in step 2 into Eqs. (3.28)–(3.35);
Step 5: Repeat steps 3–5 until the maximum deflection of the plate with edge beams converges.

3.4 Numerical Results for Static Solution

Static analysis for a rectangular normal plate with edge beams has been presented by means of the Galerkin method. Then, the proposed method is examined from numerical calculations for the following four types P1–P4 of square plates supported with four edge beams:

P1	$h_0 = 0.12\,\text{m}$	$h_{x1} = h_{x2} = h_{y1} = h_{y2} = 0.72\,\text{m}$
P2	$h_0 = 0.12\,\text{m}$	$h_{x1} = h_{x2} = h_{y1} = h_{y2} = 1.02\,\text{m}$
P3	$h_0 = 0.20\,\text{m}$	$h_{x1} = h_{x2} = h_{y1} = h_{y2} = 0.80\,\text{m}$
P4	$h_0 = 0.20\,\text{m}$	$h_{x1} = h_{x2} = h_{y1} = h_{y2} = 1.10\,\text{m}$

in which the spans l_x and l_y are 6 m and the width of the edge beams is 0.3 m. The boundary conditions at the four corners of the plates with edge beams are considered to be simply supported or clamped for bending and twisting and constrained for deflection. The difference between these boundary conditions is indicated by the initial "*S*" and "*F*" for simply supported and clamped, respectively, such as P1-S, P1-F, etc. Young's modulus E is 2.06×10^{10} N/m^2; and Poisson's ratio v is 0.17; and mass density is 2400 kg/m^3. For simplicity, the lateral loads are of two types: uniformly distributed load, p_0, and a concentrated load, p_0, acting at the midspan.

Tables 3.1, 3.2, 3.3, and 3.4 show the maximum deflections and maximum bending moments, respectively, for P2-F. Figures 3.3 and 3.4 show the distributions of the deflections and bending moments at $y = l_y/2$ for P2-F, respectively. In these figures, the solid lines represent the numerical results obtained from the proposed theory; the circles indicate results obtained by using FEM code NASTRAN. The results obtained from the present theory show relatively good agreement with the results obtained from the finite element method. Figure 3.5 shows the convergence

Table 3.1 Maximum deflection of plates, unconstrained bending and twisting of edge beams at corners [15]

Type	Uniform load			Concentrated load		
	Present theory (m)	FEM (m)	Ratio	Present theory (m)	FEM (m)	Ratio
P1-S	0.00702	0.00776	0.90	0.000720	0.000764	0.94
P2-S	0.00657	0.00624	1.05	0.000692	0.000697	0.99
P3-S	0.00220	0.00260	0.85	0.000194	0.000196	0.99
P4-S	0.00184	0.00195	0.94	0.000175	0.000176	0.99

Ratio = (present theory)/(FEM)

Table 3.2 Maximum deflection of plates, constrained bending and twisting of edge beams at corners [15]

Type	Uniform load			Concentrated load		
	Present theory (m)	FEM (m)	Ratio	Present theory (m)	FEM (m)	Ratio
P1-F	0.00680	0.00635	1.07	0.000720	0.000707	1.02
P2-F	0.00643	0.00624	1.03	0.000683	0.000678	1.01
P3-F	0.00205	0.00194	1.06	0.000185	0.000182	1.02
P4-F	0.00172	0.00150	1.15	0.000168	0.000159	1.06

Ratio = (present theory)/(FEM)

Table 3.3 Maximum bending moment of plates, unconstrained bending and twisting of edge beams at corners [15]

Type	Uniform load			Concentrated load		
	Present theory (N m/m)	FEM (N m/m)	Ratio	Present theory (N m/m)	FEM (N m/m)	Ratio
P1-S	8298	8264	1.00	2209	2157	1.02
P2-S	8040	7629	1.05	2193	2127	1.03
P3-S	10,130	11,004	0.92	2322	2286	1.02
P4-S	9258	8749	1.06	2269	2181	1.04

Ratio = (present theory)/(FEM)

characteristics of the deflections and bending moments for P2-F. The convergence is obtained by each 15 terms for m and n. Also, changing the value of the translational stiffness and rotational stiffness in the shape function used here, the theory proposed here includes a plate with classical boundary conditions, such as simply support and clamped, as shown in Appendix 2.

Table 3.4 Maximum bending moment of plates, constrained bending and twisting of edge beams at corners [15]

Type	Uniform load			Concentrated load		
	Present theory (N m/m)	FEM (N m/m)	Ratio	Present theory (N m/m)	FEM (N m/m)	Ratio
P1-F	8181	7761	1.05	2200	2137	1.03
P2-F	7945	7629	1.04	2187	2120	1.03
P3-F	9670	9090	1.06	2294	2209	1.04
P4-F	8911	8064	1.11	2248	2153	1.04

Ratio = (present theory)/(FEM)

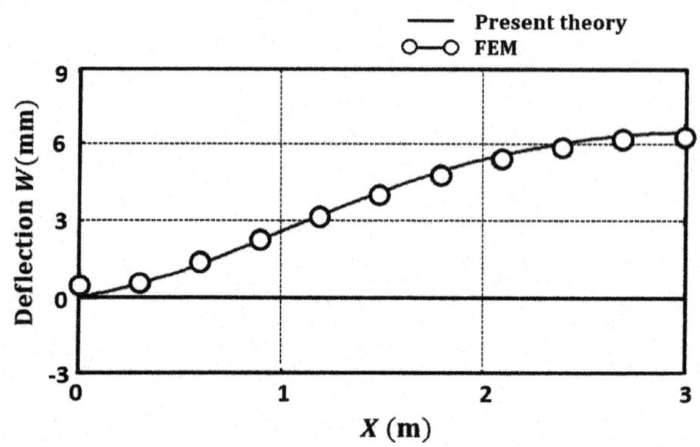

Fig. 3.3 Deflections for P2-F [15]

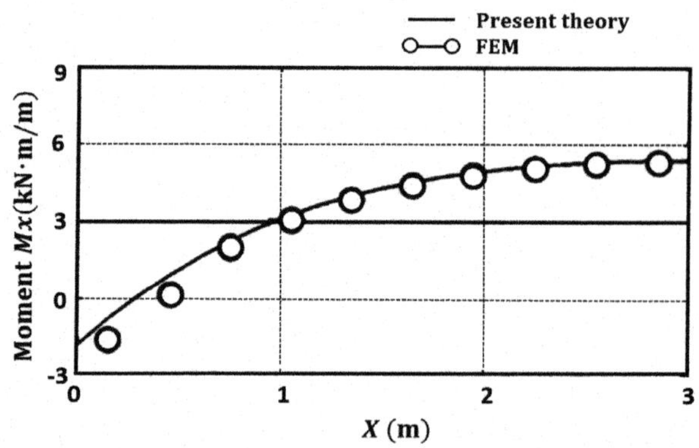

Fig. 3.4 Bending moment for P2-F [15]

3.5 Free Transverse Vibrations of a Plate with Edge Beams

For free transverse vibration of rectangular plates with edge beams, the method of separation of variables is also employed, assuming that

$$w(x, y, t) = W(x, y)\,e^{i\omega t} \tag{3.49}$$

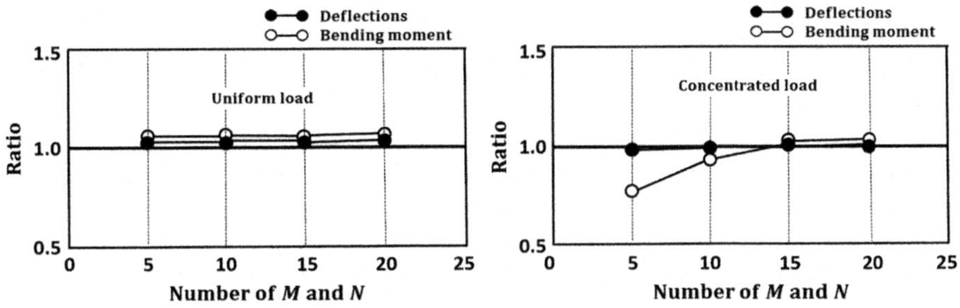

Fig. 3.5 Convergence characteristics for P2-F [15]

in which $W(x, y)$ is a function of x and y, and is a constant. The substitution of Eq. (3.49) into the equation for current free transverse vibrations becomes

$$D_0 \left(W_{,xxxx} + W_{,yyyy} + 2W_{,xxyy}\right) - \omega^2 m_0 W(x, y) = 0 \tag{3.50}$$

The natural frequencies of a rectangular plate supported with edge beams are presented by means of the Galerkin method, as presented already in Chap. 1. $W(x, y)$ is expressed as

$$W(x, y) = \sum_{m=1} \sum_{n=1} w_{mn} f_{xm}(x)\, f_{yn}(y) \tag{3.51}$$

in which w_{mn} is the unknown coefficient and $f_{xm}(x)$ and $f_{yn}(y)$ are shape functions satisfying the specified boundary conditions of the current plate in the x- and y-directions, respectively. The Galerkin equations of Eq. (3.50) can be written as

$$\delta w_{\bar{m}\bar{n}} : \sum_{m=1} \sum_{n=1} w_{mn} \left(A_{\bar{m}\bar{n}mn} - \lambda^2 B_{\bar{m}\bar{n}mn}\right) = 0 \tag{3.52}$$

in which $A_{\bar{m}\bar{n}mn}$ is given by Eq. (3.46) and $B_{\bar{m}\bar{n}mn}$ and λ are defined as

$$B_{\bar{m}\bar{n}mn} = \int_0^{l_x} \int_0^{l_y} f_{xm}(x)\, f_{yn}(y)\, f_{x\bar{m}}(x)\, f_{y\bar{n}}(y)\, \mathrm{d}x\, \mathrm{d}y \tag{3.53}$$

$$\lambda^2 = \frac{\omega^2 m_0}{D_0} \tag{3.54}$$

Equation (3.52) is a system of linear, homogeneous, simultaneous algebraic equation with respect to the unknown displacement coefficients w_{mn}. Solving Eq. (3.52) as eigenvalue problems for λ, the i-th natural frequency, ω_i, of a plate with edge beams, corresponding to the i-th value, λ_i, of λ, is determined from Eq. (3.54) as

$$\omega_i = \lambda_i \sqrt{\frac{D_0}{m_0}} \tag{3.55}$$

The natural frequencies of current plate with edge beams are now obtained from numerical computations. Then, we consider approximate expressions to obtain directly the natural frequencies. Equation (3.52) is coupled equation with respect to $\bar{m},\ \bar{n}$ and $m,\ n$. Assuming that the orthogonality condition of the shape functions used is satisfied approximately, the main behavior is dominated by only the diagonal terms of the coefficient matrices $A_{\bar{m}\bar{n}mn}$ and $B_{\bar{m}\bar{n}mn}$. Hence, Eq. (3.52) reduces to uncoupled equations. The approximate value for the i-th natural frequency is easily obtained as

$$\omega_i \approx \sqrt{\frac{A_{mnmn}}{B_{mnmn}}} \sqrt{\frac{D_0}{m_0}} \tag{3.56}$$

In the above equations, the shape functions $f_{xm}(x)$ and $f_{yn}(y)$ of the current plate are used the ones for transverse free vibration of beams supported with translational stiffness and rotational stiffness at both the ends, as shown in Appendix 2. Since the shape functions used here do not directly include the effect of the stiffness of the attached edge beam, the process to improve the shape function used, as stated already in static analysis, is also employed for the convergence of natural frequency.

3.6 Numerical Results for Natural Frequencies

The natural frequencies for a rectangular uniform plate with edge beams have been presented by means of the Galerkin method. Then, the proposed method is examined from numerical calculations for the same models used for the static problem.

Table 3.5 shows the natural frequencies for P1-F and P3-F. The numerical results show that the differences between the results obtained using Eq. (3.52) and the approximate results obtained using Eq. (3.56) are negligible for practical purposes. The results obtained from the proposed theory show excellent agreement with results obtained from FEM.

3.7 Forced Vibrations of a Plate with Edge Beams

We consider forced vibrations of a plate with edge beams. The general solution of Eq. (3.1) is assumed to be of the form

$$w(x, y, t) = \sum_{m=1}^{\infty} \sum_{n=1}^{\infty} f_{xm}(x) f_{yn}(y) \Phi_{mn}(t) \tag{3.57}$$

Table 3.5 Natural frequencies of P1-F and P3-F [15]

Type	Analytical method	First (rad/s)	Second (rad/s)	Third (rad/s)	Fourth (rad/s)
P1-F	Rigorous method	94.24	188.90	188.90	273.90
	Approximate solution	94.80	190.35	190.35	276.86
	FEM	93.76	185.04	185.05	288.28
P3-F	Rigorous method	96.56	196.10	196.10	287.90
	Approximate solution	98.29	199.59	199.59	293.10
	FEM	98.65	197.96	197.96	300.71

in which $\Phi_{mn}(t)$ are unknown functions of time t, and $f_{xm}(x)$ and $f_{yn}(y)$ are the natural functions satisfying both Eq. (3.50) and the specified boundary conditions of the current plate. Substituting Eqs. (3.57) into (3.1) and using Eq. (3.50), we have

$$\sum_{m=1}\sum_{n=1} f_{xm} f_{yn} m_0 [\ddot{\Phi}_{mn}(t) + 2h_{mn}\omega_{mn}\dot{\Phi}_{mn}(t) + \omega_{mn}^2 \Phi_{mn}(t)] = p(x, y, t) \quad (3.58)$$

in which the relationships between the damping coefficients c_{mn} and damping constants h_{mn} for the m,n-th mode shape of the current plate may be written as

$$\frac{c_{mn}}{m_0} = 2h_{mn}\omega_{mn} \quad (3.59)$$

As already stated in Sect. 1.9, for simplicity, representing Eq. (3.58) with Q, the Galerkin equation of Eq. (3.58) may be expressed in

$$\int_0^{l_x}\int_0^{l_y} \delta w \cdot Q \mathrm{d}x\mathrm{d}y = 0 \quad (3.60)$$

The variation δw is given from Eq. (3.57) by

$$\delta w = \sum_{\bar{m}=1}\sum_{\bar{n}=1} f_{x\bar{m}}(x) f_{y\bar{n}}(y)\delta\Phi_{\bar{m}\bar{n}}(t) \quad (3.61)$$

From the condition which the above equation is always applicable for arbitral variation $\delta\Phi_{\bar{m}\bar{n}}$, we have the following expressions for $\delta\Phi_{\bar{m}\bar{n}}$ with indexes $\bar{m}$ and $\bar{n}$.

$$\delta\Phi_{\bar{m}\bar{n}} : \sum_{m=1}\sum_{n=1}\left\{ K_{\bar{m}\bar{n}mn}[\ddot{\Phi}_{mn}(t) + 2h_{mn}\omega_{mn}\dot{\Phi}_{mn}(t) + \omega_{mn}^2 \Phi_{mn}(t)] \right\} = \frac{1}{m_0} Q_{\bar{m}\bar{n}}(t) \quad (3.62)$$

in which the notations $K_{\bar{m}\bar{n}mn}$ and $Q_{\bar{m}\bar{n}}(t)$ are defined as

$$K_{\bar{m}\bar{n}mn} = \int_0^{l_x}\int_0^{l_y} f_{\bar{m}\bar{n}} f_{mn}\,\mathrm{d}x\,\mathrm{d}y \tag{3.63}$$

$$Q_{\bar{m}\bar{n}}(t) = \int_0^{l_x}\int_0^{l_y} p(x, y, t) f_{\bar{m}\bar{n}}\,\mathrm{d}x\,\mathrm{d}y \tag{3.64}$$

Equation (3.62) is also obtained from multiplying both sides of Eq. (3.58) by $f_{x\bar{m}}$ and $f_{y\bar{n}}$ and integrating between 0 to l_x and 0 to l_y.

Equation (3.62) is coupled equation and can be solved by the use of suitable dynamic calculation method, such as step-by-step integration methods, based on linear acceleration method, Wilson-θ method, constant acceleration method, and Newmark-β method.

3.8 Approximate Solutions for Forced Vibrations

The dynamic analysis for a damped plate with edge beams may be solved by means of step-by-step integration methods. Since the solution of Eq. (3.62) requires dynamic analysis of simultaneous equation coupled with respect to $\bar{m}$, $\bar{n}$ and m, n, its calculation is complicated. Then, for practical use, we consider an approximate solution in closed form.

When the shape functions used have orthogonality relations, Eq. (3.62) takes on an uncoupled form. Even if the orthogonality relations of the shape functions are not realized exactly, Eq. (3.62) may be approximated an uncoupled form because the behavior of a plate with edge beams is dominated by the diagonal terms in the coefficients $K_{\bar{m}\bar{n}mn}$. Then, Eq. (3.62) becomes the following uncoupled form with respect to $\delta\Phi_{mn}$:

$$\delta\Phi_{mn} : K_{mnmn}[\ddot{\Phi}_{mn}(t) + 2h_{mn}\omega_{mn}\dot{\Phi}_{mn}(t) + \omega_{mn}^2\Phi_{mn}(t)] = \frac{1}{m_0}Q_{mn}(t) \tag{3.65}$$

The general solution of Eq. (3.65) may be obtained as

$$\Phi_{mn}(t) = \exp(-h_{mn}\omega_{mn}t)(C_1 \sin\omega_{Dmn}t + C_2\cos\omega_{Dmn}t) + \int_0^t \exp[-h_{mn}\omega_{mn}(t-\tau)]\sin\omega_{Dmn}(t-\tau)\,Q_{mn}(\tau)\,\mathrm{d}\tau \tag{3.66}$$

in which C_1 and C_2 are constants determined from the initial conditions, and ω_{mn} and ω_{Dmn} are the natural frequencies of the undamped and damped plates with edge beams, respectively. The relationship between ω_{mn} and ω_{Dmn} is

$$\omega_{Dmn} = \omega_{mn}\sqrt{1 - h_{mn}^2} \tag{3.67}$$

Thus, the dynamic deflections of the current plate are determined by substituting Eqs. (3.66) into (3.57).

The Duhamel integral in Eq. (3.66) may be calculated approximately as follows:

$$\Phi_{mn}(t) = A_{mn}(t) \sin \omega_{Dmn}t - B_{mn}(t) \cos \omega_{Dmn} t \tag{3.68}$$

in which $A_{mn}(t)$ and $B_{mn}(t)$ are expressed approximately with the incremental form as

$$A_{mn}(t) \approx A_{mn}(t - \Delta t) \exp(-h_{mn}\omega_{mn}\Delta t) + \frac{1}{\omega_{Dmn}} \int_{t-\Delta t}^{t} \exp[-h_{mn}\omega_{mn}(t - \tau)] \cos \omega_{Dmn}\tau \frac{Q_{mn}(\tau)}{m_0 K_{mnmn}} \mathrm{d}\tau \tag{3.69}$$

$$B_{mn}(t) \approx B_{mn}(t - \Delta t) \exp(-h_{mn}\omega_{mn}\Delta t) + \frac{1}{\omega_{Dmn}} \int_{t-\Delta t}^{t} \exp[-h_{mn}\omega_{mn}(t - \tau)] \sin \omega_{Dmn}\tau \frac{Q_{mn}(\tau)}{m_0 K_{mnmn}} \mathrm{d}\tau \tag{3.70}$$

Here, each first term on the right side of the above equation represents the value of A_{mn} and B_{mn} at time $t - \Delta t$, respectively.

Meanwhile, for harmonic external loads, the Duhamel integral may be written in closed-form solution, as shown in Sect. 1.10. When we consider the harmonic external load as the same as Eq. (1.104), the expressions from Eqs. (1.105) to (1.109) which are already stated in Sect. 1.10 are applicable to the closed-form solution of the current problem subject to the harmonic external load.

3.9 Numerical Results for Dynamic Responses

To examine the closed-form approximate solution proposed here for an isotropic rectangular plate with edge beams, numerical computations are carried out for four kinds of rectangular plates with edge beams, used in static problem. The damping constants, h_{mn}, are assumed to be 0.03 for all modes. Recently, human response to vibration in building slabs has become significant in design. So, the external lateral loads without the additional mass are assumed to be test loads for checking the disturbing effect of the vibration of floors caused by people walking and other

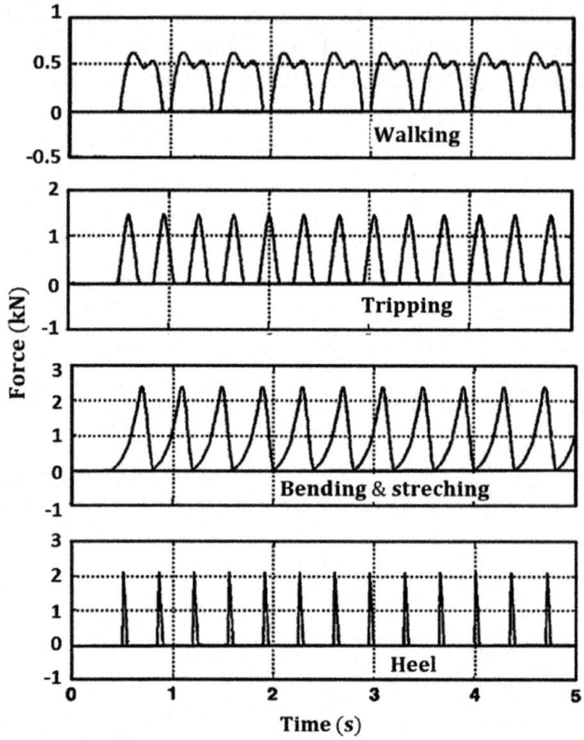

Fig. 3.6 Test loads for floors

everyday usage, as shown in Fig. 3.6. The test loads consist of four loading types: walking, tripping, bending and stretching, and heel. These data are offered from Kushida [17]. The current shape functions use natural functions obtained from the free vibration mentioned above.

Table 3.6 represents the maximum dynamic deflections at the midpoint of the plates with edge beams of P1-F, subjected to the test loads at the midpoint. Figure 3.7 shows the time history of the dynamic deflections at the midpoint of P1-F and P2-F. In this figure, the solid line indicates values obtained from the numerical computations using the linear acceleration method into Eq. (3.62); the broken line indicates values obtained from the closed-form solution given by Eq. (3.68); and the solid line with circles indicates values obtained from FEM. The difference between the solid and broken lines is too small and does not show the plot. These results show that the closed-form approximate solution proposed here is also usable.

Table 3.6 Maximum dynamic deflections of plates with edge beams for P1-F

Load type	Analytical method	Maximum deflection (μm)	Ratio
Walking	Wilson-θ method	53.80	1.07
	Closed-form solution	53.64	1.06
	FEM	50.45	
Tripping	Wilson-θ method	125.90	1.07
	Closed-form solution	125.50	1.06
	FEM	118.12	
Bending and stretching	Wilson-θ method	196.30	1.09
	Closed-form solution	195.80	1.09
	FEM	179.73	
Heel	Wilson-θ method	231.80	1.05
	Closed-form solution	231.60	1.05
	FEM	221.35	

Ratio = (calculated value)/(FEM)

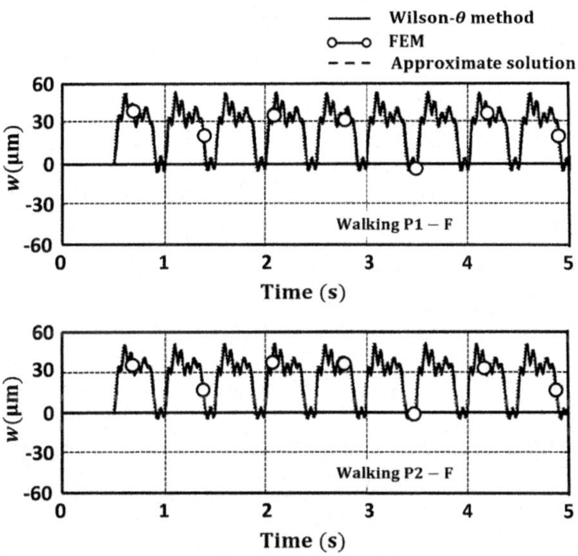

Fig. 3.7 Dynamic deflections for P1-F and P2-F, subjected to test loads for floors

3.10 Conclusions

The general analysis methods and closed-form approximate solutions for an isotropic rectangular uniform plate with edge beams have been presented by means of the Galerkin method. The exactness of the closed-form approximate solutions proposed here has been proven by comparing them with the numerical results obtained from FEM.

Appendix 1: Static Analysis of Beam with Translational Stiffness and Rotational Stiffness at Both Ends

We consider the static problem of a uniform beam supported by translational stiffness and rotational stiffness at the edges, as shown in Fig. 3.8, in which k_A, k_B, R_A, and R_B are translational stiffness and rotational stiffness of the supports, respectively. The equilibrium equation and boundary conditions of the beam with translational and rotational stiffness at both the ends are written as

$$w_{,xxxx} - \frac{p}{EI} = 0, \tag{3.71}$$

$$k_A w_a = -EIw_{,xxx}, \quad \text{at}\, x = 0, \tag{3.72}$$

$$k_B w_b = EIw_{,xxx}, \quad \text{at}\, x = l, \tag{3.73}$$

$$R_A w_a' = EIw_{,xx}, \quad \text{at}\, x = 0, \tag{3.74}$$

$$R_B w_b' = EIw_{,xx}, \quad \text{at}\, x = l, \tag{3.75}$$

The solution of Eq. (3.71) is given as

$$w = C_1\frac{x^3}{6} + C_2\frac{x^2}{2} + C_3x + C_4 + \frac{1}{EI}\iint\iint p\,\mathrm{d}x\,\mathrm{d}x\,\mathrm{d}x\,\mathrm{d}x \tag{3.76}$$

in which the integrate constants C_1–C_4 are determined from the boundary conditions given in Eqs. (3.72)–(3.75). Now, expanding the static load into Fourier series:

$$p(x) = \sum_{m=1} P_m \sin\frac{m\pi x}{l} \tag{3.77}$$

in which the Fourier coefficients P_m are given as

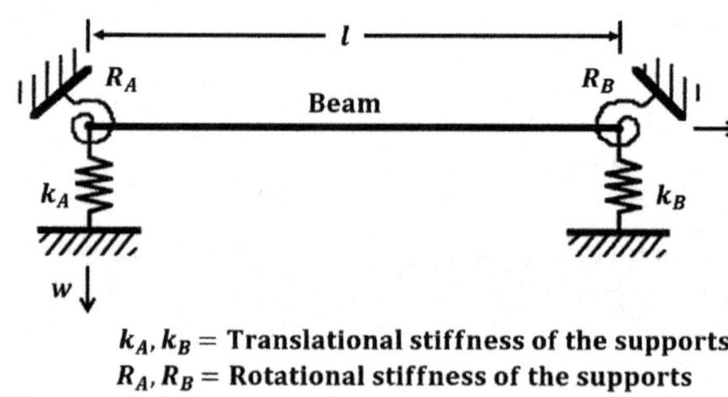

Fig. 3.8 Beam supported with translational stiffness and rotational stiffness [15]

$$P_m = \frac{2}{l}\int_0^l p(x)\sin\frac{m\pi x}{l}\mathrm{d}x \tag{3.78}$$

Now, the deflection w of the beam is expressed by

$$w(x) = \sum_{m=1} w_m f_m(x) \tag{3.79}$$

in which the shape functions $f_m(x)$ are obtained by substituting Eqs. (3.77) into (3.76) as

$$f_m = \frac{P_m}{EI\pi^4}\sin\frac{m\pi x}{l} + \frac{C_1}{6l}\left(\frac{x}{l}\right)^3 + \frac{C_2}{2l^2}\left(\frac{x}{l}\right)^2 + \frac{C_3}{l^3}\left(\frac{x}{l}\right) + \frac{C_4}{l^4}\ (for\ m = 1) \tag{3.80}$$

$$f_m = \frac{P_m}{EI(m\pi)^4}\sin\frac{m\pi x}{l} \quad (\text{for } m \geq 2) \tag{3.81}$$

The shape functions proposed here expand only the term including the external load into series expansion, while the terms including integral constants are only taken to the first term of the shape function. Therefore, the method proposed here is more general than the method of Zhou [12]. Although Zhou expands all terms on the right side of Eq. (3.76), the method is applicable to only special cases of load and boundary conditions.

Appendix 2: Free Transverse Vibrations of a Beam with Translational Stiffness and Rotational Stiffness at the Both Ends

We consider a uniform beam supported by translational stiffnesses k_A and k_B and rotational stiffnesses R_A and R_B at both the ends, as shown in Fig. 3.8. Expressing the deflection $w\,(x, t)$ with

$$w(x, t) = f(x)\mathrm{e}^{\mathrm{i}\omega t} \tag{3.82}$$

the frequency equation may be written as

$$f_{,xxxx} - k^4 f = 0 \tag{3.83}$$

in which k^2 is defined as

$$k^2 \equiv \omega\sqrt{\frac{m}{EI}} \tag{3.84}$$

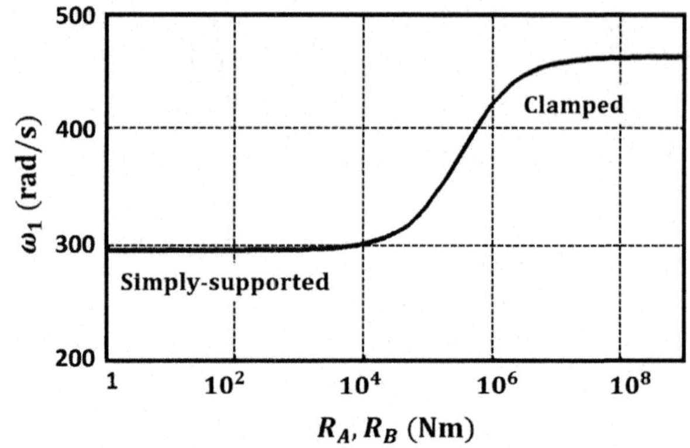

Fig. 3.9 Variation of first natural frequency of beams corresponding rotational stiffness

The solution of Eq. (3.83) is given as

$$f(x) = C_1[\cos(kx) + \cosh(kx)] + C_2[\cos(kx) - \cosh(kx)] \\ + C_3[\sin(kx) + \sinh(kx)] + C_4[\sin(kx) - \sinh(kx)] \tag{3.85}$$

in which C_1–C_4 are integral constants. Substituting Eq. (3.85) into the boundary conditions Eqs. (3.72)–(3.75), which are modified by Eq. (3.82), $f(x)$ may be obtained as

$$f(x) = k_1 \cos(kx) + k_2 \cosh(kx) + k_3 \sin(kx) + k_4 \sinh(kx) \tag{3.86}$$

in which k_1–k_4 are constants determined from boundary conditions.

Shape function $f(x)$ given in Eq. (3.86) is applicable to beams supported with boundary conditions, such as free, simply supports, clamped, elastic support, and restraint. Figure 3.9 shows the variation of the first natural frequency obtained by changing R_A and R_B from 0 to ∞ under the constraint $k_A = k_B = \infty$. It follows from Fig. 3.9 that when the rotational stiffness R_A or R_B is 10^4 Nm or less, the boundary condition for the rotation may be regarded as free. On the other hand, for the perfect constraint for the rotation, the rotational stiffness R_A or R_B is necessary 10^7 Nm and over.

References

1. Leissa, A.W.: Recent research in plate vibrations: classical theory. Shock Vib. Dig. **9**(10), 13–24 (1977)
2. Leissa, A.W.: Recent studies in plate vibrations: 1981–85 part II. Complicating effects. Shock Vib. Dig. **19**(3), 10–24 (1991)
3. Takabatake, H.: Static analyses of elastic plates with voids. Int. J. Solids Struct. **28**(2), 179–196 (1991)

4. Takabatake, H.: Dynamic analyses of elastic plates with voids. Int. J. Solids Struct. **28**(7), 879–895 (1991)
5. Takabatake, H., Morimoto, H., Fujiwara, T., Honma, T.: Simplified analysis of circular plates including voids. Comput. Struct. **58**(2), 263–275 (1996)
6. Takabatake, H., Imaizumi, T., Kunihiro, O.: Simplified analysis of rectangular plates with stepped thickness. J. Struct. Eng. ASCE **121**(1), 28–38 (1995)
7. Takabatake, H., Yanagisawa, N., Kawano, T.: A simplified analysis of rectangular cellular plates. Int. J. Solids Struct. **33**(14), 2055–2074 (1996)
8. Takabatake, H., Kajiwara, K., Takesako, R.: A simplified analysis of circular cellular plates. Comput. Struct. **61**(5), 789–804 (1996)
9. Laura, P.A.A., Grossi, R.O.: Transverse vibrations of rectangular plates with edges elastically restrained against translation and rotation. J. Sound Vib. **75**(1), 101–107 (1981)
10. Laura, P.A.A., Grossi, R.O.: Transverse vibration of a rectangular plate elastically restrained against rotation along three edges and free on the fourth edge. J. Sound Vib. **59**(3), 355–368 (1978)
11. Leissa, A.W., Laura, P.A.A., Gutierrez, R.H.: Transverse vibrations of circular plates having nonuniform edge constraints. J. Acoust. Soc. Am. **66**(1), 180–184 (1979)
12. Zhou, D.: Natural frequencies of elastically restrained rectangular plates using a set of static beam functions in the Rayleigh-Ritz method. Comput. Struct. **57**(4), 731–735 (1995)
13. Warburton, G.B., Edney, S.L.: Vibrations of rectangular plates with elastically restrained edges. J. Sound Vib. **95**(4), 537–552 (1984)
14. Gorman, D.J.: A general solution for the free vibration of rectangular plates with arbitrarily distributed lateral and rotational elastic edge support. J. Sound Vib. **174**(4), 451–459 (1994)
15. Takabatake, H., Nagareda, Y.: A simplified analysis of elastic plates with edge beams. Comput. Struct. **70**(2), 129–139 (1999)
16. Vinson, J.R.: The behavior of thin walled structures: beams, plates, and shells. Springer, Kluwer, Dordrecht, The Netherlands (1989)
17. Kushida, H.: A study of design method in consideration of vibration characteristics and optimum environment (in Japanese). J. Archit. Plann. Environ. **404** (1989)

Part II
Static and Dynamic Analyses of Various Plates

Chapter 4
Static and Dynamic Analyses of Rectangular Plates with Voids

4.1 Introduction

To guarantee lightness, structural efficiency, and enough space for equipment, plates with voids are often used in building slabs. These are called multicell slabs with transverse diaphragms, or voided slabs, or cellular slabs, depending upon the shape and size of the voids used. Most analytical methods of the previous researches for plates with voids are based on the equivalent plate analogy. With this analogy, even if a plate with voids is composed of an isotropic material, the equivalent plate becomes an orthotropic plate, because the bending rigidity and torsional rigidity are different in directions owing to the existence of the voids. A number of authors in the past various studies have proposed rigidity coefficients to enable the determination of overall effects. Crisfield and Twemlow [1] proposed a full equivalent anisotropic plate solution for cellular structures by means of the finite element method. Elliott and Clark [2] analyzed a slab with one-way circular voids. Cope et al. [3] analyzed a cellular bridge deck by means of a two-dimensional finite element solution for the equivalent shear weak slab. Holmberg [4], Sawko and Cope [5], and Elliott [6] proposed rigidity coefficients for cellular slabs.

The above-mentioned equivalent approaches have the following two faults: Firstly, since the rigidity of plates with voids is determined independently of the position of the voids, the application is restricted only to plates with many voids of the same cross section, spaced uniformly. Secondly, local variations of stress couples due to the existence of voids cannot be expressed.

On the other hand, analyses based on the finite element method for plates with voids are effective but much numerical calculation is needed. A general and simple analytical method usable in both the preliminary and final stages of the design of a plate with voids is desired from the practical use. However, as mentioned above, a general analytical method for plates with arbitrarily positioned voids has not been established.

H. Takabatake, *Simplified Analytical Methods of Elastic Plates*,
https://doi.org/10.1007/978-981-13-0086-8_4

The purpose of this chapter is to propose a general analytical methodology for plates with arbitrarily positioned voids. The discontinuous variation of rigidity and mass of plates due to the voids is expressed as a continuous function by means of an extended Dirac function. The extended Dirac function proposed here by the author is defined as a Dirac function existing continuously in a prescribed region. For the current problem, the extended Dirac function has meaning in the region where voids exist, and can express analytically the discontinuous variation in the rigidity and mass of the plates due to the voids with a continuous function. Therefore, it is effective as for a general analytical method for plates with arbitrarily positioned voids. The theory of plates with voids is formulated without modifying the rigidity and mass of the plates, as done in the equivalent plate analogy. The author [7, 8] demonstrated the effectiveness of the extended Dirac function both for bending and torsional analyses of tube structures in high-rise buildings [9, 10] and for lateral buckling of *I* beams with web stiffeners and batten plates [8].

In this chapter, the height of voids used is assumed to be relatively small compared to the plate thickness. By this assumption, the Kirchhoff-Love hypothesis can be applied to bending behavior of plates with voids. A cellular plate whose void height is much larger than the thickness of the plate will be developed in Chaps. 6 and 7. This chapter introduces how the discontinuous variation in the rigidity and mass of the plates due to the voids is expressed as a continuous function by means of the extended Dirac function, as the above-mentioned.

First, the general governing equations for rectangular plates with voids are proposed by using Hamilton's principle. Second, the static solutions for simply supported and clamped plates are presented by means of the Galerkin method. The exactness of the solution proposed for static problems is established from numerical results and experimental results of reduced model using acrylic material. Third, the natural frequencies of a rectangular plate with voids are presented by means of the Galerkin method. The proposed solutions are examined by comparing them with experimental results using acrylic plates and with results obtained from the FEM code NASTRAN for simply supported and clamped plates with voids. Forth, the forced vibrations of a damped plate with voids are presented by the use of the linear acceleration method. For practical use, the approximate solutions for static and dynamic problems of a plate with voids, subjected to general and harmonic external loads, are proposed in closed-form solution. Last, the validity of the closed-form solutions proposed here is shown by comparing them with the numerical results obtained using the linear acceleration method and using the FEM code NASTRAN. The methodology of this chapter is based on the application of an extended Dirac function to the discontinuous variation of the stiffness and mass due to the existence of voids in addition to the expression of Chap. 1. The methodology proposed in this chapter is mainly based on Refs. [11, 12] proposed by author.

4.2 Governing Equations of Plates with Voids

We consider a rectangular plate with arbitrarily positioned voids, as shown in Fig. 4.1, in which the height of voids used is assumed to be relatively small compared to the plate thickness. A Cartesian coordinate system x, y, z is employed. Assume that each void is a rectangular parallelepiped whose ridgelines are parallel to the axis x or y and which is symmetrically positioned with respect to the middle plane of the plate, as shown in Fig. 4.2. The position of the i, j-th void is indicated by the coordinate value $\left(x_i, y_j\right)$ of the midpoint of the void. The widths in the x and y directions of the void are $b_{xi,j}$ and $b_{yi,j}$, respectively, and its height is $h_{i,j}$. The size and position of each void are arbitrary except for the assumptions mentioned above.

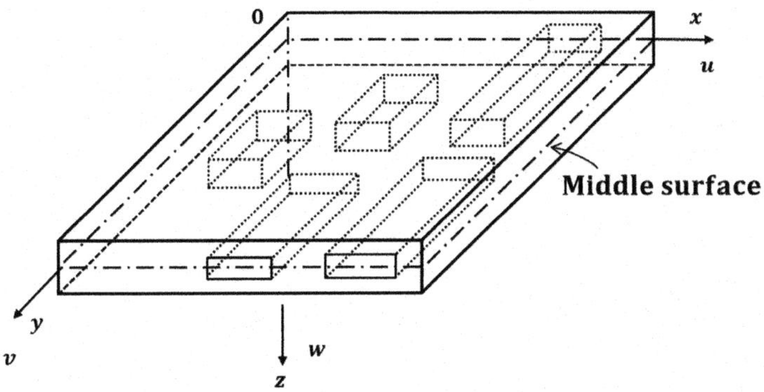

Fig. 4.1 Coordinates of a rectangular plate with voids

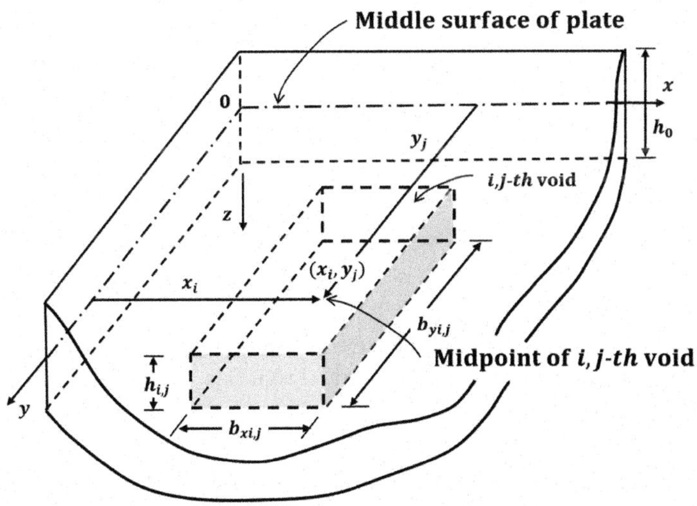

Fig. 4.2 Details of a void [11, 12]

We consider the bending of isotropic thin plates to small deformations and assume the validity of the Kirchhoff-Love plate theory for the current problem. Hence, the transverse shear deformation is neglected. The assumption used here may be effective for structures like floors, roofs, bridges because the height of the voids is relatively small.

Thus, the shape of a plate with voids is adequately defined by describing the geometry of its middle surface, which is a surface that bisects the plate thickness h_0 at each point. The governing equation of plates with voids is proposed by means of the following Hamilton's principle:

$$\delta I = \delta \int_{t_1}^{t_2} (T - U - V)\,\mathrm{d}t = 0 \tag{4.1}$$

in which T is the kinetic energy; U is the strain energy; V is the potential energy produced by the external loads; and δ is the variational operator taken during the indicated time interval from t_1 to t_2. The strain energy U for the current problem is given by

$$U = \frac{1}{2} \iint \left[M_x \kappa_x + M_y \kappa_y + 2M_{xy} \kappa_{xy}\right] \mathrm{d}x\,\mathrm{d}y \tag{4.2}$$

in which κ_x, κ_y, and κ_{xy} are the curvatures and twist of the deflected middle surface, respectively, and M_x, M_y, and M_{xy} denote the bending and twisting moments per unit width, respectively, given by

$$M_x = \int \sigma_x z\,\mathrm{d}z; \quad M_y = \int \sigma_y z\,\mathrm{d}z; \quad M_{xy} = M_{yx} = \int \tau z\,\mathrm{d}z \tag{4.3}$$

where σ_x and σ_y are the normal stress components; τ $\left(= \sigma_{xy} = \sigma_{yx}\right)$ is the shear stress; and z is the distance measured from the middle surface of the plate. The stresses σ_x, σ_y, and τ for isotropic plates can be expressed in terms of the lateral deflections w on the middle surface of the plates:

$$\sigma_x = -\frac{Ez}{1-\nu^2}(w_{,xx} + \nu w_{,yy}); \quad \sigma_y = -\frac{Ez}{1-\nu^2}(w_{,yy} + \nu w_{,xx}); \quad \tau = -2G\,z w_{,xy} \tag{4.4}$$

in which E is Young's modulus of the isotropic material, G is the shear modulus of the isotropic material, and ν is Poisson's ratio. From Eqs. (4.3) and (4.4), the bending moment M_x may be written as

$$M_x = -\frac{E}{1-\nu^2}(w_{,xx} + \nu w_{,yy}) \int z^2\,\mathrm{d}z \tag{4.5}$$

At a section where a void exists, the calculation of the above integral must be amended to exclude the void, i.e.,

$$\int z^2\,\mathrm{d}z = \int_{-h_0/2}^{h_0/2} z^2\,\mathrm{d}z - \int_{-h_1/2}^{h_1/2} z^2\,\mathrm{d}z \tag{4.6}$$

in which h_0 is the thickness of solid plates and h_1 is the height of the void. h_1 is a function of x and y. At all points in the region where the i, j-th void exists, the relation $h_1 = h_{i,j}$ is valid. Hence, $h_1(x, y)$ can generally be expressed by

$$h_1(x, y) = \sum_{i=1}^{m^*}\sum_{j=1}^{n^*} h_{i,j} D(x - x_i) D(y - y_i) \tag{4.7}$$

in which Σ is the sum for the total number of voids in the plates; m^* and n^* indicate the final numbers of voids in position counting from $i=1$ and $j=1$, respectively. $D(x - x_i)$ and $D(y - y_j)$ are the extended Dirac functions.

The extended Dirac function $D(x - x_i)$ is defined as a function where the Dirac delta function $\delta(x - \xi)$ exists continuously in the x direction through the i, j-th void, namely the region from $x_i - b_{xi,j}/2$ to $x_i + b_{xi,j}/2$, in which the supplementary variable ξ can take values continuously from $x_i - b_{xi,j}/2$ to $x_i + b_{xi,j}/2$. Similarly, $D(y - y_j)$ is a function where the Dirac delta function $\delta\,(y - \eta)$ exists continuously in the y direction through the i, j-th void, namely the region from $y_j - b_{yi,j}/2$ to $y_j + b_{yi,j}/2$, in which the supplementary variable η can take values continuously from $y_j - b_{yi,j}/2$ to $y_j + b_{yi,j}/2$. Briefly, the extended Dirac functions $D(x - x_i)$ and $D(y - y_j)$ are considered as the sum of the Dirac delta function δ distributed continuously in a prescribed region. Hence, $D(x - x_i)$ is valid for a region from $x_i - b_{xi,j}/2$ to $x_i + b_{xi,j}/2$. Also, $D(y - y_j)$ is valid for a region from $y_j - b_{yi,j}/2$ to $y_j + b_{yi,j}/2$.

The integral calculation including the extended Dirac function $D(x - x_i)$ for a given function $f(x)$ can be expressed as:

$$\begin{aligned}\int_0^{l_x} D\,(x - x_i) f\,(x)\,\mathrm{d}x &= \int_{x_i-(b_{xi,j}/2)}^{x_i+(b_{xi,j}/2)} [\delta(x - \xi) f(x)\,\mathrm{d}x]\,\mathrm{d}\xi \\ &= \int_{x_i-(b_{xi,j}/2)}^{x_i+(b_{xi,j}/2)} f(\xi)\,\mathrm{d}\xi \end{aligned} \tag{4.8}$$

in which ξ indicates a supplementary variable of x. Similarly,

$$\int_0^{l_y} D(y - y_j) f(y)\,\mathrm{d}y = \int_{y_j-(b_{yi,j}/2)}^{y_j+(b_{yi,j}/2)} f(\eta)\,\mathrm{d}\eta \tag{4.9}$$

in which η indicates a supplementary variable of y. The n-th derivatives of the extended Dirac functions can therefore be given as

$$\begin{aligned}\int_0^{l_x} D^{(n)}(x - x_i) f(x)\,\mathrm{d}x &= \int_{x_i-(b_{xi,j}/2)}^{x_i+(b_{xi,j}/2)} (-1)^n f^{(n)}(\xi)\,\mathrm{d}\xi \\ \int_0^{l_y} D^{(n)}(y - y_j) f(y)\,\mathrm{d}y &= \int_{y_j-(b_{yi,j}/2)}^{y_j+(b_{yi,j}/2)} (-1)^n f^{(n)}(\eta)\,\mathrm{d}\eta\end{aligned} \tag{4.10}$$

in which superscripts enclosed within parentheses represent the differential order.

When the conditions $b_{xi,j} \ll l_x$ and $b_{yi,j} \ll l_y$ are satisfied, the extended Dirac functions $D(x - x_i)$ and $D(y - y_j)$ are approximately related to the Dirac delta functions $\delta(x - x_i)$ and $\delta(y - y_j)$ in the following way:

$$D(x - x_i) \approx b_{xi,j}\delta(x - x_i); \quad D(y - y_i) \approx b_{yi,j}\delta(y - y_i) \tag{4.11}$$

The author [11, 12] has demonstrated the effectiveness of the extended Dirac function for beam problems. The details and employment of the Dirac delta function are given in Mikusinski and Sikorski [13] and Frýba [14], respectively.

Now, the substitution of Eq. (4.7) into Eq. (4.6) gives

$$\int z^2\,\mathrm{d}z = \frac{1}{12}\left[h_0^3 - \sum_{i=1}^{m^*}\sum_{j=1}^{n^*} (h_{i,j})^3 D(x - x_i) D(y - y_j)\right] \tag{4.12}$$

Hence, Eq. (4.5) may be written as

$$M_x = -D_0\,\mathrm{d}(x, y)\left[w_{,xx} + \nu w_{,yy}\right] \tag{4.13}$$

in which D_0 is the flexural rigidity of a solid plate neglecting voids and is given by $Eh_0^3/12(1 - \upsilon^2)$, and the coefficient of the flexural rigidity, $\mathrm{d}(x, y)$, is given by

$$\mathrm{d}(x, y) = 1 - \sum_{i=1}^{m^*}\sum_{j=1}^{n^*} \alpha_{i,j} D(x - x_i)\, D(y - y_j) \tag{4.14}$$

in which $\alpha_{i,j}$ is defined as

$$\alpha_{i,j} = \left(\frac{h_{i,j}}{h_0}\right)^3 \tag{4.15}$$

Similarly, the bending moment M_y and twisting moment M_{xy} can be expressed as

$$M_y = -D_0 \mathrm{d}(x, y)\left[w_{,yy} + \nu w_{,xx}\right] \tag{4.16}$$

$$M_{xy} = -(1-\nu) D_0 \, \mathrm{d}(x, y) w_{,xy} \tag{4.17}$$

On the other hand, the curvature changes κ_x, κ_y, and κ_{xy}, as defined in Eq. (1.8), are

$$\kappa_x = -w_{,xx}\,; \quad \kappa_y = -w_{,yy}\,; \quad \kappa_{xy} = -w_{,xy} \tag{4.18}$$

Hence, substituting Eqs. (4.13) and (4.16)–(4.18) into Eq. (4.2), the strain energy U becomes

$$U = \frac{D_0}{2} \iint \mathrm{d}(x, y)\left[(w_{,xx})^2 + (w_{,yy})^2 + 2\nu w_{,xx}\, w_{,yy} + 2(1-\nu)(w_{,xy})^2\right] \mathrm{d}x\, \mathrm{d}y \tag{4.19}$$

Next, the potential energy V produced by the external lateral loads p may be expressed as

$$V = -\iint p w \, \mathrm{d}x\, \mathrm{d}y \tag{4.20}$$

The kinetic energy T neglected the effect of rotatory inertia which is given as

$$T = \frac{1}{2} \iint \rho\, h(x, y)\, (\dot{w})^2 \, \mathrm{d}x\, \mathrm{d}y \tag{4.21}$$

in which ρ is the mass density of the plate with void. The real thickness, which deducts the height of a void from the total height, h_0, of the plate, is functions of x and y. So, $h(x, y)$ is expressed as

$$h(x, y) = h_0\, \alpha_h(x, y) \tag{4.22}$$

in which the notation $\alpha_h(x, y)$ is defined as

$$\alpha_h(x, y) = 1 - \sum_{i=1} \sum_{j=1} \left(\frac{h_{i,j}}{h_0}\right) D(x - x_i) D(y - y_i) \tag{4.23}$$

Substitution of Eqs. (4.19)–(4.21) into Eq. (4.1) yields

$$\delta I = \int_{t_1}^{t_2} \left\{ \iint F_1\, \delta w\, \mathrm{d}x\, \mathrm{d}y + \int F_2\, \delta w \Big|_0^{l_x} \mathrm{d}y - \int F_3\, \delta w_{,x} \Big|_0^{l_x} \mathrm{d}y \right.$$
$$\left. + \int F_4 \delta w \Big|_0^{l_y} \mathrm{d}x - \int F_5 \delta\, w_{,y} \Big|_0^{l_y} \mathrm{d}x - 2(1-\nu) F_6 \delta\, w \Big|_0^{l_x} \Big|_0^{l_y} \right\} \mathrm{d}t = 0 \quad (4.24)$$

in which $F_1 - F_6$ are given by the LHS of Eq. (4.25) and each second equation of (4.26)–(4.29). Here, l_x and l_y are the span lengths in the x and y directions of the plate, respectively.

From Eq. (4.24), and adding the viscus damping term with a damping coefficient c, the differential equation of motion can be given as

$$\rho h(x,y)\ddot{w} + c\dot{w} + D_0\big[\mathrm{d}(x,y)w_{,xx} + \nu \mathrm{d}w_{,yy}\big]_{,xx}$$
$$+ D_0\big[\mathrm{d}(x,y)w_{,yy} + \nu \mathrm{d}(x,y)w_{,xx}\big]_{,yy} + 2(1-\nu)D_0\big[\mathrm{d}(x,y)w_{,xy}\big]_{,xy} = p(x,y,t) \quad (4.25)$$

together with the associated boundary conditions

$$w = 0 \quad \text{or} \quad D_0[\mathrm{d}w_{,xx}]_{,x} + \nu D_0[\mathrm{d}w_{,yy}]_{,x} + 2(1-\nu)D_0[\mathrm{d}w_{,xy}]_{,y} = 0 \quad (4.26)$$
$$w_{,x} = 0 \quad \text{or} \quad D_0[\mathrm{d}w_{,xx} + \nu \mathrm{d}w_{,yy}] = 0 \quad (4.27)$$

at $x = 0$ and l_x; and

$$w = 0 \quad \text{or} \quad D_0[\mathrm{d}w_{,yy}]_{,y} + \nu D_0[\mathrm{d}w_{,xx}]_{,y} + 2(1-\nu)D_0[\mathrm{d}w_{,xy}]_{,x} = 0 \quad (4.28)$$
$$w_{,y} = 0 \quad \text{or} \quad D_0[\mathrm{d}w_{,yy} + \nu \mathrm{d}w_{,xx}] = 0 \quad (4.29)$$

at $y = 0$ and l_y; and

$$w = 0 \quad \text{or} \quad D_0 \mathrm{d}w_{,xy} = 0 \quad (4.30)$$

at the corners.

For solid plates without voids, the coefficient $\mathrm{d}(x, y)$ becomes 1 and the coefficient $h(x, y)$ is h_0 so that the governing equations proposed here reduce to the general Eq. (1.87) for rectangular solid (normal) plates.

4.3 Static Analyses to Rectangular Plates with Voids

In the preceding section, the governing equations for rectangular plates with voids have been proposed in general form. Then, we consider the static solutions for simply supported and clamped plates by means of the Galerkin method. If a suitable shape function for arbitral boundary conditions is used, the methodology demonstrated in this chapter is applicable to a rectangular thin plate with voids supported by general boundary conditions. The deflections $w(x, y)$ can be expressed by a power series expansion as follows:

$$w(x, y) = \sum_{m=1} \sum_{n=1} w_{mn} f_{mn}(x, y) \tag{4.31}$$

in which f_{mn} are shape functions satisfying the specified boundary conditions. The following functions represent f_{mn} for simply supported and clamped plate

$$\left.\begin{array}{l} f_{mn}(x, y) = \sin \dfrac{m\pi x}{l_x} \sin \dfrac{n\pi x}{l_y} \quad \text{(for simply supported plates)} \\ f_{mn}(x, y) = \sin \dfrac{\pi x}{l_x} \sin \dfrac{m\pi x}{l_x} \sin \dfrac{\pi y}{l_y} \sin \dfrac{n\pi y}{l_y} \quad \text{(for clamped plates)} \end{array}\right\} \tag{4.32}$$

As stated in Chap. 1, the Galerkin equation for the static problems can be expressed as

$$\int_0^{l_x} \int_0^{l_y} Q \cdot \delta w \, \mathrm{d}x \, \mathrm{d}y = 0 \tag{4.33}$$

in which the notation Q indicates the equation neglecting the inertia term in Eq. (4.25). Substituting Eq. (4.31) into Eq. (4.33), the Galerkin equations are obtained as

$$\begin{aligned} \delta w_{\bar{m}\bar{n}} : \sum_{m=1} \sum_{n=1} w_{mn} \int_0^{l_x} \int_0^{l_y} & \{[\mathrm{d} f_{mn,xx}]_{,xx} + [\mathrm{d} f_{mn,yy}]_{,yy} + \nu[\mathrm{d} f_{mn,yy}]_{,xx} \\ & + \nu[\mathrm{d} f_{mn,xx}]_{,yy} + 2(1-\nu)[\mathrm{d} f_{mn,xy}]_{,xy}\} f_{\bar{m}\bar{n}} \, \mathrm{d}x \, \mathrm{d}y = \int_0^{l_x} \int_0^{l_y} \frac{p}{D_0} f_{\bar{m}\bar{n}} \, \mathrm{d}x \, \mathrm{d}y \end{aligned} \tag{4.34}$$

To simplify, we assume the lateral loads p are a uniform load p_0. Substituting Eq. (4.32) into Eq. (4.34), the Galerkin equations reduce to a system of linear algebraic equations with respect to the displacement coefficients w_{mn}, viz.

$$\delta w_{\bar{m}\bar{n}} : \sum_{m=1} \sum_{n=1} w_{mn} A_{\bar{m}\bar{n}mn} = B_{\bar{m}\bar{n}} \tag{4.35}$$

In the above system of linear algebraic equations, the row is given by the subscripts $\bar{m}$ and $\bar{n}$ and the column by the subscripts m and n.

By substituting Eq. (4.32) for Eq. (4.34) and calculating it concretely, $A_{\bar{m}\bar{n}mn}$ and $B_{\bar{m}\bar{n}}$ are obtained as follows: For simply supported plates with voids, $A_{\bar{m}\bar{n}mn}$ and $B_{\bar{m}\bar{n}}$ are given by

$$\begin{aligned}
A_{\bar{m}\bar{n}mn} &= \pi^4\left[m^2+\left(\frac{n}{\alpha}\right)^2\right]^2\delta_{m\bar{m}}\delta_{n\bar{n}} - \sum_{i=1}\sum_{j=1} 4\pi^4\alpha_{i,j}\Big[F_{xss}(m,\bar{m};i)F_{yss}(n,\bar{n};j) \\
&\times\left[m^2+\left(\frac{n}{\alpha}\right)^2\right]^2 - 2m\left[m^2+\left(\frac{n}{\alpha}\right)^2\right][mF_{xss}(m,\bar{m};i)-\bar{m}F_{xcc}(m,\bar{m};i)]F_{yss}(n,\bar{n};j) \\
&-2\left[\left(\frac{n}{\alpha}\right)^3+m^2\frac{n}{\alpha}\right]\left[\frac{n}{\alpha}F_{yss}(n,\bar{n};j)-\frac{\bar{n}}{\alpha}F_{ycc}(n,\bar{n};j)\right]F_{xss}(m,\bar{m};i) \\
&+\left[m^2+\nu\left(\frac{n}{\alpha}\right)^2\right]\Big\{[m^2+\bar{m}^2]F_{xss}(m,\bar{m};i)-2m\bar{m}F_{xcc}(m,\bar{m};i)\Big\}F_{yss}(n,\bar{n};j) \\
&+\left[\left(\frac{n}{\alpha}\right)^2+\nu m^2\right]\left\{\left[\left(\frac{n}{\alpha}\right)^2+\left(\frac{\bar{n}}{\alpha}\right)^2\right]F_{yss}(n,\bar{n};j)-2\frac{n}{\alpha}\frac{\bar{n}}{\alpha}F_{ycc}(n,\bar{n};j)\right\}F_{xss}(m,\bar{m};i) \\
&+2(1-\nu)m\frac{n}{\alpha}[mF_{xss}(m,\bar{m};i)-\bar{m}F_{xcc}(m,\bar{m};i)] \\
&\times\left[\frac{n}{\alpha}F_{yss}(n,\bar{n};j)+\frac{\bar{n}}{\alpha}F_{ycc}(n,\bar{n};j)\right]\Big]
\end{aligned} \tag{4.36}$$

$$B_{\bar{m}\bar{n}} = \begin{cases} \frac{16}{\bar{m}\bar{n}\pi^2}\left(\frac{p_0 l_x^4}{D_0}\right) & \text{for odd } \bar{m},\bar{n} \\ 0 & \text{otherwise} \end{cases} \tag{4.37}$$

in which $\delta_{m\bar{m}}$ and $\delta_{n\bar{n}}$ are the Kronecker deltas; α is the aspect ratio $\alpha = l_y/l_x$ of the span lengths; and the notations $F_{xss}(m,\bar{m};i)$ and $F_{xcc}(m,\bar{m};i)$ are defined as

$$\begin{aligned}
\begin{Bmatrix} F_{xss}(m,\bar{m};i) \\ F_{xcc}(m,\bar{m};i) \end{Bmatrix} &= \frac{1}{l_x}\int_{x_i-(b_{xi,j}/2)}^{x_i+(b_{xi,j}/2)} \begin{Bmatrix} \sin\left(\frac{m\pi x}{l_x}\right)\sin\left(\frac{\bar{m}\pi x}{l_x}\right) \\ \cos\left(\frac{m\pi x}{l_x}\right)\cos\left(\frac{\bar{m}\pi x}{l_x}\right) \end{Bmatrix} dx \\
&= \frac{1}{(m-\bar{m})\pi}\cos\frac{(m-\bar{m})\pi x_i}{l_x}\sin\frac{(m-\bar{m})\pi b_{xi,j}}{2l_x}(1-\delta_{m\bar{m}}) \\
&+\frac{1}{2}\left(\frac{b_{xi,j}}{l_x}\right)\delta_{m\bar{m}} \mp \frac{1}{(m+\bar{m})\pi}\cos\frac{(m+\bar{m})\pi x_i}{l_x}\sin\frac{(m+\bar{m})\pi b_{xi,j}}{2l_x}
\end{aligned} \tag{4.38}$$

The notations $F_{yss}(n,\bar{n};j)$ and $F_{ycc}(n,\bar{n};j)$ are obtained by transforming $m \to n$, $\bar{m} \to \bar{n}$, $x_i \to y_j$, $b_{xi,j} \to b_{yi,j}$, and $l_x \to l_y$ in Eq. (4.38).

On the other hand, the expressions for $A_{\bar{m}\bar{n}mn}$ and $B_{\bar{m}\bar{n}}$ for clamped plates are expressed as

$$
\begin{aligned}
A_{\bar{m}\bar{n}mn} = {} & \pi^4\left[\alpha F_{m\bar{m}}(4,0)F_{n\bar{n}}(0,0) + \frac{2}{\alpha}F_{m\bar{m}}(2,0)F_{n\bar{n}}(2,0) + \left(\frac{1}{\alpha}\right)^3 F_{m\bar{m}}(0,0)F_{n\bar{n}}(4,0)\right] \\
& - \sum_{i=1}\sum_{j=1}\pi^4\alpha_{i,j}\Bigg[\alpha F_{m\bar{m}}(4,0;i)F_{n\bar{n}}(0,0;j) + \frac{2}{\alpha}F_{m\bar{m}}(2,0;i)F_{n\bar{n}}(2,0;j) \\
& + \left(\frac{1}{\alpha}\right)^3 F_{m\bar{m}}(0,0;i)F_{n\bar{n}}(4,0;j) - 2\alpha[F_{m\bar{m}}(4,0;i) + F_{m\bar{m}}(3,1;i)]F_{n\bar{n}}(0,0;j) \\
& - 2\left(\frac{1}{\alpha}\right)^3[F_{n\bar{n}}(4,0;j) + F_{n\bar{n}}(3,1;j)]F_{m\bar{m}}(0,0;i) \\
& + \alpha[F_{m\bar{m}}(4,0;i) + 2F_{m\bar{m}}(3,1;i) + F_{m\bar{m}}(2,2;i)]F_{n\bar{n}}(0,0;j) \\
& + \left(\frac{1}{\alpha}\right)^3[F_{n\bar{n}}(4,0;j) + 2F_{n\bar{n}}(3,1;j) + F_{n\bar{n}}(2,2;j)]F_{m\bar{m}}(0,0;i) \\
& - \frac{2}{\alpha}\{[F_{m\bar{m}}(2,0;i) + F_{m\bar{m}}(1,1;i)]F_{n\bar{n}}(2,0;j) \\
& + [F_{n\bar{n}}(2,0;j) + F_{n\bar{n}}(1,1;j)]F_{m\bar{m}}(2,0;i)\} \\
& + \frac{2(1-\nu)}{\alpha}[F_{m\bar{m}}(2,0;i) + F_{m\bar{m}}(1,1;i)][F_{n\bar{n}}(2,0;j) + F_{n\bar{n}}(1,1;j)] \\
& + \frac{\nu}{\alpha}[F_{m\bar{m}}(2,0;i) + 2F_{m\bar{m}}(1,1;i) + F_{m\bar{m}}(0,2;i)]F_{n\bar{n}}(2,0;j) \\
& + [F_{n\bar{n}}(2,0;j) + 2F_{n\bar{n}}(1,1;j) + F_{n\bar{n}}(0,2;j)]F_{m\bar{m}}(2,0;i)\Bigg]
\end{aligned}
\tag{4.39}
$$

$$
B_{\bar{m}\bar{n}} = \frac{1}{4}\delta_{\bar{m}1}\delta_{\bar{n}1}\alpha\left[\frac{P_0 l_x^4}{D_0}\right] \tag{4.40}
$$

in which $F_{m\bar{m}}(0,0)$, $F_{m\bar{m}}(2,0)$, …, $F_{m\bar{m}}(0,0;i)$… are expressed in general form by

$$
\left.
\begin{aligned}
F_{m\bar{m}}(k_1,k_2) &= (l_x)^{k_1+k_2-1}\int_0^{l_x} f_{xm}^{(k_1)} f_{x\bar{m}}^{(k_2)}\,\mathrm{d}x \\
F_{n\bar{n}}(k_1,k_2) &= (l_y)^{k_1+k_2-1}\int_0^{l_y} f_{yn}^{(k_1)} f_{y\bar{n}}^{(k_2)}\,\mathrm{d}y \\
F_{m\bar{m}}(k_1,k_2;i) &= (l_x)^{k_1+k_2-1}\int_0^{l_x} D(x-x_i) f_{xm}^{(k_1)} f_{x\bar{m}}^{(k_2)}\,\mathrm{d}x \\
&= (l_x)^{k_1+k_2-1}\int_{x_i-(b_{xi,j})/2}^{x_i+(b_{xi,j})/2} f_{xm}^{(k_1)}(\xi) f_{x\bar{m}}^{(k_2)}(\xi)\,\mathrm{d}\xi \\
F_{n\bar{n}}(k_1,k_2;j) &= (l_y)^{k_1+k_2-1}\int_{y_j-(b_{yi,j})/2}^{y_j+(b_{yi,j})/2} f_{yn}^{(k_1)}(\eta) f_{y\bar{n}}^{(k_2)}(\eta)\,\mathrm{d}\eta
\end{aligned}
\right\}
\tag{4.41}
$$

in which f_{xm} and f_{yn} are x and y components of the shape function given in Eq. (4.32), viz.

$$f_{xm} = \sin\frac{\pi x}{l_x}\sin\frac{m\pi x}{l_x}; \quad f_{yn} = \sin\frac{\pi y}{l_y}\sin\frac{n\pi y}{l_y} \tag{4.42}$$

Thus, solving Eq. (4.35) for the unknown displacement coefficients w_{mn} and substituting them into Eq. (4.31), the deflections w are obtained. The integrals involving the extended Dirac functions in Eqs. (4.36) and (4.39) have been rigorously calculated on the basis of Eqs. (4.8), (4.9), and (4.10).

However, when the width of each void is small compared with the corresponding span length, the integral calculation is rapidly simplified by the use of the approximate relations given in Eq. (4.11). For example, Eqs. (4.38) and (4.41) can be simplified as follows:

$$\left.\begin{aligned} F_{xss}(m,\bar{m};i) &\approx b_{xi,j}\sin\left(\tfrac{m\pi x_i}{l_x}\right)\sin\left(\tfrac{\bar{m}\pi x_i}{l_x}\right) \\ F_{xcc}(m,\bar{m};i) &\approx b_{xi,j}\cos\left(\tfrac{m\pi x_i}{l_x}\right)\cos\left(\tfrac{\bar{m}\pi x_i}{l_x}\right) \\ F_{m\bar{m}}(k_1,k_2;i) &\approx b_{xi,j}(l_x)^{k_1+k_2-1} f_{xm}^{(k_1)}(x_i) f_{x\bar{m}}^{(k_2)}(x_i) \\ F_{n\bar{n}}(k_1,k_2;j) &\approx b_{yi,j}(l_y)^{k_1+k_2-1} f_{yn}^{(k_1)}(y_j) f_{y\bar{n}}^{(k_2)}(y_j) \end{aligned}\right\} \tag{4.43}$$

Next, we consider uncoupled approximate solutions from the point of the practical use. Although the behavior of plates with voids is affected by all the terms of the square matrix $A_{\bar{m}\bar{n}mn}$, the behavior is considered to be dominated by the diagonal terms in the matrix $A_{\bar{m}\bar{n}mn}$. This effect becomes prominent by using an appropriate shape function. Hence, taking into consideration only the diagonal terms of $A_{\bar{m}\bar{n}mn}$, Eq. (4.35) becomes of uncoupled form. Thus, the approximate solutions of w_{mn} are given as

$$w_{mn} \approx \frac{B_{mn}}{A_{mnmn}} \tag{4.44}$$

The bending moments M_x and M_y and twisting moment M_{xy} are given by substituting the deflections w into Eqs. (4.13), (4.16), and (4.17), respectively. The transverse shearing forces Q_x and Q_y and vertical edge forces V_x and V_y per unit length are given by

$$Q_x = M_{x,x} + M_{xy,y}; \quad Q_y = M_{y,y} + M_{xy,x} \tag{4.45}$$

$$V_x = Q_x + M_{xy,y}; \quad V_y = Q_y + M_{yx,x} \tag{4.46}$$

Here, the differential $M_{x,x}$ is calculated as

$$M_{x,x} = -D_0\left\{\mathrm{d}(x,y)_{,x}\left[w_{,xx} + \nu w_{,yy}\right] + \mathrm{d}(x,y)\left[w_{,xxx} + \nu w_{,xyy}\right]\right\} \tag{4.47}$$

From Eq. (4.14), the differential of d(x, y) with respect to x is expressed as

$$\mathrm{d}(x, y)_{,x} = -\sum_{i=1}\sum_{j=1}\alpha_{i,j} D(x - x_i)_{,x} D(y - y_j) \tag{4.48}$$

From Sinozaki et al. [15], the integration involving the differential of the Dirac function is expressed by

$$\int \delta(x - x_i)_{,x} f(x)\,\mathrm{d}x = -\int \delta(x - x_i) f(x)_{,x}\,\mathrm{d}x \tag{4.49}$$

Differentiating the above equation with respect to x yields

$$\delta(x - x_i)_{,x} f(x) = -\delta(x - x_i) f(x)_{,x} \tag{4.50}$$

For the extended Dirac function, Eq. (4.50) may be extended as

$$D(x - x_i)_{,x} f(x) = -D(x - x_i) f(x)_{,x} \tag{4.51}$$

Similarly,

$$D(y - y_j)_{,y} f(y) = -D(y - y_j) f(y)_{,y} \tag{4.52}$$

The substitution of Eqs. (4.14), (4.48), and (4.51) into Eq. (4.47) results in

$$M_{x,x} = -D_0\left[w_{,xxx} + \nu w_{,xyy}\right] \tag{4.53}$$

The result is not affected directly by the extended Dirac functions. Similarly, the following relations can be obtained as

$$\left.\begin{aligned} M_{y,y} &= -D_0\left[w_{,yyy} + \nu w_{,xxy}\right] \\ M_{xy,x} &= -(1 - \nu) D_0 w_{,xxy} \\ M_{xy,y} &= -(1 - \nu) D_0 w_{,xyy} \end{aligned}\right\} \tag{4.54}$$

Thus, the transverse shearing forces Q_x and Q_y and the vertical edge forces V_x and V_y become

$$\left.\begin{aligned} Q_x &= -D_0\left[w_{,xxx} + w_{,xyy}\right]; \quad Q_y = -D_0\left[w_{,yyy} + w_{,xxy}\right] \\ V_x &= -D_0\left[w_{,xxx} + 2w_{,xyy} - \nu w_{,xyy}\right]; \quad V_y = -D_0\left[w_{,yyy} + 2w_{,xxy} - \nu w_{,xxy}\right] \end{aligned}\right\} \tag{4.55}$$

4.4 Numerical Results

In the preceding section, the static solutions for simply supported and clamped plates with voids have been presented by means of the Galerkin method. All external terms $B_{\bar{m}\bar{n}}$ given in Eqs. (4.37) and (4.40) have the same dimension, $p_0 l_x^4/D_0$. Hence, the displacements w, stress couples M_x, M_y, and M_{xy}, stress resultants Q_x and Q_y, V_x, and V_y can be expressed in nondimensioned forms by taking $p_0 l_x^4/D_0$, $p_0 l_x^2$, and $p_0 l_x$ as the units, respectively.

Then, in order to examine the effectiveness of the proposed solutions, numerical calculations are carried out for three cases as shown in Table 4.1, in which Poisson's ratio is 0.17. Figures 4.3, 4.4, 4.5, 4.6, 4.7 and 4.8 show the deflections and bending moments M_y for the three cases of simply supported and clamped plates with voids, respectively. It is shown from the numerical results that the differences between the rigorous solutions based on Eq. (4.35) and the approximate solutions based on Eq. (4.44) are negligible in practice. The results obtained from the Galerkin method show good agreement with the results obtained from the finite element method. The finite element method used here is based on isotropic and rectangular plate elements due to Adini-Clough-Melosh, as given by Rao [16] and Ugural [17], in which an element with voids includes the effect of the voids, and is independent of FEM based on equivalent orthotropic plate theory as given by Hinton and Owen [18]. In addition, the numerical results obtained from the equivalent plate analogy proposed by Crisfield and Twemlow [1] are close to the numerical results of the Galerkin method. However, it is clear that the equivalent plate analogy cannot give good results for all cases and that, especially, the values obtained for the bending moments M_y indicate mean values including the effect of the local rigidity due to voids. This point must be taken into consideration in designs using the equivalent plate analogy.

4.5 Relationships Between Theoretical and Experimental Results

In order to experimentally examine the theory proposed here, static experiments for acrylic plates with voids were carried out for simply supported and clamped plates. The experimental equipment is demonstrated, in outline, in Fig. 4.9, in which the span lengths $l_x = l_y = 30$ cm. Although the positions of the voids in the specimens are the same as the voided plates shown in Table 4.1, used in the numerical calculations mentioned above, the thickness and the ratios of void size, $h_{i,j}/h_0$, $b_{xi,j}/l_x$, and $b_{yi,j}/l_y$, take the following values:

Type 0: $h_0 = 0.6$ cm
Type 1: $h_0 = 0.6$ cm, $h_{i,j}/h_0 = 0.33$, $b_{xi,j}/l_x = 0.1$, $b_{yi,j}/l_y = 0.5$
Type 2: $h_0 = 0.6$ cm, $h_{i,j}/h_0 = 0.33$, $b_{xi,j}/l_x = 0.1$, $b_{yi,j}/l_y = 1.0$
Type 3: $h_0 = 0.6$ cm, $h_{i,j}/h_0 = 0.33$, $b_{xi,j}/l_x = 0.1$, $b_{yi,j}/l_y = 0.1$.

Table 4.1 Lists of isotropic rectangular plates with voids [11]

Type	Plane	Section	$\frac{h_{i,j}}{h_0}$	$\frac{b_{xi,j}}{l_x}$	$\frac{b_{yi,j}}{l_y}$	$\alpha = \frac{l_y}{l_x}$
1	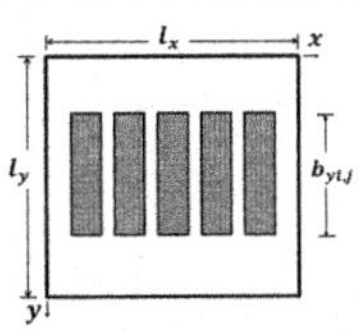	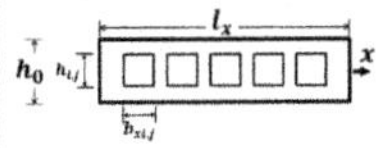	0.5	0.1	0.5	1.0
2	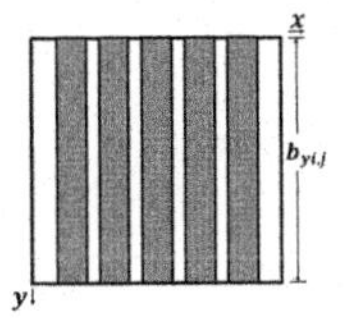	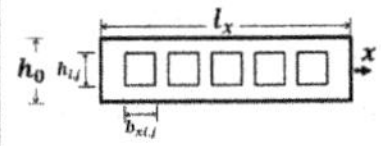	0.5	0.1	1.0	1.0
3	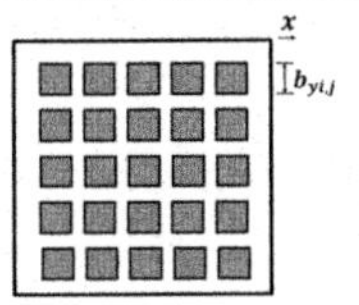	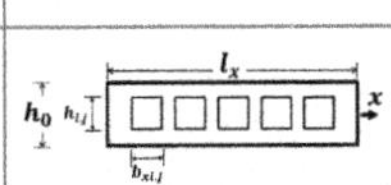	0.5	0.1	0.1	1.0

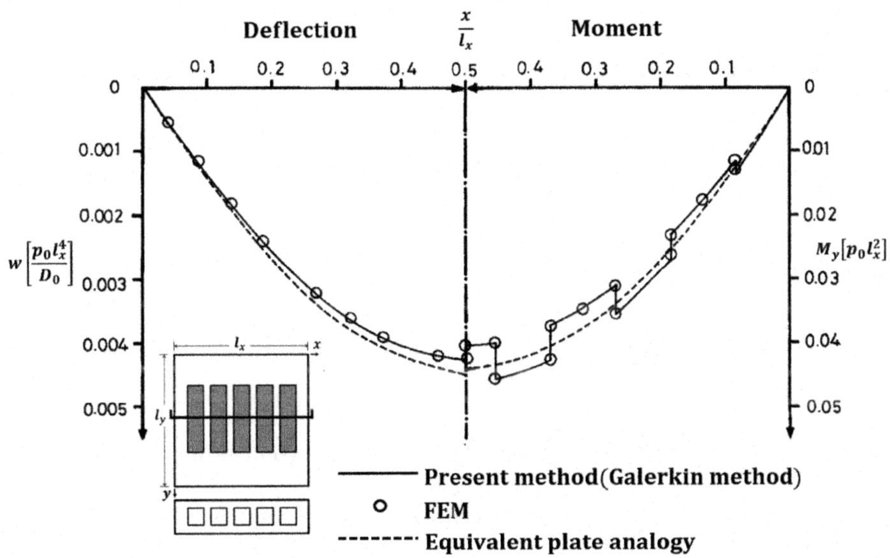

Fig. 4.3 w and M_y for a simply supported plate with voids (Type 1) [11]

Young's modulus and Poisson's ratio of the acrylic plates used are $E = 320{,}700$ N/cm^2 and $\nu = 0.34$, respectively. Experiments for plates without voids, called Type 0, were carried out in order to examine the experimental equipment used. The experimental results showed good agreement with the theoretical results in the

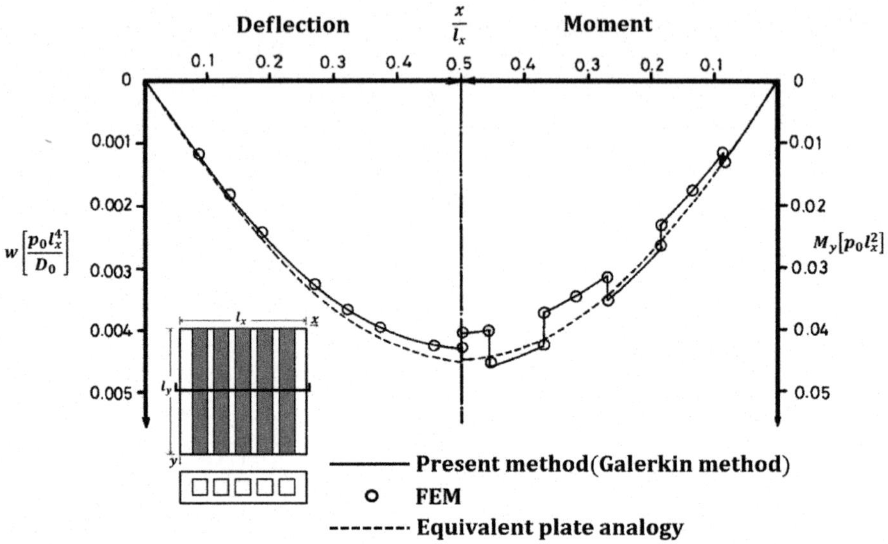

Fig. 4.4 w and M_y for a simply supported plate with voids (Type 2) [11]

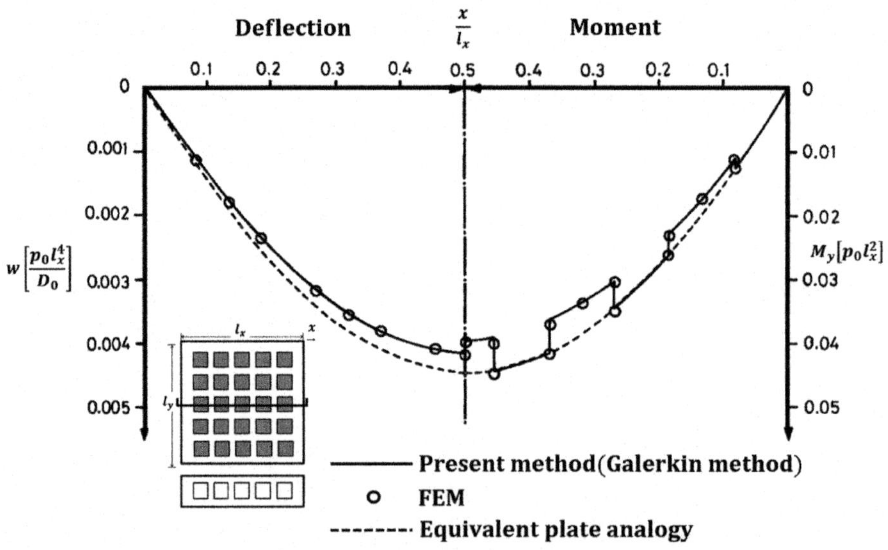

Fig. 4.5 w and M_y for a simply supported plate with voids (Type 3) [11]

linear region, as shown in Fig. 4.10. The relationships between the deflections at the midpoint of the specimens and the lateral uniform load per unit area are shown in Figs. 4.11, 4.12, and 4.13. It follows from these figures that the theory proposed here shows strong agreement in the linear region. Thus, it is demonstrated that the theory proposed here can be applied practically to plates with voids.

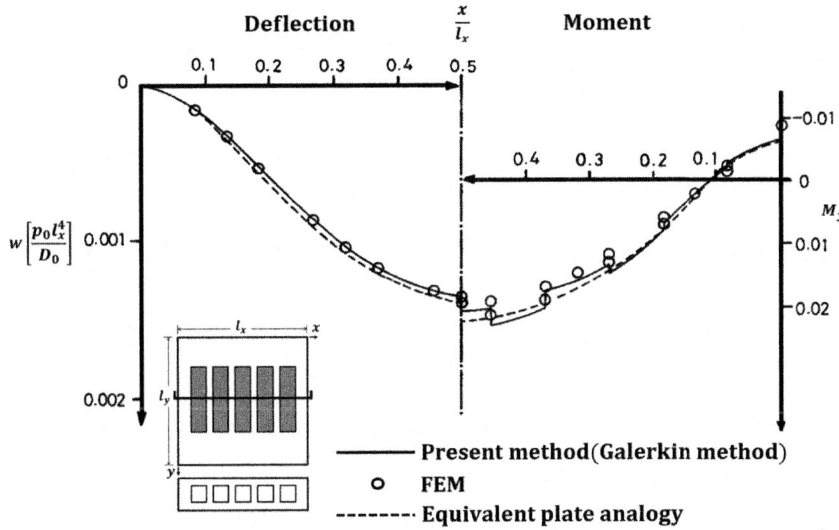

Fig. 4.6 w and M_y for a clamped plate with voids (Type 1) [11]

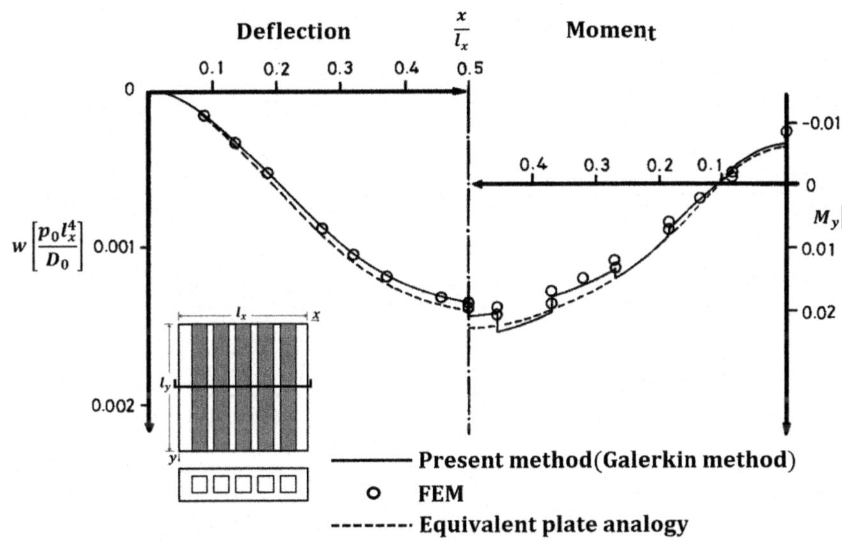

Fig. 4.7 w and M_y for a clamped plate with voids (Type 2) [11]

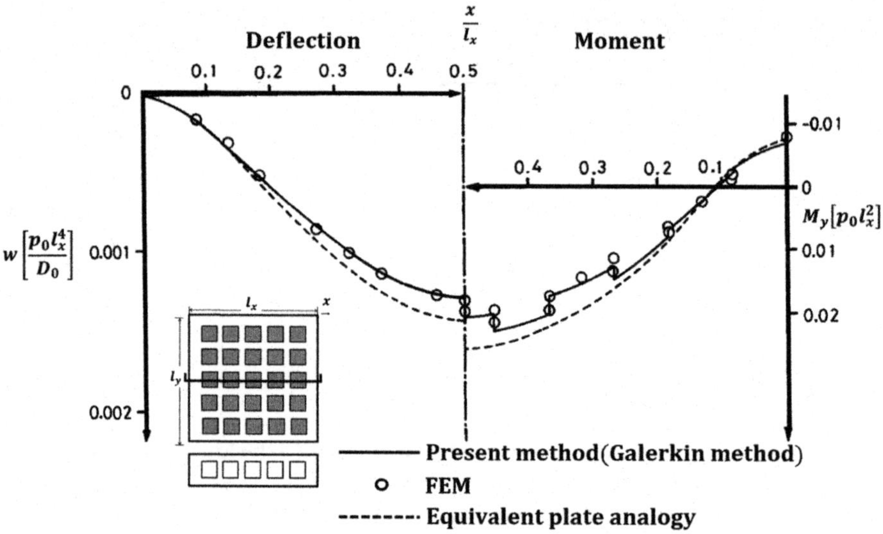

Fig. 4.8 w and M_y for a clamped plate with voids (Type 3) [11]

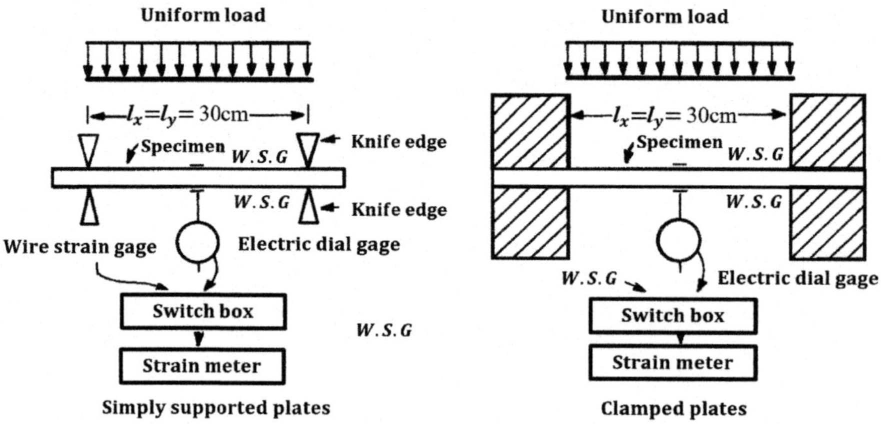

Fig. 4.9 Outline of the experimental equipment [11]

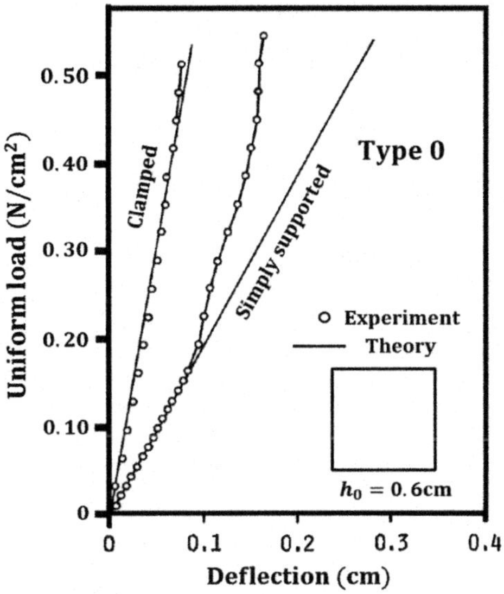

Fig. 4.10 Relationship between the lateral load and deflection (Type 0) (normal plate) [11]

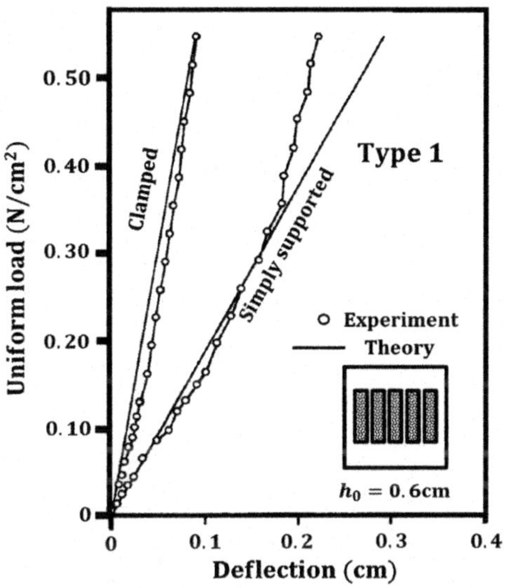

Fig. 4.11 Relationship between the lateral load and deflection (Type 1) [11]

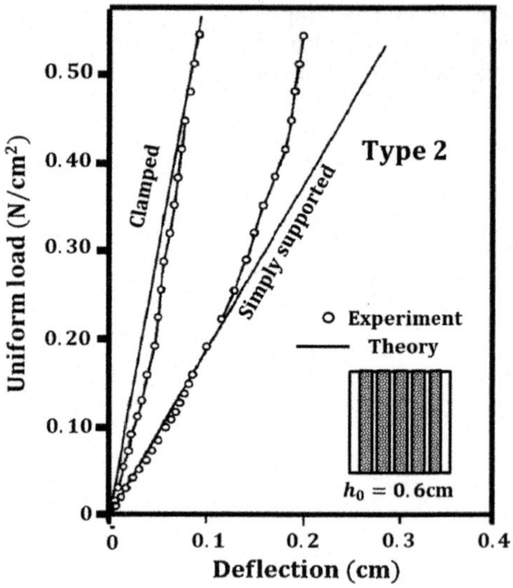

Fig. 4.12 Relationship between the lateral load and deflection (Type 2) [11]

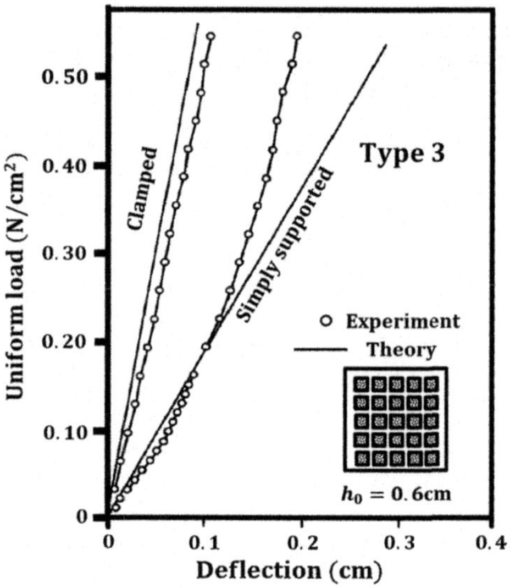

Fig. 4.13 Relationship between the lateral load and deflection (Type 3) [11]

4.6 Free Transverse Vibrations of a Plate with Voids

We consider the free transverse vibrations of rectangular plates with voids. The method of separation of variables for the dynamic deflections w is also employed, assuming that

$$w(x, y, t) = W(x, y)\,\Phi(t) \tag{4.56}$$

in which $W(x, y)$ is a function of x and y, and $\Phi(t)$ is a function of time t. Applying Eq. (4.56) to the equation for free transverse vibrations obtained from Eq. (4.25), the two equations satisfied by $W(x, y)$ and $\Phi(t)$ can be expressed as follows:

$$\ddot{\Phi}(t) + \omega^2\Phi(t) = 0 \tag{4.57}$$

$$\begin{aligned}&[\mathrm{d}W_{,xx}]_{,xx} + [\mathrm{d}W_{,yy}]_{,yy} + \nu[\mathrm{d}W_{,yy}]_{,xx} + \nu[\mathrm{d}W_{,xx}]_{,yy}\\&+ 2(1-\nu)[\mathrm{d}W_{,xy}]_{,xy} - \frac{\omega^2 m_0\alpha_h(x, y)}{D_0} = 0\end{aligned} \tag{4.58}$$

in which ω is a constant. We also use the relation $\rho h(x, y) = m_0\alpha_h(x, y)$, because the mass is variable due to the existence of voids.

The natural frequencies of a plate with voids are presented by means of the Galerkin method. $W(x, y)$ is expressed by a power series expansion as follows:

$$W(x, y) = \sum_{m=1}\sum_{n=1} W_{mn} f_{mn}(x, y) \tag{4.59}$$

in which W_{mn} are the unknown coefficients, and $f_{mn}(x, y)$ are functions satisfying the specified boundary conditions of the plates. The Galerkin equations of Eq. (4.58) can be written as

$$\delta W_{\bar{m}\bar{n}} : \sum_{m=1}\sum_{n=1} W_{mn}\left[A_{\bar{m}\bar{n}mn} - \lambda^2 B_{\bar{m}\bar{n}mn}\right] = 0 \tag{4.60}$$

in which the notations $A_{\bar{m}\bar{n}mn}$, $B_{\bar{m}\bar{n}mn}$ and λ are defined as

$$\begin{aligned}A_{\bar{m}\bar{n}mn} = &\int_0^{l_x}\int_0^{l_y}\Big\{\left[\mathrm{d}f_{mn,xx}\right]_{,xx} + \left[\mathrm{d}f_{mn,yy}\right]_{,yy} + \nu\left[\mathrm{d}f_{mn,yy}\right]_{,xx} + \nu\left[\mathrm{d}f_{mn,xx}\right]_{,yy}\\&+2(1-\nu)\left[\mathrm{d}f_{mn,xy}\right]_{,xy}\Big\} f_{\bar{m}\bar{n}}\,\mathrm{d}x\,\mathrm{d}y\end{aligned} \tag{4.61}$$

$$B_{\bar{m}\bar{n}mn} = \frac{1}{l_x^4}\int_0^{l_x}\int_0^{l_y} \alpha_h(x, y) f_{mn} f_{\bar{m}\bar{n}}\,\mathrm{d}x\,\mathrm{d}y \tag{4.62}$$

$$\lambda^2 = \frac{\omega^2 m_0 l_x^4}{D_0} \tag{4.63}$$

where the integral calculations including the extended Dirac functions, $D(x - x_i)$ and $D(y - y_i)$, are given in Eqs. (4.8)–(4.11). Equation (4.60) is a system of linear, homogeneous, simultaneous algebraic equations with respect to the unknown displacement coefficients W_{mn}. The coefficients $B_{\bar{m}\bar{n}mn}$ appear in diagonal and nondiagonal terms due to the existence of the voids. Solving Eq. (4.60) as eigenvalue problems for λ, the i-th natural frequency ω_i of a plate with voids, corresponding to the i-th value λ_i of λ, is determined from Eq. (4.63) as

$$\omega_i = \lambda_i \frac{1}{l_x^2} \sqrt{\frac{D_0}{m_0}} \tag{4.64}$$

The natural frequencies of a plate with voids are now obtained from the numerical calculation for Eq. (4.60) having the coupled form. Then, we consider the approximate expressions to directly obtain the natural frequencies. Although the natural frequencies are affected by the diagonal and nondiagonal terms in the square matrices $A_{\bar{m}\bar{n}mn}$ and $B_{\bar{m}\bar{n}mn}$, the main behavior is considered to be dominated by only the diagonal terms. Hence, considering only the diagonal terms in $A_{\bar{m}\bar{n}mn}$ and $B_{\bar{m}\bar{n}mn}$, Eq. (4.60) becomes of an uncoupled form. Thus, the approximate values of λ_i for the i-th mode may be obtained as

$$\lambda_i = \sqrt{\frac{B_{mnmn}}{A_{mnmn}}} \tag{4.65}$$

The approximate value for the i-th natural frequency ω_i is obtained by substituting this into Eq. (4.64).

Next, in order to specifically obtain the natural frequencies with respect to simply supported plates with voids, the coefficients $A_{\bar{m}\bar{n}mn}$ and $B_{\bar{m}\bar{n}mn}$ will be concretely shown. For simply supported plates, the following natural functions are used:

$$f_{mn}(x, y) = \sin \frac{m\pi x}{l_x} \sin \frac{n\pi y}{l_y} \tag{4.66}$$

in which the coefficients $A_{\bar{m}\bar{n}mn}$ and $B_{\bar{m}\bar{n}mn}$ are

$$
\begin{aligned}
A_{\bar{m}\bar{n}mn} &= \pi^4\left[m^2+\left(\frac{n}{\alpha}\right)^2\right]^2\delta_{m\bar{m}}\delta_{n\bar{n}} \\
&\quad -\sum_{i=1}\sum_{j=1}4\pi^4\alpha_{i,j}\left[F_{xss}(m,\bar{m};i)F_{yss}(n,\bar{n};j)\left[m^2+\left(\frac{n}{\alpha}\right)^2\right]^2\right. \\
&\quad -2m\left[m^2+\left(\frac{n}{\alpha}\right)^2\right][mF_{xss}(m,\bar{m};i)-\bar{m}F_{xcc}(m,\bar{m};i)]F_{yss}(n,\bar{n};j) \\
&\quad -2\left[\left(\frac{n}{\alpha}\right)^3+m^2\frac{n}{\alpha}\right]\left[\frac{n}{\alpha}F_{yss}(n,\bar{n};j)-\frac{n}{\alpha}F_{ycc}(n,\bar{n};j)\right]F_{xss}(m,\bar{m};i) \\
&\quad +\left[m^2+\nu\left(\frac{n}{\alpha}\right)^2\right]\left\{\left[m^2+\bar{m}^2\right]F_{xss}(m,\bar{m};i)-2m\bar{m}F_{xcc}(m,\bar{m};i)\right\}F_{yss}(n,\bar{n};j) \\
&\quad +\left[\left(\frac{n}{\alpha}\right)^2+\nu m^2\right]\left\{\left[\left(\frac{n}{\alpha}\right)^2+\left(\frac{\bar{n}}{\alpha}\right)^2\right]F_{yss}(n,\bar{n};j)-2\frac{n}{\alpha}\frac{\bar{n}}{\alpha}F_{ycc}(n,\bar{n};j)\right\}F_{xss}(m,\bar{m};i) \\
&\quad +2(1-\nu)m\frac{n}{\alpha}[mF_{xss}(m,\bar{m};i)-\bar{m}F_{xcc}(m,\bar{m};i)] \\
&\quad \left.\times\left[\frac{n}{\alpha}F_{yss}(n,\bar{n};j)+\frac{\bar{n}}{\alpha}F_{ycc}(n,\bar{n};j)\right]\right]
\end{aligned}
\tag{4.67}
$$

$$
B_{\bar{m}\bar{n}mn} = \delta_{\bar{m}m}\delta_{\bar{n}n}-4\sum_{i=1}\sum_{j=1}\left(\frac{h_{i,j}}{h_0}\right)F_{xss}(m,\bar{m};i)F_{yss}(n,\bar{n};j) \tag{4.68}
$$

where $\delta_{m\bar{m}}$ and $\delta_{n\bar{n}}$ are the Kronecker deltas, and the notations $F_{xss}(m,\bar{m};i)$ and $F_{xcc}(m,\bar{m};i)$ are defined as

$$
\begin{aligned}
\left\{\begin{array}{l} F_{xss}(m,\bar{m};i) \\ F_{xcc}(m,\bar{m};i)\end{array}\right\} &= \frac{l}{l_x}\int_{x_i-(b_{xi,j}/2)}^{x_i+(b_{xi,j}/2)}\left\{\begin{array}{l}\sin\left(\frac{m\pi x}{l_x}\right)\sin\left(\frac{\bar{m}\pi x}{l_x}\right) \\ \cos\left(\frac{m\pi x}{l_x}\right)\cos\left(\frac{\bar{m}\pi x}{l_x}\right)\end{array}\right\}\mathrm{d}x \\
&= \frac{1}{(m-\bar{m})\pi}\cos\frac{(m-\bar{m})\pi x_i}{l_x}\sin\frac{(m-\bar{m})\pi b_{xi,j}}{2l_x}(1-\delta_{m\bar{m}}) \\
&\quad +\frac{1}{2}\left(\frac{b_{xi,j}}{l_x}\right)\delta_{m\bar{m}}\mp\frac{1}{(m+\bar{m})\pi}\cos\frac{(m+\bar{m})\pi x_i}{l_x}\sin\frac{(m+\bar{m})\pi b_{xi,j}}{2l_x}
\end{aligned}
\tag{4.69}
$$

The notations $F_{yss}(n,\bar{n};j)$ and $F_{ycc}(n,\bar{n};j)$ are obtained by making the substitutions $m\rightarrow n$, $\bar{m}\rightarrow\bar{n}$, $x_i\rightarrow y_j$, $b_{xi,j}\rightarrow b_{yi,j}$, and $l_x\rightarrow l_y$ in Eq. (4.69).

On the other hand, the natural frequencies for clamped rectangular plates with voids can be obtained similarly by using suitable natural functions for clamped solid plates.

4.7 Numerical Results for Natural Frequencies

The natural frequencies for a plate with voids have been presented by means of the Galerkin method. The effectiveness of the natural frequencies proposed here is examined from numerical computations carried out for three cases, as shown in Table 4.1. All the voids have the same cross section, spaced equally. Figures 4.14 and 4.15 represent the first natural frequencies for simply supported and clamped plates with voids, respectively, in which Poisson's ratio ν is 0.17; the width–span ratios $b_{xi,j}/l_x$ and $b_{yi,j}/l_y$ of the i, j-th void take the values 0.05, 0.1, and 0.15; the ratios $h_{i,j}/h_0$ change from 0 to 0.9; and the aspect ratio $\alpha = l_y/l_x$ is always 1. The value at $h_{i,j}/h_0 = 0$ represents the value for normal solid plates without voids. In these figures, each first natural frequency, ω_1, is divided by the first natural frequency, ω_{01}, of the solid plates. The numerical results show that the differences between the results obtained using Eq. (4.60) and the approximate results obtained using Eq. (4.65) are too small to plot and are negligible in practical use. The results obtained from the proposed theory show excellent agreement when compared with results obtained from the FEM code NASTRAN, in which 22×22 finite elements are used. However, the ratio $h_{i,j}/h_0$ in these figures must be restricted to be smaller than 0.6 due to the use of the Kirchhoff-Love assumption made here. In this calculation, the natural functions for clamped plates with voids use, for the sake of computational simplicity, the following approximate expressions:

$$f_{mn}(x, y) = \sin\left(\frac{\pi x}{l_x}\right)\sin\left(\frac{m\pi x}{l_x}\right)\sin\left(\frac{\pi y}{l_y}\right)\sin\left(\frac{n\pi y}{l_y}\right) \tag{4.70}$$

If the well-known exact natural functions for normal plates are used, the results of the numerical computation for clamped plates with voids will be improved.

Figure 4.16 shows the convergence of the finite elements with mesh refinement on the natural frequencies of the simply supported plate with voids, named Type 3, and the solid plate. It follows that the 22×22 elements used in the above comparison are appropriate for good convergence.

For higher natural frequencies, the proposed theory also shows good agreement with the results obtained from NASTRAN. The higher natural frequencies increase in proportion to the increase in the ratio $h_{i,j}/h_0$. The behavior is the same as the first natural frequency for $h_{i,j}/h_0$ smaller than 0.6, except for the second, fourth, etc., natural frequencies of simply supported plates and clamped plates named Type 1. These differences are caused when the natural modes of Type 1 are asymmetric with respect to a line given by $y = l_y/2$ which passes through the midspan, as shown in Fig. 4.17. For these modes of Type 1, the position of the voids is such that they occur on a part of, but not throughout, the span of the plate with respect to the line $y = l_y/2$. Therefore, the effects of the voids on these natural modes cancel each other, and the increment of these natural frequencies due to the voids is less than for the other modes of Type 1 and other types. Figure 4.18 shows the second natural frequency for simply supported plates with voids.

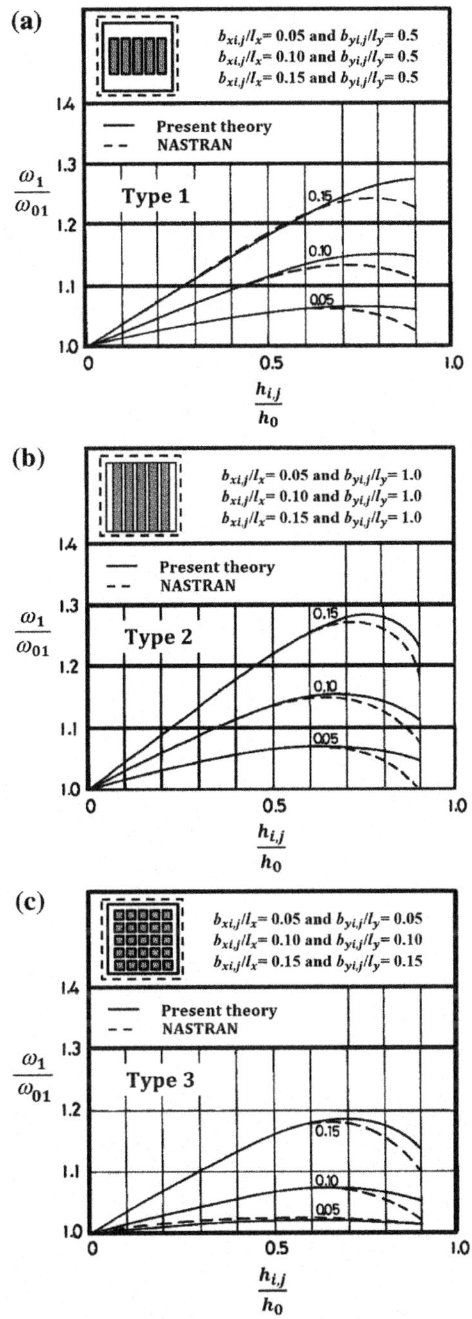

Fig. 4.14 First natural frequency of a simply supported plate with voids. **a** Type 1, **b** Type 2, and **c** Type 3 [12]

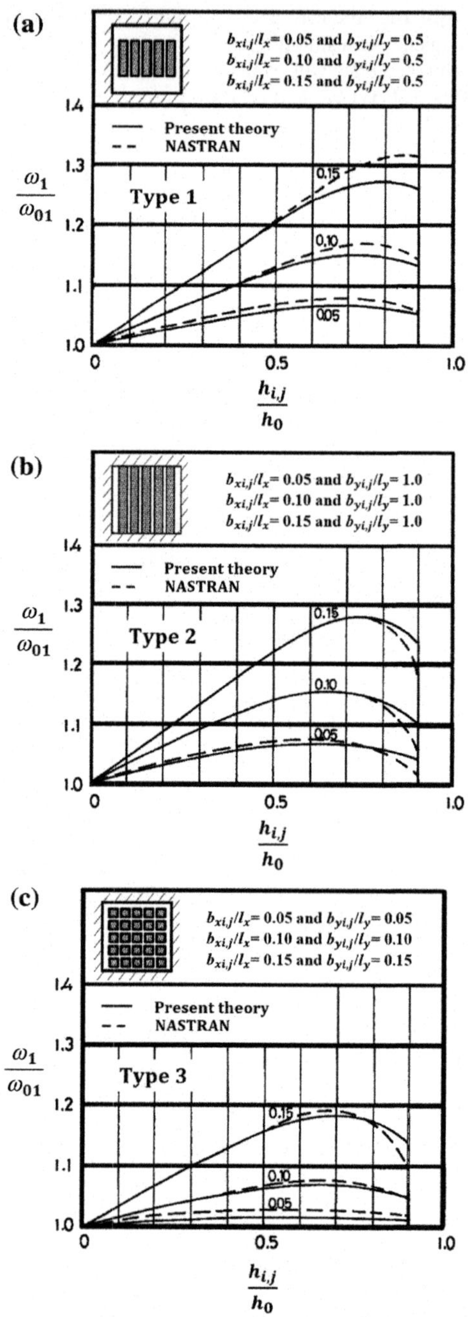

Fig. 4.15 First natural frequency of a clamped plate with voids. **a** Type 1, **b** Type 2, and **c** Type 3 [12]

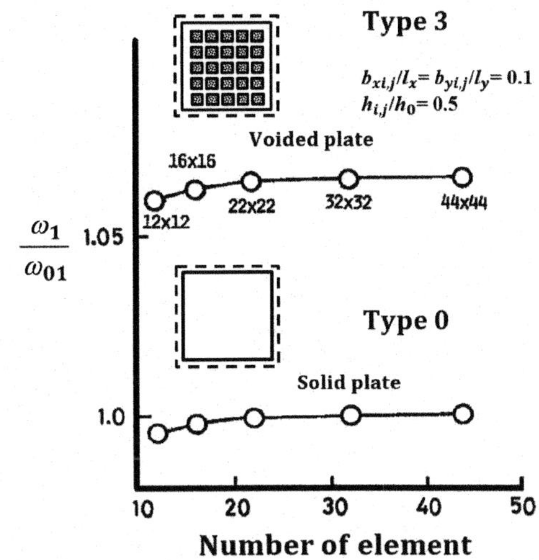

Fig. 4.16 Convergence of finite elements with mesh refinement [12]

4.8 Relationships Between Theoretical Results and Experimental Results for Natural Frequencies

In order to examine experimentally the theory proposed here, experiments of the free vibration for acrylic plates with voids were carried out for both simply supported and clamped boundary conditions.

The experimental equipment is demonstrated schematically in Fig. 4.19. The span lengths $l_x = l_y = 30\,\text{cm}$. The material constants are Young's modulus $E = 320{,}700\ \text{N/cm}^2$; Poisson's ratio $\nu = 0.34$; and the mass density $\rho = 11.76\ \text{kg/cm}^3$. The natural frequencies are calculated from the dynamical deflections at the mid-point of the specimens. The relationships for the first natural frequency between the experimental results and the theoretical results proposed here are shown in Table 4.2. It shows that the proposed theory agrees well with the experimental results. Although the specimens are insufficient in number, it is shown that the proposed theory can apply in practice to plates with voids.

4.9 Forced Vibrations of Plates with Voids

In the preceding sections, the free transverse vibrations of a plate with voids have been presented. Next, we consider forced vibrations of plates with voids. Since the dynamic deflection w is a function of x, y, and t, the general solution of Eq. (4.25) is also assumed to be of the form

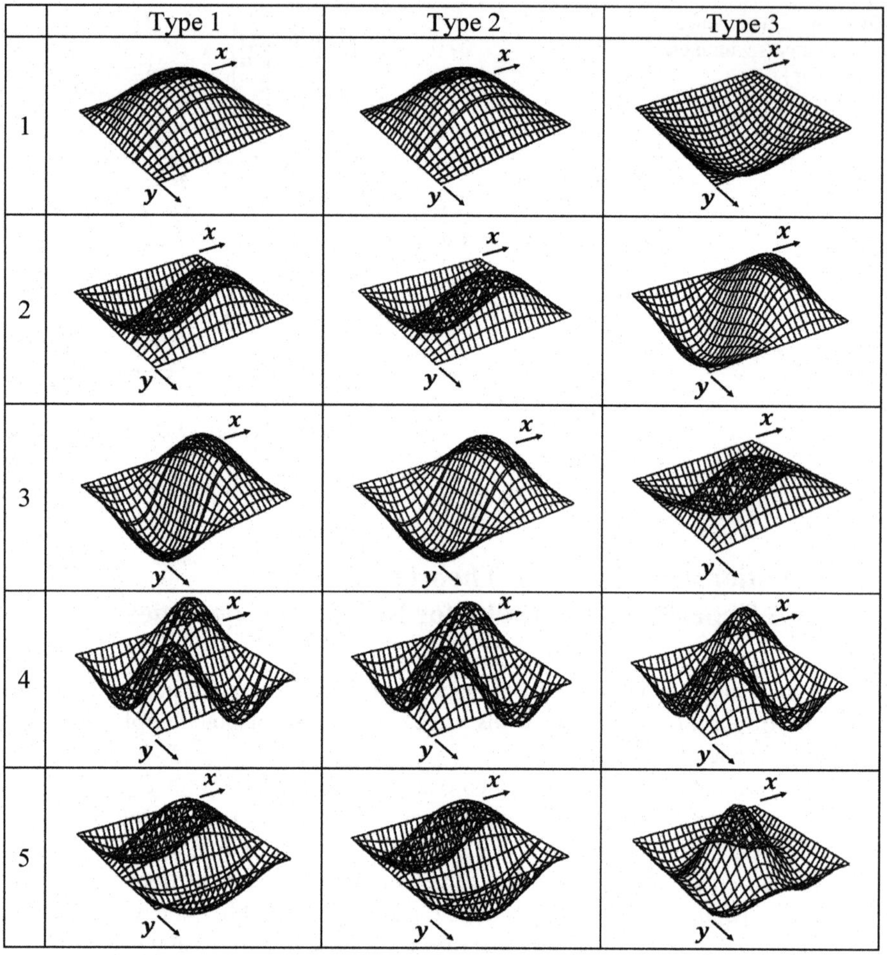

Fig. 4.17 Frequency modes of simply supported plates with voids [12]

$$w(x, y, t) = \sum_{m=1}^{\infty} \sum_{n=1}^{\infty} f_{mn}(x, y)\, \Phi_{mn}(t) \tag{4.71}$$

in which $\Phi_{mn}(t)$ are unknown functions of time t, and the functions $f_{mn}(x, y)$ are the natural functions satisfying both Eq. (4.58) and the specified boundary conditions at the ends of the plate. Substituting Eq. (4.71) into Eq. (4.25) and using the Eq. (4.58), we have

$$\sum_{m=1}^{\infty} \sum_{n=1}^{\infty} f_{mn}(x, y)\left[\rho h(x, y)\ddot{\Phi}_{mn}(t) + c\, \dot{\Phi}_{mn}(t) + \rho h(x, y)\, \omega_{mn}^2\, \Phi_{mn}(t)\right] = p(x, y, t) \tag{4.72}$$

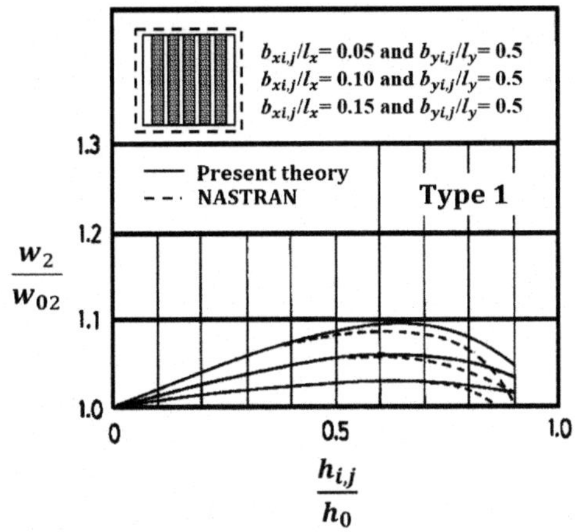

Fig. 4.18 Second natural frequency of a simply supported plate with voids (Type 1) [12]

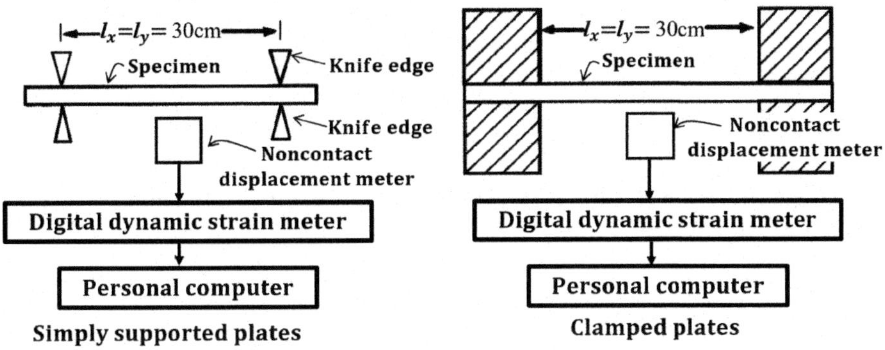

Fig. 4.19 Schematic diagram of the experimental equipment [12]

Since the thickness $h(x, y)$ is a function of x and y, Eq. (4.72) cannot be transformed into an uncoupled form by means of the orthogonality relations for natural functions. Therefore, Eq. (4.72) will be solved by either numerical computation based on the step-by-step integration methods in Sect. 4.10 or the closed-form approximate solution in Sect. 4.11.

4.10 Dynamic Analyses Based on the Step-by-Step Integration Method

Since Eq. (4.72) is in a coupled form, the solution can be obtained from the numerical computations. We have the relation $\rho h(x, y) = m_0\alpha_h(x, y)$. For simplicity, we

Table 4.2 Relationships between the theoretical and experimental results [12]

Type	$h_{i,j}/h_0$	$b_{xi,j}/l_x$	$b_{yi,j}/l_y$	First natural frequencies		
				Analytical method	Simply supported plate	Clamped plate
0				Experiment	720.8 rad/s	1296.3 rad/s
	0	0	0	Theory	666.5 rad/s	1257.8 rad/s
				Error	7.5%	2.9%
				Experiment	746.3 rad/s	1344.8 rad/s
1	0.33	0.1	0.5	Theory	738.2 rad/s	1439.4 rad/s
				Error	1.0%	−7.0%
				Experiment	755.7 rad/s	1320.9 rad/s
2	0.33	0.1	1.0	Theory	758.5 rad/s	1464.7 rad/s
				Error	−0.4%	−10.8%
				Experiment	729.0 rad/s	1399.0 rad/s
3	0.33	0.1	0.1	Theory	723.9 rad/s	1423.2 rad/s
				Error	0.7%	−1.7%

Error = (Experiment − Theory)/(experiment); $\alpha = l_y/l_x = 1.0$; $l_x = 30$ cm; and $h_0 = 0.6$ cm

assume the following relation between the damping coefficient c and the damping constant h_{mn} as a whole:

$$\frac{c}{\rho h(x, y)} \approx 2h_{mn}\omega_{mn} \tag{4.73}$$

Substituting Eq. (4.73) into Eq. (4.72) and multiplying both sides of Eq. (4.72) by $f_{\bar{m}\bar{n}}(x, y)$; and integrating between 0 to l_x and 0 to l_y, respectively, Eq. (4.72) may be translated with respect to indexes $\bar{m}$ and $\bar{n}$,

$$\sum_{m=1}^{\infty}\sum_{n=1}^{\infty} K_{\bar{m}\bar{n}mn}\left[\ddot{\Phi}_{mn} + 2h_{mn}\omega_{mn}\dot{\Phi}_{mn} + \omega_{mn}^2\Phi_{mn}\right] = \frac{Q_{\bar{m}\bar{n}}(t)}{m_0} \tag{4.74}$$

in which the notations $K_{\bar{m}\bar{n}mn}$ and $Q_{\bar{m}\bar{n}}(t)$ are defined as

$$K_{\bar{m}\bar{n}mn} = a_{mn}\delta_{m\bar{m}}\delta_{n\bar{n}} - \sum_{i=1}\sum_{j=1}\left(\frac{h_{i,j}}{h_0}\right)F_x(m, \bar{m}; i)F_y(n, \bar{n}; i) \tag{4.75}$$

$$Q_{\bar{m}\bar{n}}(t) = \int_0^{l_x}\int_0^{l_y} p(x, y, t) f_{\bar{m}\bar{n}}(x, y)\,\mathrm{d}x\,\mathrm{d}y \tag{4.76}$$

in which a_{mn}, $F_x(m, \bar{m}; i)$ and $F_y(n, \bar{n}; j)$ are defined as

$$a_{mn} = \int_0^{l_x}\int_0^{l_y} f_{mn}(x, y) f_{mn}(x, y)\,\mathrm{d}x\,\mathrm{d}y \tag{4.77}$$

in which the nondimensional quantities $F_x(m, \bar{m}; i)$ and $F_y(n, \bar{n}; j)$ are defined as

$$F_x(m, \bar{m}; i) = \int_0^{l_x} D(x - x_i) f_{xm} f_{x\bar{m}}\,\mathrm{d}x \tag{4.78}$$

$$F_y(n, \bar{n}; j) = \int_0^{l_y} D(y - y_j) f_{yn} f_{y\bar{n}}\,\mathrm{d}y \tag{4.79}$$

The shape functions f_{xm} and f_{yn} are the x and y components of f_{mn}, respectively. Equation (4.74) can be solved by using the step-by-step integration methods. The current natural functions $f_{mn}(x, y)$ take Eq. (4.66) for a simply supported plate with voids and Eq. (4.70) for a clamped plate with voids.

4.11 Closed-Form Approximate Solutions for Forced Vibrations

In the preceding section, the dynamic analysis for a damped plate with voids has been presented to be solvable by means of the step-by-step integration methods. Then, for practical use, we consider an approximation of the closed-form solution for a damped plate with voids.

The dynamical response of a plate with voids is obtained by solving Eq. (4.74). Since the coefficients $K_{\bar{m}\bar{n}mn}$ in Eq. (4.74) have nondiagonal terms $m \neq \bar{m}$ and/or $n \neq \bar{n}$ due to the effect of the voids, Eq. (4.74) takes on a coupled form. For simplicity, assuming that the behavior of a plate with voids is dominated by only the diagonal terms in the coefficients $K_{\bar{m}\bar{n}mn}$, Eq. (4.74) becomes of an uncoupled form. The effectiveness of the assumption used here has been numerically demonstrated in the presentation of the approximate solutions for natural frequencies of a plate with voids. Thus, Eq. (4.74) can be approximated as

$$\ddot{\Phi}_{mn}(t) + 2h_{mn}\omega_{mn}\dot{\Phi}_{mn}(t) + \omega_{mn}^2\Phi_{mn}(t) = \frac{Q_{mn}}{m_0 K_{mnmn}} \tag{4.80}$$

The general solution of Eq. (4.80) can be obtained as

$$\Phi_{mn}(t) = \exp(-h_{mn}\omega_{mn}t)[C_1 \sin\omega_{Dmn}t + C_2\cos\omega_{Dmn}t]$$
$$+ \frac{1}{m_0 K_{mnmn}\omega_{Dmn}} \int_0^t \exp[-h_{mn}\omega_{mn}(t - \tau)] \sin\omega_{Dmn}(t - \tau) Q_{mn}(\tau)\,\mathrm{d}\tau \tag{4.81}$$

in which ω_{mn} are the natural frequencies of the undamped plates with voids, and ω_{Dmn} are the natural frequencies of the damped plates with voids. The relationship between ω_{mn} and ω_{Dmn} has the well-known relation

$$\omega_{Dmn} = \omega_{mn}\sqrt{1 - h_{mn}^2} \tag{4.82}$$

Hence, the dynamic deflections of plates with voids are determined by substituting Eq. (4.81) into Eq. (4.71).

Next, let us consider the dynamic solutions of a plate with voids, subjected to the following harmonic external load:

$$p(x, y, t) = p_{xy}(x, y)\sin\omega_p t \tag{4.83}$$

in which $p_{xy}(x, y)$ is a function of the external loads, and ω_p is the frequency of the external loads. Then, the notation $Q_{mn}(t)$ can be written as

$$Q_{mn}(t) = \sin\omega_p t\ Q_p(m, n) \tag{4.84}$$

in which the notation $Q_p(m, n)$ is defined as

$$Q_p(m, n) = \int_0^{l_x}\int_0^{l_y} p_{xy}(x, y) f_{mn}(x, y)\,\mathrm{d}x\,\mathrm{d}y \tag{4.85}$$

If at $t = 0$ the following condition is prescribed

$$w\,(x, y, 0) = 0,\ \dot{w}\,(x, y, 0) = 0 \tag{4.86}$$

then $\Phi_{mn}(t)$ become the following simplified expression:

$$\begin{aligned}\Phi_{mn}(t) = {} & \frac{Q_p(m, n)}{m_0\,K_{mnmn}\,\omega_{Dmn}}\frac{1}{2}\left\{\frac{h_{mn}\omega_{mn}\cos(\omega_p t) + (\omega_p + \omega_{Dmn})\sin(\omega_p t)}{(h_{mn}\omega_{mn})^2 + (\omega_p + \omega_{Dmn})^2}\right.\\ & - \frac{h_{mn}\omega_{mn}\cos(\omega_p t) + (\omega_p - \omega_{Dmn})\sin(\omega_p t)}{(h_{mn}\omega_{mn})^2 + (\omega_p - \omega_{Dmn})^2}\\ & + \exp(-h_{mn}\omega_{mn}t)\left[\frac{(\omega_p + \omega_{Dmn})\sin(\omega_{Dmn}t) - h_{mn}\omega_{mn}\cos(\omega_{Dmn}t)}{(h_{mn}\omega_{mn})^2 + (\omega_p + \omega_{Dmn})^2}\right.\\ & \left.\left. + \frac{(\omega_p - \omega_{Dmn})\sin(\omega_{Dmn}t) - h_{mn}\omega_{mn}\cos(\omega_{Dmn}t)}{(h_{mn}\omega_{mn})^2 + (\omega_p - \omega_{Dmn})^2}\right]\right\}\end{aligned} \tag{4.87}$$

Thus, the closed-form approximate solutions for a damped plate with voids subjected to harmonic external loads are presented in Eqs. (4.81) and (4.87). These general approximate solutions are also applicable to both undamped plates with voids and normal plates. The closed-form solution for undamped plates with voids is

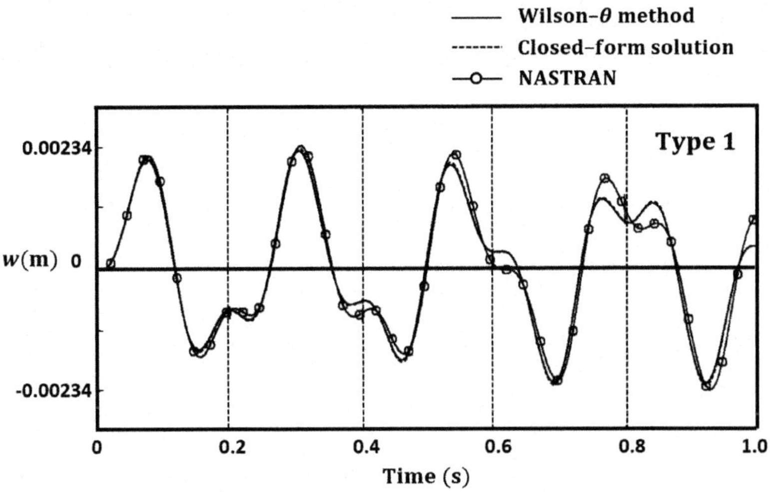

Fig. 4.20 Dynamic deflections w for a simply supported plate with voids (Type 1) [12]

easily given by changing, as $h_{mn} \to 0$, $\omega_{Dmn} \to \omega_{mn}$ and $\exp(-h_{mn}\omega_{mn} t) \to 1$ in Eqs. (4.81) and (4.87), while the solution for solid plates without voids is obtained by replacing K_{mnmn} with 1. The reduced equation is equal to Eq. (1.108) presented in Chap. 1.

4.12 Numerical Results for Dynamical Responses

In this section, in order to examine the approximate solution proposed here for an isotropic rectangular plate with voids, numerical computations are carried out for three cases, as shown in Table 4.1. The data used are as follows: Young's modulus $E = 2.06 \times 10^{10}$ N/m^2; Poisson's ratio $\nu = 0.17$; thickness $h_0 = 0.3$ m; span lengths $l_x = l_y = 10$ m; voided ratio $h_{i,j}/h_0 = 0.5$; and the mass density $\rho = 2400$ kg/m^3. The external lateral load is assumed to be a harmonic and uniformly distributed force given in Eq. (4.83), in which $p_{xy}(x, y) = 1.47\,\mathrm{k\,N m^{-2}}$, and $\omega_p = \omega_{01}/2$. Here, ω_{01} is the first natural frequency for solid plates without voids.

Figures 4.20, 4.21 and 4.22 show time histories of the dynamic deflections at the midpoint of simply supported plates with voids of Types 1–3, respectively, while Figs. 4.23, 4.24, and 4.25 show time histories of the dynamic deflections at the midpoint of clamped plates with voids of Types 1–3, respectively. In these figures, the solid lines represent values obtained from the numerical computations of coupled Eq. (4.73) using Wilson-θ method; the broken lines indicate values obtained from the closed-form approximate solution; and the solid lines with circles indicate values obtained from the FEM code NASTRAN. The numerical results demonstrate that

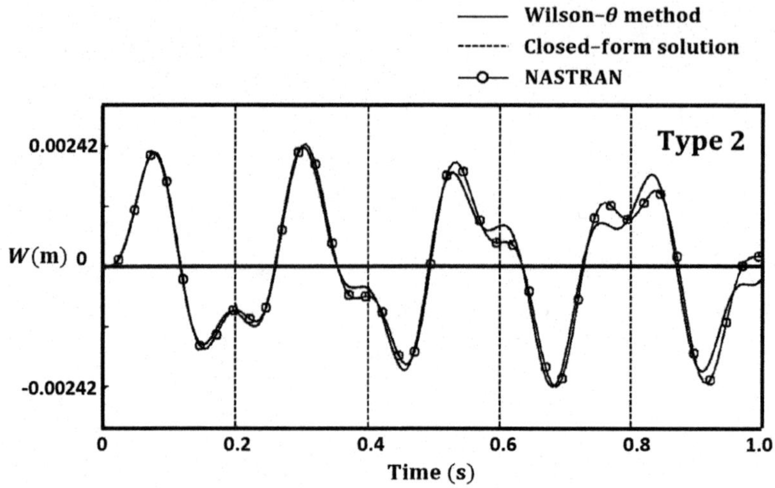

Fig. 4.21 Dynamic deflections w for a simply supported plate with voids (Type 2) [12]

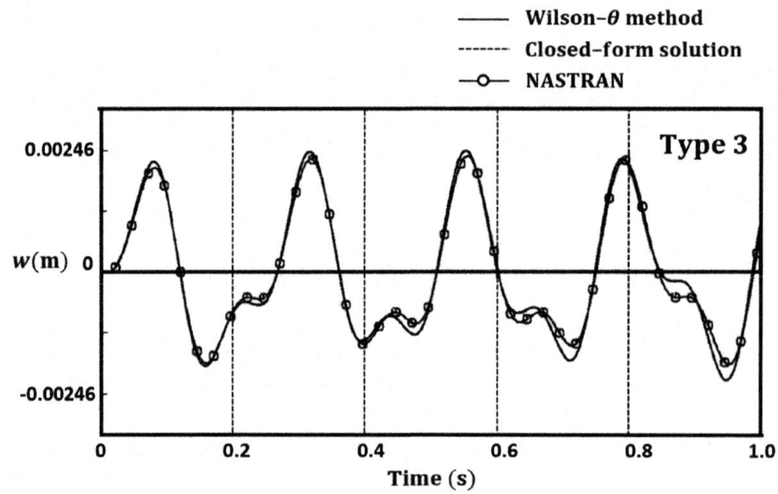

Fig. 4.22 Dynamic deflections w for a simply supported plate with voids (Type 3) [12]

the closed-form approximate solution proposed here is applicable to the dynamic analyses of plates with voids, in practical use.

The limitation of the ratio $h_{i,j}/h_0$ in the proposed theory on natural frequencies has already been stated as being smaller than 0.6. This limitation is also effective for dynamic plates with voids, subjected to forced vibrations, because the inaccuracy of the natural frequencies results in an inexact dynamical response.

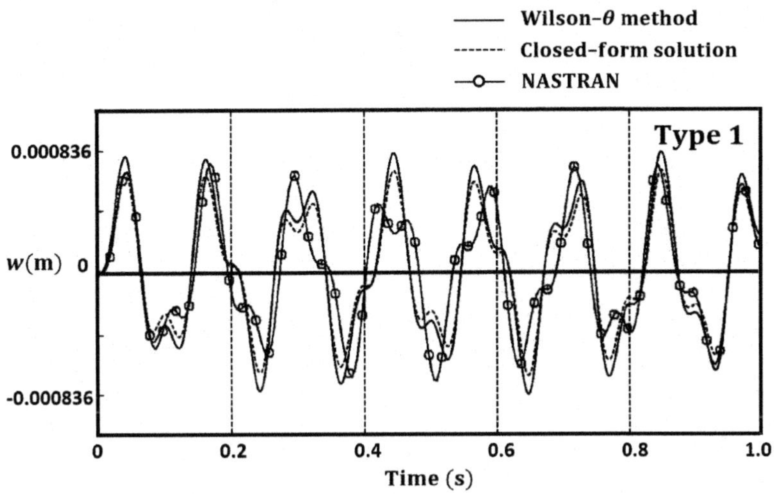

Fig. 4.23 Dynamic deflections w for a clamped plate with voids (Type 1) [12]

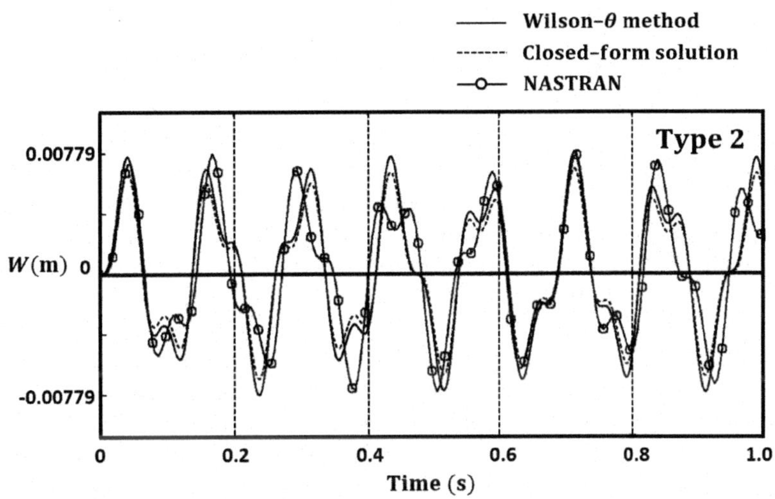

Fig. 4.24 Dynamic deflections w for a clamped plate with voids (Type 2) [12]

On the other hand, the limitation of the ratios b_{xi}/l_x and b_{yj}/l_y is not clear. They are restricted by the shear forces and the validity of the Kirchhoff-Love plate theory. West [19] suggested that cell distortion of a plate with voids should be taken into account when the void area exceeds 60% of the total cross section.

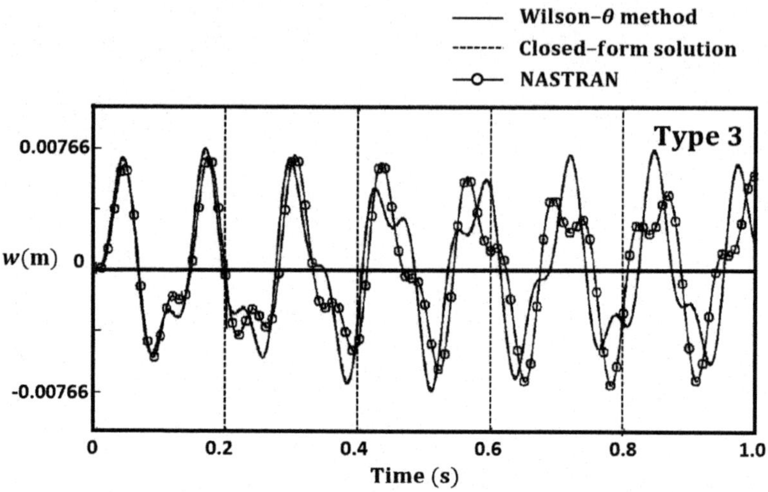

Fig. 4.25 Dynamic deflections w for a clamped plate with voids (Type 3) [12]

4.13 Conclusions

The general analysis methods and closed-form approximate solutions for static, free vibration and forced vibrations problems of an isotropic rectangular plate with arbitrarily disposed voids have been proposed by means of both the extended Dirac function and the Galerkin method. The closed-form solutions proposed here have been validated by comparing them with the numerical results obtained from more rigorous coupled equations and the FEM code NASTRAN and with the experimental results.

For the sake of simplicity, this chapter disregards the transverse shear deformation and the local deformation of the top and bottom platelets of the void. It will be necessary to consider these deformations when the cross section and/or number of voids increase. However, in practice, the occurrence of the local deformation should be restricted for building slabs. The transverse shear deformation can be considered by replacing the Kirchhoff-Love plate hypothesis with the Mindlin-Reissner hypothesis.

Each void was assumed to be a rectangular parallelepiped for simplicity's sake, but it is relatively easy to extend the proposed theory to a void with circular or symmetric cross section. The advanced theory will be presented in Chap. 6.

References

1. Crisfield, M.A., Twernlow, R.P.: The equivalent plate approach for the analysis of cellular structures. In: Civil Engineering and Public Works. March, 259–263 (1971)
2. Elliott, G., Clark, L.A.: Circular voided concrete slab stiffness. J. Struct. Div. ASCE **108**(11), 2379–2393 (1982)
3. Cope, R.J., Harris, G., Sawko, F.: A new approach to the analysis of cellular bridge decks. Analysis of structural systems for torsion. ACI **35**, 185–210 (1973)
4. Holmberg, A.: Shear-Weak Beams an Elastic Foundation, p. 10. International Association for Bridge and Structural Engineering (IABSE), Zurich (1960)
5. Sawko, F., Cope, R.J.: Analysis of multi-cell bridges without transverse diaphragms—a finite element approach. Struct Eng **47**(11), 455–460 (1969)
6. Elliott, G.: Partial Loading on Orthotropic Plates, vol 42, p. 519. Cement and Concrete Association Technical Report, London (1978)
7. Takabatake, H.: Bending and torsional analyses of tube systems (in Japanese). In: Symposium Proceedings on Computational Methods in Structural Engineering and Related Fields (JSSC), vol 11, pp. 205–210 (1987)
8. Takabatake, H.: Lateral buckling of I beams with web stiffeners and batten plates. Int. J. Solids Struct. **24**(10), 1003–1019 (1988)
9. Takabatake, H., Muaki, H., Hirano, T.: Doubly symmetric tube structures. I: static analysis. J. Struct. Eng. **119**(7), 1981–2001 (1983)
10. Takabatake, H., Muaki, H., Hirano, T.: Doubly symmetric tube structures. II: dynamic analysis. J. Struct. Eng. **119**(7), 2002–2016 (1983)
11. Takabatake, H.: Static analyses of elastic plates with voids. Int. J. Solids Struct. **28**(2), 179–196 (1991)
12. Takabatake, H.: Dynamic analyses of elastic plates with voids. Int. J. Solids Struct. **28**(7), 879–895 (1991)
13. Mikusiński, J., Sikorski, R.: The elementary theory of distributions, I. Państwowe Wydawnictwo Naukowe, Dissertationes Mathematicae, vol. 12. Warszawa (1957)
14. Frýba, L.: Vibration of solids and structures under moving loads. Noordhoff International Publishing, Groningen, The Netherlands (1972)
15. Sinozaki, H., Matsumori, T., Matsuura, T.: Introduction to the Delta Function. Gendaikougakusha, Tokyo (1983)
16. Rao, S.S.: The Finite Element Method in Engineering. Pergamon Press, Oxford (1982)
17. Ugural, A.C.: Stresses in Plates and Shells. McGraw-Hill, New York (1981)
18. Hinton, E., Owen, D.R.J.: Finite Element Software for Plates and Shells. Pineridge Press, Swansea, UK (1984)
19. West, R.: Recommendations on the Use of Grillage Analysis for Slabs and Pseudo-Slab Bridge Decks. Cement and Concrete Association (1973)

Chapter 5
Static and Dynamic Analyses of Circular Plates with Voids

5.1 Introduction

In the preceding chapter, we have presented the static and dynamic problems of rectangular plates with voids, in which the size of voids considered is assumed to be relatively small compared to the thickness of the plate. This chapter discusses the same problem of circular plates with voids by converting from the rectangular plates to circular ones.

For reasons of lightness and structural efficiency, circular plates with voids are often used. Most analytical methods of the previous studies for the analysis of plates with voids are based on the equivalent plate analogy. A number of researchers [1–8] have proposed stiffness coefficients to enable the determination of overall effects. However, since in the equivalent approaches the stiffness and mass of plates with voids are determined independently of the position of the voids, the application of the theory is restricted only to a rectangular plate with many voids of the same cross section, spaced uniformly. The irregular variation in stiffness and mass of the plate due to voids is more pronounced in the circular plates than in the rectangular plates. Therefore, it cannot apply to a circular plate with irregularly spaced voids and/or with voids of different cross section.

On the other hand, although analyses based on the finite element method for a plate with voids are effective, much numerical calculation is needed. In Chap. 4, a general and simple analytical method for static and dynamic problems of rectangular plates with arbitrarily positioned voids under the Kirchhoff-Love hypotheses has been presented by the use of the extended Dirac function. The discontinuous variation of stiffness and mass of such a rectangular plate, due to voids, was expressed as a continuous function by means of an extended Dirac function. This theory is usable in both the preliminary and final stages of the design of a rectangular plate with voids.

This chapter proposes a general analytical methodology for a circular plate with arbitrarily positioned voids. The discontinuous variation of the bending stiffness and

H. Takabatake, *Simplified Analytical Methods of Elastic Plates*,
https://doi.org/10.1007/978-981-13-0086-8_5

mass is also expressed as a continuous function in a distribution sense by means of extended Dirac functions, as shown in Chap. 4.

The general governing equation for a circular plate with voids is derived by using Hamilton's principle. Then the static and dynamic solutions are presented from the proposed governing equations by the use of the Galerkin method, as the same approach stated in Chap. 4. Finally, the exactness of the proposed solutions is established from numerical results by comparing the results obtained from the proposed theory for simply supported and clamped plates with the results obtained from the FEM. The methodology proposed in this chapter is mainly based on Ref. [9].

5.2 Governing Equations of a Circular Plate with Voids

We consider a circular plate with arbitrarily positioned voids, as shown in Fig. 5.1. A polar coordinate system r, ϕ, z is employed, as presented in Chap. 2. The height of voids does not exceed half the thickness of plates, as stated in Chap. 4. Each void is assumed to be a fan parallelepiped whose ridgelines are parallel to be the axis r or ϕ and which is symmetrically positioned with respect to the middle plane of the plate. The position of the i-, j-th void is indicated by the coordinate value (r_i, ϕ_j) of the midpoint of the void; the widths in the r and ϕ directions of the void are $b_{ri,j}$ and $b_{\phi i,j}$, respectively, and its height is $h_{i,j}$. The size and position of each void are arbitrary, except for the assumption mentioned above.

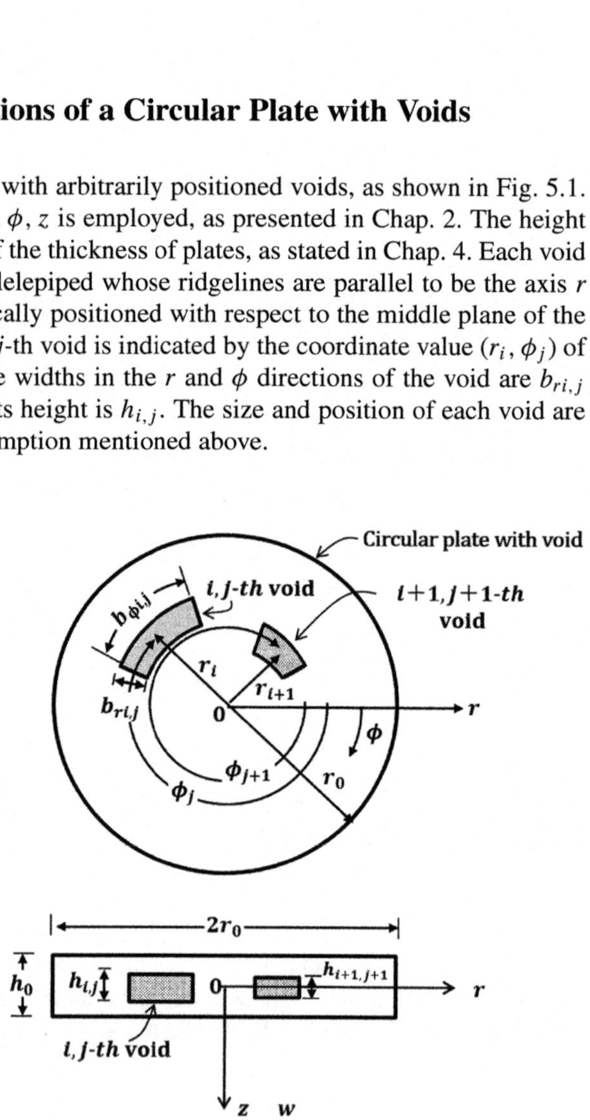

Fig. 5.1 Coordinates of a circular plate with voids

We consider the linear theory of bending of isotropic circular plates under the Kirchhoff-Love hypotheses. The transverse shear deformation, including the deformation of a frame, which is composed of top and bottom platelets and partitions of each void, is neglected, because the height of voids is assumed not to exceed half that of the slab height. Usual plates are subjected to lateral loads on the top platelet, and the occurrence of the local deformation on the top platelet is undesirable in practical use. Hence, the lateral loads are assumed to act on the middle surface. Thus, the shape of a circular plate with voids is adequately defined by describing the geometry of its middle surface that bisects the plate thickness h_0 at each point.

The displacement components U, V, and W of a general point of the circular plate with voids are assumed to be

$$U(r, \phi, z) = -z w_{,r} \tag{5.1}$$

$$V(r, \phi, z) = -z \frac{1}{r} w_{,\phi} \tag{5.2}$$

$$W(r, \phi, z) = w \tag{5.3}$$

where $w(x, y, t)$ indicates the displacement component in the z-direction on the middle surface and z is the distance measured from the middle surface of the plate. The displacement is defined as positive when it points toward the positive direction of the coordinate axis z.

The governing equation for the circular plate with voids is also derived by means of Hamilton's principle

$$\delta I = \delta \int_{t_1}^{t_2} (T - U - V)\mathrm{d}t = 0 \tag{5.4}$$

where T is the kinetic energy; U is the strain energy; V is the potential energy produced by external loads. δ indicates the variation operator.

Using the linear strain–displacement relations and linear stress–displacement relations of isotropic circular plates, the strain energy U may be expressed as

$$U = \frac{1}{2} \int_0^{2\pi} \int_0^{r_0} (M_r \kappa_r + M_\phi \kappa_\phi + 2M_{r\phi} \kappa_{r\phi}) \mathrm{d}r \, r \, \mathrm{d}\phi \tag{5.5}$$

where

$$\kappa_r = -w_{,rr}; \quad \kappa_\phi = -\frac{1}{r} w_{,r} - \frac{1}{r^2} w_{,\phi\phi}; \quad \kappa_{r\phi} = -\frac{1}{r} w_{,r\phi} + \frac{1}{r^2} w_{,\phi} \tag{5.6}$$

are the curvatures and twist of the middle surface, respectively. On the other hand, M_r, M_ϕ, and $M_{r\phi}$ are the bending and twisting moments per unit width, respectively, and they are given as

$$M_r = -D_0\, d(r, \phi)\left[w_{,rr} + \nu\left(\frac{1}{r}w_{,r} + \frac{1}{r^2}w_{,\phi\phi}\right)\right] \tag{5.7}$$

$$M_\phi = -D_0\, d(r, \phi)\left(\frac{1}{r}w_{,r} + \frac{1}{r^2}w_{,\phi\phi} + \nu w_{,rr}\right) \tag{5.8}$$

$$M_{r\phi} = -(1-\nu)D_0\, d(r, \phi)\left(\frac{1}{r}w_{,r\phi} - \frac{1}{r^2}w_{,\phi}\right) \tag{5.9}$$

Here D_0 is the well-known flexural rigidity of a solid plate neglecting voids, given by

$$D_0 = \frac{Eh_0^3}{12(1-\nu^2)} \tag{5.10}$$

where E is Young's modulus; ν is Poisson's ratio; and h_0 is the thickness of the solid plate. On the other hand, the coefficient of the flexural rigidity $d(r, \phi)$ is defined as

$$d(r, \phi) \equiv 1 - \left(\frac{h_{i,j}}{h_0}\right)^3 D(r - r_i)D(\phi - \phi_j) \tag{5.11}$$

in which $D(r - r_i)$ and $D(\phi - \phi_j)$ are the extended Dirac functions corresponding to the radial and circumferential directions, respectively.

The calculations including the extended Dirac functions in Polar coordinate system are similar to in Cartesian coordinate system. Specifically, the integral calculation including the extended Dirac function $D(r - r_i)$ can be expressed as

$$\begin{aligned}\int_0^{r_0} D(r - r_i)f(r)\mathrm{d}r &= \int_{r_i-(b_{ri,j}/2)}^{r_i+(b_{ri,j}/2)} \left[\int_0^{r_0} \delta(r-\xi)f(r)\mathrm{d}r\right]\mathrm{d}\xi \\ &= \int_{r_i-(b_{ri,j}/2)}^{r_i+(b_{ri,j}/2)} f(\xi)\mathrm{d}\xi\end{aligned} \tag{5.12}$$

in which ξ indicates a supplementary variable of r. Similarly,

$$\begin{aligned}\int_0^{2\pi} D(\phi - \phi_j)f(\phi)\,\mathrm{d}\phi &= \int_{\phi_j-(b_{\phi i,j}/2)}^{\phi_j+(b_{\phi i,j}/2)} \left[\int_0^{2\pi} \delta(\phi-\eta)f(\phi)\,\mathrm{d}\phi\right]\mathrm{d}\eta \\ &= \int_{\phi_j-(b_{\phi i,j}/2)}^{\phi_j+(b_{\phi i,j}/2)} f(\eta)\mathrm{d}\eta\end{aligned} \tag{5.13}$$

in which η is a supplementary variable of ϕ. The n-th derivatives of the extended Dirac function can be transformed

$$\left.\begin{aligned}\int_0^{r_0} D^{(n)}(r-r_i)f(r)\mathrm{d}r &= \int_{r_i-(b_{ri,j}/2)}^{r_i+(b_{ri,j}/2)} (-1)^n f^{(n)}(\xi)\mathrm{d}\xi \\ \int_0^{2\pi} D^{(n)}(\phi-\phi_j)f(\phi)\mathrm{d}\phi &= \int_{\phi_j-(b_{\phi i,j}/2)}^{\phi_j+(b_{\phi i,j}/2)} (-1)^n f^{(n)}(\eta)\mathrm{d}\eta\end{aligned}\right\} \tag{5.14}$$

in which the superscripts enclosed within parentheses represent the differential order.

When the conditions $b_{ri,j} \ll r_0$ and $b_{\phi i,j} \ll 2\pi$ are satisfied, the extended Dirac functions $D(r-r_i)$ and $D(\phi-\phi_j)$ are approximately related to the following relations by the use of the Dirac functions $\delta(r-r_i)$ and $\delta(\phi-\phi_j)$, prescribed on the midpoint of the i-, j-th void.

$$\left.\begin{aligned}D(r-r_i) &\approx b_{ri,j}\delta(r-r_i) \\ D(\phi-\phi_i) &\approx b_{\phi i,j}\delta(\phi-\phi_j)\end{aligned}\right\} \tag{5.15}$$

Substituting Eqs. (5.7)–(5.9) into Eq. (5.5), the strain energy U can be obtained as

$$U = \frac{1}{2} = \int_0^{2\pi}\int_0^{r_0} D_0 d[(\kappa_r)^2 + (\kappa_\phi)^2 + 2\nu\kappa_r\kappa_\phi + 2(1-\nu)(\kappa_{r\phi})^2] r\,\mathrm{d}\phi\,\mathrm{d}r \tag{5.16}$$

Next, the variation of the potential of the external lateral loads is given as

$$\delta V = -\int_0^{2\pi}\int_0^{r_0} (p - c\dot{w})\delta w\, r\,\mathrm{d}r\,\mathrm{d}\phi + \int_{c^*} \left[m_r^*\,\delta w_{,\phi} + v_r^*\delta w\right]_0^{r_0} r\,\mathrm{d}\phi \tag{5.17}$$

where p is external lateral load per unit area; c is the damping coefficient, such as structural damping; v_r^* and m_r^* are external vertical forces and external moments prescribed on the mechanical boundary edges c^* at $r = r_0$, respectively.

The kinetic energy, T, may be written as

$$T = \frac{1}{2}\int_0^{2\pi}\int_0^{r_0} m_0\alpha_h(\dot{w})^2 \mathrm{d}r\, r\,\mathrm{d}\phi \tag{5.18}$$

where the dot indicates differentiation with respect to time; and the notations α_h and m_0 are defined as

$$\alpha_h(r,\phi) \equiv 1 - \left(\frac{h_{i,j}}{h_0}\right) D(r-r_i)D(\phi-\phi_j) \tag{5.19}$$

$$m_0 = \rho h_0 \tag{5.20}$$

in which ρ is the mass density of the plate. The notation $\alpha_h(r, \phi)$ represents the coefficient of mass.

Substituting Eqs. (5.16)–(5.18) into Eq. (5.4), the differential equation of motion for circular plates with voids can be obtained as

$$\frac{m_0\alpha_h\ddot{w}}{D_0} + \frac{c\dot{w}}{D_0} - \frac{p}{D_0} + \frac{1}{r}(rdw_{,rr})_{,rr} - \frac{1}{r}\left[d\left(\frac{1}{r}w_{,r} + \frac{1}{r^2}w_{,\phi\phi}\right)\right]_{,r}$$
$$+ \left[\frac{d}{r^2}\left(\frac{1}{r}w_{,r} + \frac{1}{r^2}w_{,\phi\phi}\right)\right]_{,\phi\phi} - \frac{\nu}{r}(dw_{,rr})_{,r} + \nu\left(d\frac{1}{r^2}w_{,rr}\right)_{,\phi\phi}$$
$$+ \frac{2(1-\nu)}{r}\left[d\left(\frac{1}{r}w_{,r\phi} - \frac{1}{r^2}w_{,\phi}\right)\right]_{,\phi r} + 2(1-\nu)\left[\frac{d}{r^2}\left(\frac{1}{r}w_{,r\phi} - \frac{1}{r^2}w_{,\phi}\right)\right]_{,\phi} = 0 \tag{5.21}$$

together with the associated boundary conditions

$$w = 0 \quad \text{or}$$
$$-\frac{1}{r}(rD_0\,dw_{,rr})_{,r} + D_0 d\frac{1}{r}\left(\frac{1}{r}w_{,r} + \frac{1}{r^2}w_{,\phi\phi}\right)$$
$$+ \nu D_0 d\frac{1}{r}w_{,rr} - \frac{\nu}{r}\left[D_0 d\left(w_{,r} + \frac{1}{r}w_{,\phi\phi}\right)\right]_{,r}$$
$$- 2(1-\nu)\left[D_0\frac{d}{r}\left(\frac{1}{r}w_{,r\phi} - \frac{1}{r^2}w_{,\phi}\right)\right]_{,\phi} + v_r^* = 0 \tag{5.22}$$

$$w_{,r} = 0 \quad \text{or}$$
$$D_0 d\,w_{,rr} + \nu D_0 d\left(\frac{1}{r}w_{,r} + \frac{1}{r^2}w_{,\phi\phi}\right) + m_r^* = 0 \tag{5.23}$$

at $r = r_0$, in which v_r^* and m_r^* are external vertical force and moment prescribed at the boundary $r = r_0$, respectively.

Equation (5.21) is the general equation of motion for a circular thin plate with voids. Since for solid plates the coefficients α_h and d for the coefficients of mass and stiffness, respectively, become all 1, the governing equations proposed here reduce to the general equations for a circular solid plate, as presented in Chap. 2.

5.3 Static Analysis

In the preceding section, the governing equations for a circular plate with voids have been proposed in general by means of the extended Dirac function, in order to express rationally the discontinuous variation of the plate's stiffness and mass. This section indicates general analytical method of the static problem by the use of the Galerkin method. The methodology is the same as the approach described in

Chap. 2, except for treatments including the extended Dirac functions. The deflections w are expressed by

$$w(r, \phi) = \sum_{m=1} \sum_{n=0} w_{mn} f_{rm}(r) f_{\phi n}(\phi) \tag{5.24}$$

in which w_{mn} represent unknown displacement coefficients; and f_{rm} and $f_{\phi n}$ indicate shape functions in the radial and circumferential directions, respectively, satisfying the specified boundary conditions.

The Galerkin equations for the static problem of Eq. (5.21) can be expressed as

$$\delta w_{\bar{m}\bar{n}}\colon \sum_{m=1} \sum_{n=0} w_{mn} A_{mn\bar{m}\bar{n}} = B_{\bar{m}\bar{n}} \tag{5.25}$$

in which the coefficients are obtained as follows:

$$\begin{aligned}
A_{mn\bar{m}\bar{n}} = {} & F_{\phi n\bar{n}}(0, 0; 0)[2F_{rm\bar{m}}(3, 0, 0; 0) + F_{rm\bar{m}}(4, 0, 1; 0) \\
& \qquad + F_{rm\bar{m}}(1, 0, -2; 0) - F_{rm\bar{m}}(2, 0, -1; 0)] \\
& + 2F_{\phi n\bar{n}}(2, 0; 0)\,[2F_{rm\bar{m}}(0, 0, -3; 0) - F_{rm\bar{m}}(1, 0, -2; 0) + F_{rm\bar{m}}(2, 0 - 1; 0)] \\
& + F_{\phi n\bar{n}}(4, 0; 0) F_{rm\bar{m}}(0, 0, -3; 0) \\
& + \sum \left(\frac{h_{i,j}}{h_0}\right)^3 [F_{\phi n\bar{n}}(0, 0; j)\{-F_{rm\bar{m}}(2, 2, 1; i) - F_{rm\bar{m}}(1, 1, -1; i) \\
& - \nu[F_{rm\bar{m}}(2, 1, 0; i) + F_{rm\bar{m}}(1, 2, 0; i)]\} \\
& + F_{\phi n\bar{n}}(2, 0; j)\{-F_{rm\bar{m}}(0, 1, -2; i) + \nu[-8F_{rm\bar{m}}(0, 0, -3; i) \\
& + 4F_{rm\bar{m}}(1, 0, -2; i) + 4F_{rm\bar{m}}(0, 1, -2; i) - F_{rm\bar{m}}(0, 2, -1; i)]\} \\
& + 2(1 - \nu) F_{\phi n\bar{n}}(1, 1; j)[F_{rm\bar{m}}(0, 1, -2; i) - F_{rm\bar{m}}(1, 1, -1; i) \\
& + F_{rm\bar{m}}(1, 0, -2; i) - F_{rm\bar{m}}(0, 0, -3; i)] \\
& + F_{\phi n\bar{n}}(0, 2; j)\,[-F_{rm\bar{m}}(1, 0, -2; i) - \nu F_{rm\bar{m}}(2, 0, -1; i)] \\
& - F_{\phi n\bar{n}}(2, 2; j)\, F_{rm\bar{m}}(0, 0, -3; i)]
\end{aligned} \tag{5.26}$$

$$B_{\bar{m}\bar{n}} = \int_0^{\phi_0} \int_0^{r_0} \frac{p}{D_0} f_{r\bar{m}}\, f_{\phi\bar{n}}\, r \,\mathrm{d}r\, \mathrm{d}\phi \tag{5.27}$$

where $F_{rm\bar{m}}(k_1, k_2, k_3; 0)$, $F_{rm\bar{m}}(k_1, k_2, k_3; i)$, $F_{\phi n\bar{n}}(k_1, k_2; 0)$, and $F_{\phi n\bar{n}}(k_1, k_2; j)$ are defined in general form as

$$F_{rm\bar{m}}(k_1, k_2, k_3; 0) = \int_0^{r_0} f_{rm}^{(k_1)} f_{r\bar{m}}^{(k_2)} r^{k_3} \mathrm{d}r \tag{5.28}$$

$$F_{rm\bar{m}}(k_1, k_2, k_3; i) = \int_0^{r_0} f_{rm}^{(k_1)} f_{r\bar{m}}^{(k_2)} r^{k_3} D(r - r_i) \mathrm{d}r \tag{5.29}$$

$$F_{\phi n\bar{n}}(k_1, k_2; 0) = \int_0^{2\pi} f_{\phi n}^{(k_1)} f_{\phi \bar{n}}^{(k_2)} \mathrm{d}\phi \tag{5.30}$$

$$F_{\phi n\bar{n}}(k_1, k_2; j) = \int_0^{2\pi} f_{\phi n}^{(k_1)} f_{\phi \bar{n}}^{(k_2)} D(\phi - \phi_j) \mathrm{d}\phi \tag{5.31}$$

The integral calculations including the extended Dirac function are already described in Eqs. (5.12)–(5.15). The unknown displacement coefficients w_{mn} can be obtained from solving a set of algebraic linear equations given by Eq. (5.25).

Since Eq. (5.25) has coupled form given in a set of algebraic linear equations, the solution is based on the numerical calculation by the use of computer. Therefore, for practical use, we present an approximate solution in uncoupled form. Although the behavior of circular plates with voids is affected by all the terms of the square matrix $A_{\bar{m}\bar{n}mn}$, the behavior is considered to be dominated by only the diagonal terms in the matrix $A_{\bar{m}\bar{n}mn}$. Hence, taking into consideration only the diagonal terms of $A_{\bar{m}\bar{n}mn}$, Eq. (5.25) becomes of uncoupled form. Thus, the approximate solutions of w_{mn} are obtained as

$$w_{mn} \approx \frac{B_{mn}}{A_{mnmn}} \tag{5.32}$$

5.4 Numerical Results for Static Problems

In the preceding section, a statically loaded circular plate with voids has been analyzed by means of the Galerkin method. Then, the exactness of the proposed method is examined from numerical calculations carried out for two types of plates: simply supported and clamped plates, as shown in Fig. 5.2. Data used are as follows: Young's modulus $E = 2.06 \times 10^{10}$ N/m^2; Poisson's ratio $\nu = 0.17$; slab's height $h_0 = 0.8$ m; span radius $r_0 = 8$ m void's height $h_{i,j} = 0.4$ m; and the mass density $\rho = 2400$ kg/m^3. For simplicity, the lateral load is assumed to be uniformly distributed load $p = 9.807$ kNm^{-2}.

The shape functions f_{mn} and $f_{\phi n}$ for the simply supported plates use the following functions for solid plates subjected to uniform load:

$$f_{rm}(r) = \left[\left(\frac{r}{r_0}\right)^4 - a_1\left(\frac{r}{r_0}\right)^2 + a_2\right]^m \quad (m = 1, 3, 5 \ldots) \tag{5.33}$$

$$f_{\phi n}(\phi) = \cos n\phi \quad (n = 0) \tag{5.34}$$

in which for current problem the value of n takes 0 and the value of m takes the odd number and the constants a_1 and a_2 are given by

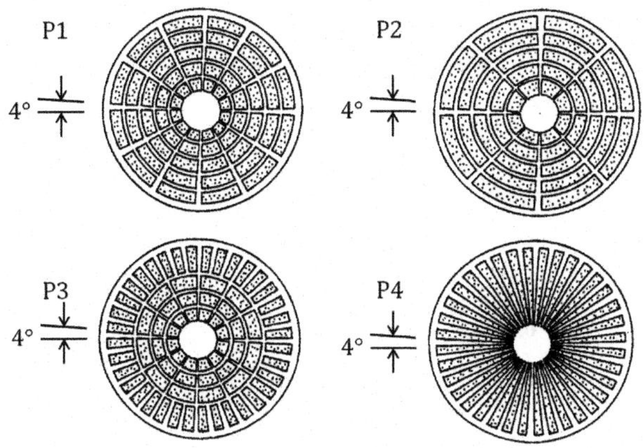

Fig. 5.2 Isotropic circular plates with voids [9]

$$a_1 = 2\frac{3+\nu}{1+\nu}; \quad a_2 = \frac{5+\nu}{1+\nu} \tag{5.35}$$

Meanwhile, the shape functions for the clamped plates use

$$f_{rm}(r) = \left[1 - \left(\frac{r}{r_0}\right)^2\right]^{2m} \quad (m = 1, 3, 5, \ldots) \tag{5.36}$$

$$f_{\phi n}(\phi) = \cos n\phi \quad (n = 0) \tag{5.37}$$

Figures 5.3 and 5.4 show the distribution of the deflections and bending moments for the clamped plates with voids, respectively. In these figures, the solid lines represent the numerical results obtained from the proposed theory given by Eqs. (5.25) and (5.32), the circles represents the numerical results obtained by using the FEM code POPLAS/FEM5. The numerical results obtained from Eq. (5.25) agree with the one obtained from Eq. (5.32). It follows from these figures that the results obtained from the present theory show excellent agreement with the results obtained from the finite element method. The finite element method used here is based on isotropic rectangular or triangular plate elements, as shown in Fig. 5.5. The plate element that includes a void is replaced with a solid plate element having the equivalent bending stiffness. This modeling is also examined by the results obtained from solid elements divided into three sections in the thickness wise.

The bending and torsional moments are given by Eqs. (5.7)–(5.9), respectively. These values at $r=0$ can be calculated by means of L'Hopital's rule as follows:

$$-\frac{1}{r}w_{,r} - \frac{1}{r^2}w_{,\phi\phi} = -\lim_{r\to 0}\frac{w_{,rr}}{r_{,r}} - \lim_{r\to 0}\frac{w_{,\phi\phi rr}}{r^2_{,rr}} \tag{5.38}$$

For current uniformly distributed loads, the order of m for f_{rm} used Eqs. (5.33) and (5.36) as for the shape functions is completely convergent only $m=1$ and the effect of the higher order is meaningless. Hence the static solution for $m=1$ becomes

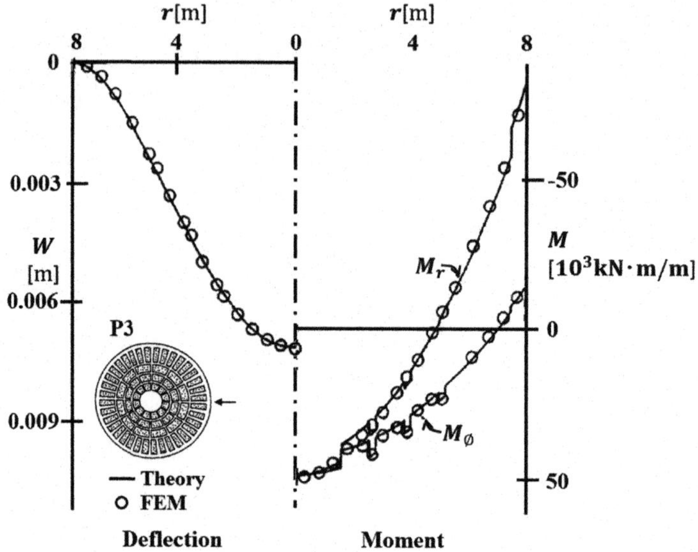

Fig. 5.3 w and M_y for a clamped plate with voids, P3 [9]

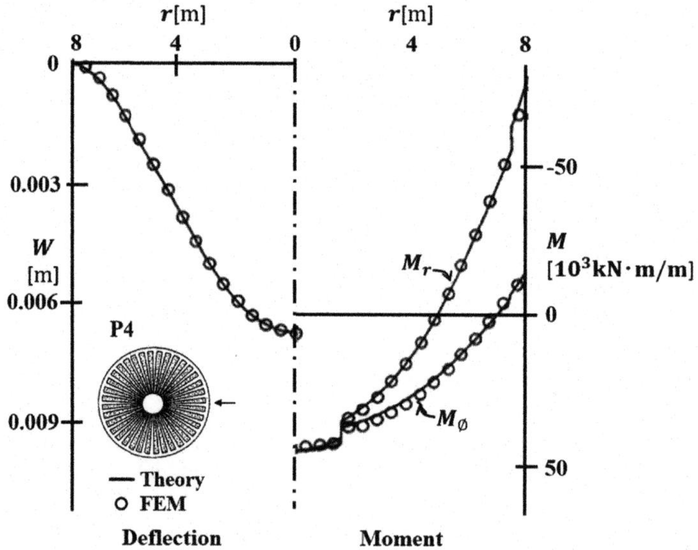

Fig. 5.4 w and M_y for a clamped plate with voids, P4 [9]

the closed-form solution of a circular plates with voids subject to a uniform load, as given by Eq. (5.32).

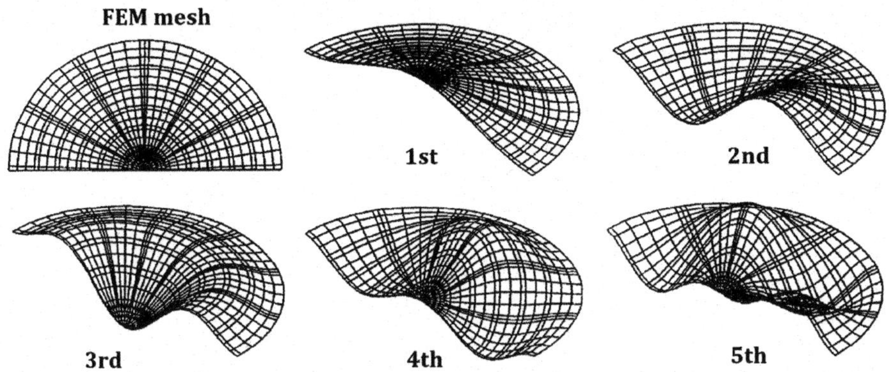

Fig. 5.5 FEM mesh and natural modes for P2 [9]

5.5 Free Transverse Vibrations of Plate with Voids

We consider the free transverse vibrations of circular plates with voids. The method of separation of variables is also employed, assuming that

$$w(r, \phi, t) = W(r, \phi)\mathrm{e}^{i\omega t} \tag{5.39}$$

in which $W(r, \phi)$ is a function of r and ϕ; and ω is a constant. Substitution of Eq. (5.39) into the equation for free transverse vibrations obtained from Eq. (5.21) becomes

$$\begin{aligned}
&\tfrac{1}{r}(d\,r\,W_{,rr})_{,rr} - \tfrac{1}{r}\left[d\left(\tfrac{1}{r}W_{,r} + \tfrac{1}{r^2}W_{,\phi\phi}\right)\right]_{,r} + \tfrac{1}{r^2}\left[d\left(\tfrac{1}{r}W_{,r} + \tfrac{1}{r^2}W_{,\phi\phi}\right)\right]_{,\phi\phi} - \tfrac{\nu}{r}(d\,W_{,rr})_{,r} \\
&+\nu\left(d\tfrac{1}{r^2}W_{,rr}\right)_{,\phi\phi} + \tfrac{\nu}{r}\left[d\left(W_{,r} + \tfrac{1}{r}W_{,\phi\phi}\right)\right]_{,rr} + \tfrac{2(1-\nu)}{r}\left[d\left(\tfrac{1}{r}W_{,r\phi} - \tfrac{1}{r^2}W_{,\phi}\right)\right]_{,\phi r} \\
&+2(1-\nu)\tfrac{1}{r^2}\left[d\left(\tfrac{1}{r}W_{,r\phi} - \tfrac{1}{r^2}W_{,\phi}\right)\right]_{,\phi} - \omega^2\tfrac{m_0\alpha_h}{D_0}W(r, \phi) = 0
\end{aligned} \tag{5.40}$$

The natural frequencies of a circular plate with voids are also computed by means of the Galerkin method. $W(r, \phi)$ is expressed as

$$W(r, \phi) = \sum_{m=1}\sum_{n=0} w_{mn} f_{rm}(r) f_{\phi n}(\phi) \tag{5.41}$$

in which w_{mn} are unknown coefficients; and $f_{rm}(r)$ and $f_{\phi n}(\phi)$ are functions satisfying the specified boundary conditions of the plates. The Galerkin equations of Eq. (5.40) can be expressed as

$$\delta w_{\bar{m}\bar{n}}\colon \sum_{m=1}\sum_{n=0} w_{mn}(A_{\bar{m}\bar{n}mn} - \lambda^2 B_{\bar{m}\bar{n}mn}) = 0 \tag{5.42}$$

where the coefficients $A_{\bar{m}\bar{n}mn}$ are given by Eq. (5.26) in the static problem; and the coefficients $B_{\bar{m}\bar{n}mm}$ and λ are defined as

$$B_{\bar{m}\bar{n}mn} = \int_0^{2\pi} \int_0^{r_0} \alpha_h f_{rm} f_{\phi n} f_{r\bar{m}} f_{\phi\bar{n}}\, r\, \mathrm{d}\phi\, \mathrm{d}r \tag{5.43}$$

$$\lambda^2 = \frac{\omega^2 m_0}{D_0} \tag{5.44}$$

Equations (5.42) are system of linear, homogeneous, simultaneous algebraic equations with respect to the unknown displacement coefficients w_{mn}. The coefficients $B_{\bar{m}\bar{n}mm}$ have nonzero values in diagonal and non-diagonal terms due to the existence of the voids. Solving Eq. (5.42) as eigenvalue problems for λ, the i-th natural frequency ω_i of a plate with voids, corresponding to the i-th value λ_i of λ, is determined from Eq. (5.44) as

$$\omega_i = \lambda_i \sqrt{\frac{D_0}{m_0}} \tag{5.45}$$

As above-mentioned, the natural frequencies of a circular plate with voids are now obtained from numerical computations using Eq. (5.42). Then, for practical use, we consider approximate expressions to directly obtain the natural frequencies. Although the natural frequencies are affected by the diagonal and non-diagonal terms in the square matrices $A_{\bar{m}\bar{n}mn}$ and $B_{\bar{m}\bar{n}mn}$, the main behavior is considered to be dominated by only the diagonal terms. Hence, taking only the diagonal terms in $A_{\bar{m}\bar{n}mn}$ and $B_{\bar{m}\bar{n}mn}$ into consideration, Eq. (5.42) becomes of an uncoupled form. Thus, the i-th approximate value λ_i is directly obtained as

$$\lambda_i \approx \sqrt{\frac{A_{mnmn}}{B_{mnmn}}} \tag{5.46}$$

The approximate value for the i-th natural frequency is easily obtained by substituting this into Eq. (5.45).

5.6 Numerical Results for Natural Frequencies

The natural frequencies for a circular plate with voids have been presented by means of the Galerkin method. Then, the exactness of the proposed analytical methodologies is examined from numerical calculations carried out for the same models used at the static problem, as shown in Fig. 5.2. The boundary conditions at $r = r_0$ are assumed to be two cases: simply supported and clamped. The following shape functions are used:

Tables 5.1 Natural frequencies of simply supported and clamped plates with voids for P3 [9]

No.	Simply supported plate				Clamped plate			
	Theory (rad/s)	FEM (rad/s)	Ratio	Mode	Theory (rad/s)	FEM (rad/s)	Ratio	Mode
1	58.12	58.79	0.989	(0,1)	126.26	126.33	0.999	(0,1)
2	172.20	173.99	0.990	(1,1)	272.42	271.60	1.003	(1,1)
3	323.14	324.21	0.997	(2,1)	449.50	448.64	1.002	(2,1)
4	355.58	358.56	0.992	(0,2)	485.24	486.25	0.998	(0,2)
5	503.20	506.78	0.993	(3,1)	652.28	656.05	0.994	(3,1)
6	606.15	608.12	0.997	(1,2)	768.55	770.93	0.997	(1,2)
7	718.02	722.40	0.994	(4,1)	890.14	896.10	0.993	(4,1)
8	893.50	891.69	1.002	(0,3)	1088.91	1088.31	1.001	(0,3)
9	896.14	894.22	1.002	(2,2)	1094.30	1088.56	1.005	(2,2)
10	966.21	969.08	0.997	(5,1)	1161.60	1166.41	0.996	(5,1)

Ratio = theory/FEM

$$f_{\phi n}(\phi) = \cos n\phi \quad \text{(for both plates)} \tag{5.47}$$

$$f_{rm}(r) = \sin\left[\frac{m\pi}{2}\left(1 - \frac{r}{r_0}\right)\right] \quad \text{(for simply supported plate)} \tag{5.48}$$

$$f_{rm}(r) = \sin\left[\frac{m\pi}{2}\left(1 - \frac{r}{r_0}\right)\right]\sin\left[\frac{\pi}{2}\left(1 - \frac{r}{r_0}\right)\right] \quad \text{(for clamped plate)} \tag{5.49}$$

in which the values of m take odd number for $n = 0$ and even number for $n \geq 1$.

Tables 5.1 and 5.2 show the numerical results of natural frequencies from first to tenth order for P3 and P4. It follows from the comparison by each analytical method that the differences between the results obtained using Eq. (5.42) and the approximate results obtained using Eq. (5.46) are negligible in practical use. Additionally, the results obtained from the proposed theory show excellent agreement when compared with results obtained from the FEM. Figure 5.5 shows the used FEM mesh and the natural modes for the clamped plate named P2.

5.7 Forced Vibrations of Plates with Voids

We consider forced vibrations of a circular plate with voids. The general solution of Eq. (5.21) for dynamic deflection $w(r, \phi, t)$ is also assumed to be of the form

$$w(r, \phi, t) = \sum_{m=1}^{\infty}\sum_{n=0}^{\infty} f_{rm}(r) f_{\phi n}(\phi) \Phi_{mn}(t) \tag{5.50}$$

Tables 5.2 Natural frequencies of simply supported and clamped plates with voids for P4 [9]

No.	Simply supported plate				Clamped plate			
	Theory (rad/s)	FEM (rad/s)	Ratio	Mode	Theory (rad/s)	FEM (rad/s)	Ratio	Mode
1	56.59	57.18	0.990	(0,1)	123.02	123.16	0.999	(0,1)
2	167.15	169.12	0.988	(1,1)	264.49	263.83	1.003	(1,1)
3	314.02	314.85	0.997	(2,1)	436.96	435.51	1.003	(2,1)
4	346.69	349.63	0.992	(0,2)	473.98	474.22	0.999	(0,2)
5	488.83	492.28	0.993	(3,1)	634.25	636.99	0.996	(3,1)
6	587.05	590.52	0.994	(1,2)	745.25	748.50	0.996	(1,2)
7	697.05	701.67	0.993	(4,1)	865.25	870.42	0.994	(4,1)
8	868.58	867.09	1.002	(0,3)	1062.11	1059.57	1.002	(0,3)
9	872.94	873.17	1.000	(2,2)	1066.06	1063.52	1.002	(2,2)
10	937.05	942.13	0.995	(5,1)	1128.90	1153.05	0.979	(5,1)

Ratio = theory/FEM

in which $\Phi_{mn}(t)$ are unknown functions of time t; and $f_{rm}(r)$ and $f_{\phi n}(\phi)$ indicate the natural functions satisfying both Eq. (5.40) and the specified boundary conditions of the circular plate. Substituting Eq. (5.50) into Eq. (5.21) and using Eq. (5.40), we can obtain the following equation for $\Phi_{mn}(t)$

$$\sum_{m=1}\sum_{n=0} f_{rm} f_{\phi n} m_0 \alpha_h [\ddot{\Phi}_{mn}(t) + 2h^*_{mn}\omega_{mn}\dot{\Phi}_{mn}(t) + \omega^2_{mn}\Phi_{mn}(t)] = p(r,\phi,t) \quad (5.51)$$

in which h^*_{mn} is defined by

$$\frac{c}{m_0\alpha_h} = 2h^*_{mn}\omega_{mn} \quad (5.52)$$

Multiplying both sides of Eq. (5.51) by $f_{r\bar{m}} f_{\phi\bar{n}}$ and integrating between 0 to r_0 and 0–2π, respectively, and approximating that h^*_{mn} at each point of current plate can be replaced with the usual damping constants h_{mn} for current plate as a whole, we have the following coupled algebraic equations

$$\delta\Phi_{\bar{m}\bar{n}}: \sum_{m=1}\sum_{n=0} K_{\bar{m}\bar{n}mn}[\ddot{\Phi}_{mn}(t) + 2h_{mn}\omega_{mn}\dot{\Phi}_{mn}(t) + \omega^2_{mn}\Phi_{mn}(t)] = \frac{1}{m_0} Q_{\bar{m}\bar{n}}(t) \quad (5.53)$$

where the notations $K_{\bar{m}\bar{n}mn}$ and $Q_{\bar{m}\bar{n}}(t)$ are defined as

$$K_{\bar{m}\bar{n}mn} = \int_0^{2\pi}\int_0^{r_0} \alpha_h(r,\phi) f_{rm} f_{r\bar{m}} f_{\phi n} f_{\phi\bar{n}} r \, d\phi \, dr \quad (5.54)$$

$$Q_{\bar{m}\bar{n}}(t) = \int_0^{2\pi}\int_0^{r_0} p(r, \phi, t) f_{r\bar{m}}\, f_{\phi\bar{n}}\, r\, d\phi\, dr \tag{5.55}$$

Since the mass coefficient α_h is a function of r and ϕ, Eq. (5.53) cannot be transformed into an uncoupled form by means of the orthogonality relations for natural functions. So Eq. (5.53) can be solved by using step-by-step integration methods, such as the Wilson-θ method. Neglecting the effects of voids, Eq. (5.53) may be easily changed to the equations for a circular normal plate without voids subjected to forced vibrations, which is mentioned in Chap. 2.

5.8 Closed-Form Approximate Solutions for Forced Vibrations

The dynamic analysis for a damped circular plate with voids has been presented to be solvable by applying the step-by-step integration method to Eq. (5.53). Then, for practical use, we consider the closed-form approximate solution for a damped circular plate with voids.

In order to formulate the closed-form approximate solution it is necessary to convert Eq. (5.53) from coupled form to uncoupled one. So, assuming that the behavior of a plate with voids is now dominated by the diagonal terms in the coefficients $K_{\bar{m}\bar{n}mn}$, Eq. (5.53) can be changed to an uncoupled form. The effectiveness of the assumption used here has been numerically demonstrated in the presentation of the approximate solutions for natural frequencies of a plate with voids. Then, Eq. (5.53) can be approximated as

$$\delta\Phi_{mn}\colon\ K_{mnmn}[\ddot{\Phi}_{mn}(t) + 2h_{mn}\omega_{mn}\dot{\Phi}_{mn}(t) + \omega_{mn}^2\Phi_{mn}(t)] = \frac{1}{m_0}Q_{mn}(t) \tag{5.56}$$

The general solution of Eq. (5.56) can be expressed as

$$\Phi_{mn}(t) = \exp(-h_{mn}\omega_{mn}t)[C_1 \sin\omega_{Dmn}t + C_2\cos\omega_{Dmn}t] + \frac{1}{K_{mnmn}m_0\omega_{Dmn}}\int_0^t \exp[-h_{mn}\omega_{mn}(t-\tau)]\sin\omega_{Dmn}(t-\tau)\, Q_{mn}(\tau)d\tau \tag{5.57}$$

in which C_1 and C_2 are constants determined from the initial conditions; and ω_{mn} and ω_{Dmn} are the natural frequencies of the undamped and damped plates with voids, respectively. The well-known relationship between ω_{mn} and ω_{Dmn} is also effective.

$$\omega_{Dmn} = \omega_{mn}\sqrt{1-h_{mn}^2} \tag{5.58}$$

The Duhamel integral in Eq. (5.57) may be calculated approximately by means of Paz [10] or Takabatake [11, 12], as already explained in Eqs. (3.68)–(3.70) in Chap. 3. The dynamic deflections of a circular plate with voids are determined by substituting $\Phi_{mn}(t)$ obtained from Eq. (5.57) into Eq. (5.50).

Next, we represent the dynamic solutions of a circular plate with voids, subjected to the following harmonic external load:

$$p(r, \phi, t) = p^*(r, \phi) \sin \omega_p t \tag{5.59}$$

where $p^*(r, \phi)$ is a function with respect to r and ϕ and indicates the distribution of external load; and ω_p is the frequency of the external loads. Then, $Q_{mn}(t)$ are expressed as

$$Q_{mn}(t) = Q_{mn}^* \sin \omega_p t \tag{5.60}$$

in which the notations Q_{mn}^* are defined as

$$Q_{mn}^* = \int_0^{2\pi} \int_0^{r_0} p^*(r, \phi) f_{mn}\, r\, \mathrm{d}\phi\, \mathrm{d}r \tag{5.61}$$

If the initial displacement and initial velocity at $t = 0$ are zero, then $\Phi_{mn}(t)$ can be written as

$$\begin{aligned}\Phi_{mn} = &\frac{1}{K_{mnmn} m_0 \omega_{Dmn}} \frac{Q_{mn}^*}{2} \left\{ \frac{h_{mn}\omega_{mn} \cos \omega_p t + (\omega_p + \omega_{Dmn}) \sin \omega_p t}{(h_{mn}\omega_{mn})^2 + (\omega_p + \omega_{Dmn})^2} \right. \\ &- \frac{h_{mn}\omega_{mn} \cos \omega_p t + (\omega_p - \omega_{Dmn}) \sin \omega_p t}{(h_{mn}\omega_{mn})^2 + (\omega_p - \omega_{Dmn})^2} \\ &+ \exp(-h_{mn}\omega_{mn} t) \left[\frac{(\omega_p + \omega_{Dmn}) \sin \omega_{Dmn} t - h_{mn}\omega_{mn} \cos \omega_{Dmn} t}{(h_{mn}\omega_{mn})^2 + (\omega_p + \omega_{Dmn})^2} \right. \\ &\left.\left. + \frac{(\omega_p - \omega_{Dmn}) \sin \omega_{Dmn} t + h_{mn}\omega_{mn} \cos \omega_{Dmn} t}{(h_{mn}\omega_{mn})^2 + (\omega_p - \omega_{Dmn})^2} \right] \right\}\end{aligned} \tag{5.62}$$

Thus, the closed-form approximate solution for a damped circular plate with voids has been presented. Since Eq. (5.62) has high expandability, it can be applied to various circular plates with voids subject to the action of a general harmonic external load. Letting $h_{mn} \to 0$, $\omega_{Dmn} \to \omega_{mn}$, and $\exp(-h_{mn}\omega_{mn} t) \to 1$ in Eq. (5.62), the closed-form solution for an undamped plate with voids becomes

$$\Phi_{mn} = \frac{Q_{mn}^*}{K_{mnmn} m_0 \omega_{mn}} \frac{\omega_p \sin \omega_{mn} t - \omega_{mn} \sin \omega_p t}{\omega_p^2 - \omega_{mn}^2} \tag{5.63}$$

While the solution for a circular solid plate without voids is obtained by replacing K_{mnmn} with 1. This solution corresponds to circular normal plates subjected to a harmonic external load, mentioned in Chap. 2.

5.9 Numerical Results for Dynamic Responses: Discussions

In order to examine the closed-form approximate solution proposed here for an isotropic circular plate with voids, numerical computations were carried out for four kinds of circular plates with voids, as shown in Fig. 5.2. The damping constants h_{mn} are assumed to be 0.03 for all modes. The following two kinds of external lateral loads without the additional mass are assumed: firstly, a harmonic and uniformly distributed force given in Eq. (5.59), in which $p^* = 9806$ N/m and $\omega_p = 62.83$ rad/s; secondly, test loads checking the disturbing effect of the vibration of floors caused by people walking and other everyday usage, as shown in Fig. 5.6. The test loads consist of four loading types: walking, tripping, bending and stretching, and heel. The current natural functions $f_{rm}(r)$ take Eq. (5.48) for a simply supported plate with voids and Eq. (5.49) for a clamped plate with voids.

Table 5.3 shows the maximum dynamic deflections at the midspan of the plates with voids, subjected to the harmonic and uniformly distributed load. The rations in this table are defined as the value obtained from the present theory to FEM. Figure 5.7 shows the time history of the dynamic deflections at the midpoint of clamped plates with voids of Type P1. In this figure, the solid line presents the numerical results obtained from Eq. (5.53) using Wilson-θ method; the broken line indicates values obtained from the closed-form approximate solution Eq. (5.62); and the solid line

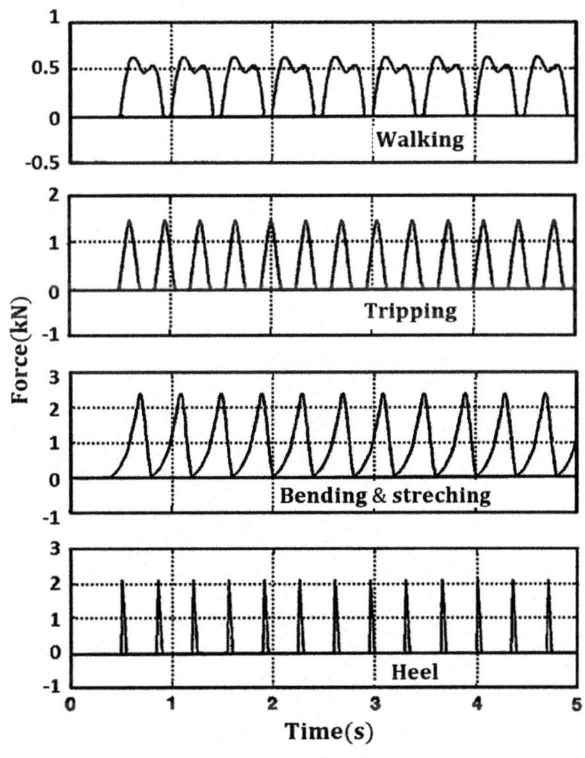

Fig. 5.6 Test loads for floors [9]

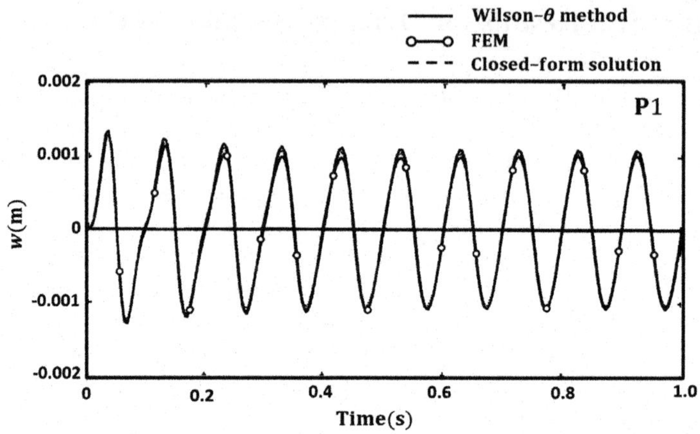

Fig. 5.7 Dynamic deflections for a clamped plate with voids for P1, subjected to harmonic and uniform load [9]

with circles indicates values obtained from FEM. The difference between the solid and broken lines is too small to plot. Thus, from the numerical results the closed-form approximate solution proposed here is examined to be applicable to the dynamic analysis of plates with voids, in practical use.

On the other hand, Table 5.4 shows the time history of the maximum dynamic deflections at the midpoint of the clamped plates with voids of Type P1, subjected to the test loads of harmonic and uniform loads at the midpoint. Figure 5.8 shows the time history of the dynamic deflections at the midpoint of the clamped plates with

Table 5.3 Maximum dynamic deflection of clamped plates with voids [9]

Type	Analytical method	Maximum dynamic deflection (m)	Ratio
P1	Wilson-θ method	0.001366	1.060
	Closed-form solution	0.001303	1.010
	FEM	0.001289	
P2	Wilson-θ method	0.001369	1.060
	Closed-form solution	0.001305	1.045
	FEM	0.001291	
P3	Wilson-θ method	0.001321	1.035
	Closed-form solution	0.001291	1.012
	FEM	0.001276	
P4	Wilson-θ method	0.001357	1.056
	Closed-form solution	0.001285	1.000
	FEM	0.001285	

Table 5.4 Maximum dynamic deflection of clamped plates with voids for P4 [9]

Load type	Analytical method	Maximum deflection (μm)	Ratio
Walking	Closed-form solution	0.9546	0.986
	FEM	0.9680	
Tripping	Closed-form solution	2.2310	0.973
	FEM	2.2930	
Bending and stretching	Closed-form solution	3.6530	0.982
	FEM	3.7210	
Heel	Closed-form solution	5.2280	1.000
	FEM	5.2280	

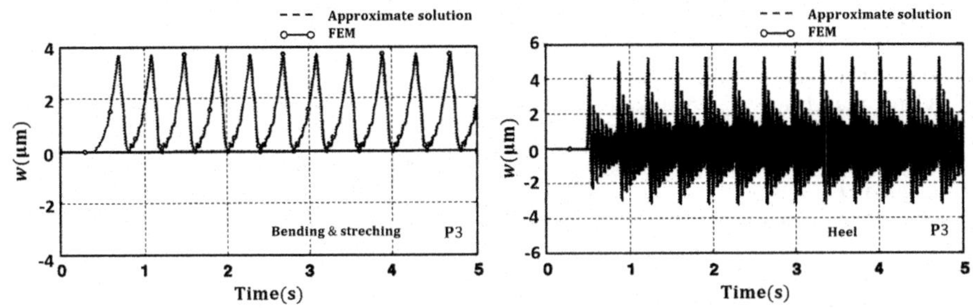

Fig. 5.8 Dynamic deflections for a clamped plate with voids for P3, subjected to test loads for floors [9]

voids of Type P3, subjected to the test loads floors. These results demonstrate that the closed-form approximate solution proposed here is also usable.

West [13] suggested that cell distortion of a plate with voids should be taken into account when the void area exceeds 60% of the total cross section. The author [14] presented the effectiveness of this limitation to the ratio $h_{i,j}/h_0$ from numerical computations for free and forced vibrations of rectangular plates with voids. This criterion is considered to be effective for current circular plates with voids.

5.10 Conclusions

The general analysis methods and closed-form approximate solutions for an isotropic circular plate with arbitrarily disposed voids have been proposed by means of the extended Dirac function. The effectiveness of the closed-form approximate solutions proposed here has been proven by comparing them with the numerical results obtained from FEM.

For the sake of simplicity, this chapter disregards the transverse shear deformation, including frame deformation and the local deformation of the top and bottom platelets of the void. It will be necessary to consider these deformations when the cross section and number of void increase. This treatment will be discussed in Chap. 7. Each void was also assumed to be a fan parallelepiped for simplicity's sake, but it is relatively easy to extend the proposed theory to a void with circular symmetric cross section.

References

1. Crisfield, M.A., Twernlow, R.P.: The equivalent plate approach for the analysis of cellular structures. Civil Engineering and Public Works, March, pp. 259–263 (1971)
2. Holmberg, A.: Shear-Weak Beams an Elastic Foundation, vol. 10. International Association for Bridge and Structural Engineering (IABSE), Zurich (1960)
3. Sawko, F., Cope, R.J.: Analysis of multi-cell bridges without transverse diaphragms—a finite element approach. Struct. Eng. **47**(11), 455–460 (1969)
4. Cope, R.J., Harris, G., Sawko, F.: A new approach to the analysis of cellular bridge decks. Anal. Struct. Syst. Torsion ACI **35**, 185–210 (1973)
5. Elliott, G.: Partial loading on orthotropic plates. Cement Concrete Assoc. Tech. Rep. **42** (London), 519 (1978)
6. Elliott, G., Clark, L.A.: Circular voided concrete slab stiffness. J. Struct. Division ASCE **108**(11), 2379–2393 (1982)
7. Szilard, R.: Theory and Analysis of Plates-Classical and Numerical Methods. Prentice-Hall, Englewood Cliffs (1974)
8. Cope, R.J., Clark, L.A.: Concrete slabs—snalysis and dsign. Elsevier Applied Science, London (1984)
9. Takabatake, H., Morimoto, H., Fujiwara, T., Honma, T.: Simplified analysis of circular plates including voids. Comput. Struct. **58**(2), 263–275 (1996)
10. Paz, M.: Structural Dynamics, Theory and Computation, 3rd edn., pp. 74–75. Van Nostrand Reinhold, New York (1985)
11. Takabatake, H., Mukai, H., Hirano, T: Doubly symmetric tube structures. I: static analysis. J. Struct. Eng., ASCE, **119**, 1981–2001 (1993)
12. Takabatake, H., Mukai, H., Hirano, T.: Doubly symmetric tube structures. II: dynamic analysis. J. Struct. Eng., ASCE, **119**, 2002–2016 (1993)
13. West, R.: Recommendations on the Use of Grillage Analysis for Slabs and Pseudo-slab Bridge Decks. Cement and Concrete Association (1973)
14. Takabatake, H.: Dynamic analyses of elastic plates with voids. Int. J. Solids Struct. **28**(7), 879–895 (1991)

Chapter 6
Static and Dynamic Analyses of Rectangular Cellular Plates

6.1 Introduction

Chapter 4 presented an analytical method for the static and dynamic analyses of rectangular plates with voids using the Kirchhoff-Love hypothesis for the case where the size of voids is less than half the thickness of the plate. This chapter extends to rectangular plates with voids of large size by changing from the Kirchhoff-Love hypothesis to the Mindlin-Reissner one.

Recently, cellular plates have been used as building slabs with long spans. The deformation is characterized by the frame deformation, as shown in Fig. 6.1. This is the flexural deformation of a frame, composed of top and bottom platelets and of partitions surrounding a void. Since the deformation of cellular plates is dominated by the frame deformation, the theory based on the validity of the Kirchhoff-Love hypothesis, as already presented in Chap. 4, cannot apply to cellular plates.

In the previous works, most analytical methods for the analysis of cellular plates are based on an equivalent plate analogy. Most of the researcher is an evaluation of

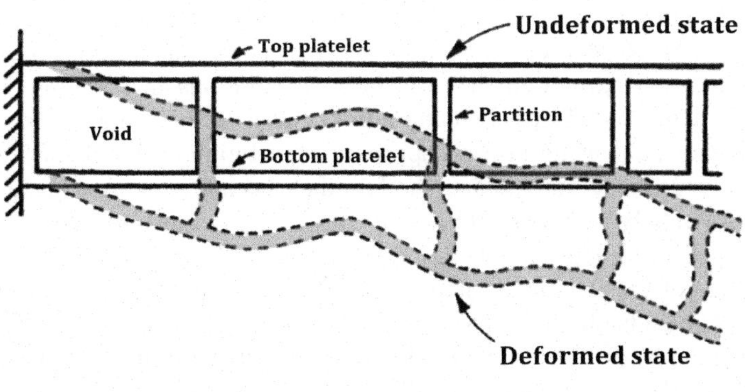

Fig. 6.1 Frame deformation of cellular plate

H. Takabatake, *Simplified Analytical Methods of Elastic Plates*,
https://doi.org/10.1007/978-981-13-0086-8_6

equivalent rigidity, for example, Crisfield and Twemlow [1], Cope et al. [2], Holmberg [3], Sawko and Cope [4], and Elliott [5]. However, the above equivalent approaches cannot apply to a plate with irregularly spaced voids and/or with voids of different cross sections, because the stiffness and mass depend on the size and position of voids used.

On the other hand, although the finite element method is effective for a cellular plate, it needs a computer of great capacity and is both costly and time-consuming for computation. So, in the preliminary stage of design, the use of the finite element method is not practical.

As a solution to these problems, the author [6, 7] presented a general and simple analytical methodology for static and dynamic analyses of plates with arbitrarily positioned voids. In this theory, the discontinuous variation of stiffness and mass of such a plate was expressed as a continuous function by means of the extended Dirac function, as already introduced in Chap. 4. Furthermore, the frame deformation is considered by an equivalent transverse shear stiffness prescribed for each void, in addition to the Mindlin-Reissner hypothesis.

Reissner [8] and Mindlin [9] presented theories including the transverse shear deformation for a rectangular solid plate, in which the bending rigidity and transverse shear stiffness were constant. However, since a cellular plate with arbitrarily positioned and arbitrarily opening voids is dominated by the frame deformation, the bending and transverse shear stiffnesses vary discontinuously. So, it is desired, for practical use, to formulate a general and simple analytical method for such a cellular plate. Takabatake et al. [10] demonstrated a general analytical method for the static and dynamic problems of rectangular cellular plates.

This chapter proposes a general and simple analytical method for a rectangular cellular plate with arbitrarily positioned voids, in which bending and the transverse shear deformations, along with the frame deformation, are considered. The discontinuous variation of the bending and transverse shear stiffnesses and of mass is treated as a continuous function by means of an extended Dirac function, as stated in Chaps. 4 and 5 [6, 7, 11]. This treatment is independent of the previous equivalent plate analogy for rectangular plates.

First, the general governing equations for a rectangular cellular plate with the transverse shear deformation along with the frame deformation are formulated by means of Hamilton's principle. Second, the static solutions are proposed by the use of the Galerkin method. Third, the natural frequencies are presented by means of the Galerkin method, and a closed-form approximate solution for the natural frequency is proposed. Fourth, the analytical method for the forced vibrations is presented to be solvable numerically by the use of the step-by-step integration method. For practical use, the closed-form approximate solution is proposed for general external loads. The exactness of the analytical methods proposed here is examined by comparing the numerical results obtained from the proposed theories with the results obtained from FEM code NASTRAN. The methodology proposed in this chapter is mainly based on Ref. [10].

6.2 Governing Equations of a Cellular Plate Including Transverse Shear Deformations Along with Frame Deformation

We consider a rectangular cellular plate with arbitrarily positioned voids. A Cartesian coordinate system *x, y, z* is also employed. Each void is assumed to be a rectangular parallelepiped whose ridgelines are parallel to the axis *x* or *y* and which is symmetrically positioned with respect to the middle plane of the cellular plate, as shown in Fig. 6.1. The midpoint, widths, and height of the *i,j*-th void are indicated by (x_i, y_j), $b_{xi,j}$, $b_{yi,j}$, and $h_{i,j}$, respectively, as shown in Fig. 6.2. The size and position of each void are arbitrary except for the above-mentioned assumptions.

This chapter considers the bending of an isotropic cellular plate subject to small deformations, including the transverse shear deformation along with the frame deformation. The deformation of cellular plates is assumed to be adequately defined by describing the geometry of its middle surface, which is a surface that bisects the plate thickness h_0 at each point. Then, neglecting the effect of shear lag and considering the in-plane mean transverse deformation, the displacement components $\bar{U}$, $\bar{V}$, and $\bar{W}$ on a general point of the cellular plate are given by deformation variables defined on the middle surface of the cellular plate as

$$\bar{U}(x, y, z, t) = z\theta_x(x, y, t) \tag{6.1}$$

$$\bar{V}(x, y, z, t) = z\theta_y(x, y, t) \tag{6.2}$$

$$\bar{W}(x, y, z, t) = w(x, y, t) \tag{6.3}$$

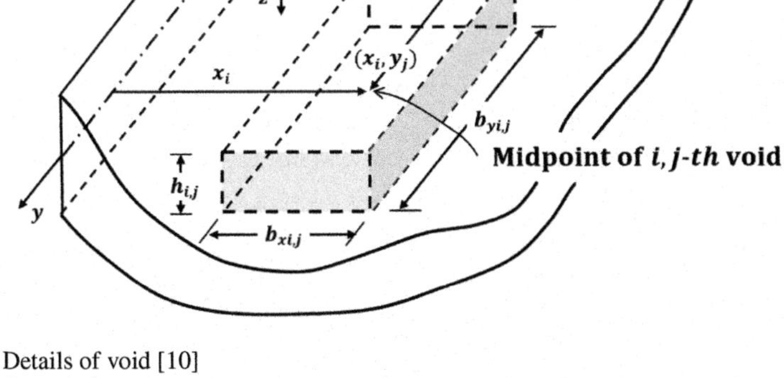

Fig. 6.2 Details of void [10]

in which $w(x, y, t)$ is the lateral displacement on the middle surface, and θ_x and θ_y are the rotational angles about y- and x-axes on the middle surface, respectively, and z is the distance measured from the middle surface of the plate. These displacement and rotational angles are considered positive when they point toward the positive direction of the coordinate axis z and clockwise with respect to the y- and x-coordinate axes, respectively.

The strain energy U of the cellular plate is given by

$$U = \frac{1}{2}\iint [M_x\theta_{x,x} + M_y\theta_{y,y} + M_{xy}(\theta_{x,y} + \theta_{y,x}) + Q_x(w_{,x} + \theta_x) + Q_y(w_{,y} + \theta_y)]\mathrm{d}x\,\mathrm{d}y \quad (6.4)$$

in which M_x, M_y, and M_{xy} are the bending and twisting moments per unit length, respectively; Q_x and Q_y indicate the transverse shearing forces per unit length. The bending and twisting moments are given by

$$M_x = D_0\mathrm{d}(x, y)(\theta_{x,x} + \nu\theta_{y,y}) \quad (6.5)$$

$$M_y = D_0\mathrm{d}(x, y)(\theta_{y,y} + \nu\theta_{x,x}) \quad (6.6)$$

$$M_{xy} = \frac{1-\nu}{2}D_0\mathrm{d}(x, y)(\theta_{x,y} + \theta_{y,x}) \quad (6.7)$$

in which the flexural stiffness $D_0 = Eh_0^3/[12(1-\nu^2)]$ for solid plate without voids. Here, E is Young's modulus and ν is Poisson's ratio. On the other hand, the stiffness coefficient $\mathrm{d}(x, y)$ indicating the variation of stiffness due to voids is defined as

$$\mathrm{d}(x, y) = 1 - \alpha_{i,j}D(x - x_i)D(y - y_j) \quad (6.8)$$

in which $\alpha_{i,j}$ is defined as

$$\alpha_{i,j} = \left(\frac{h_{i,j}}{h_0}\right)^3 \quad (6.9)$$

and $D(x - x_i)$ and $D(y - y_j)$ are the extended Dirac functions, as already used in Chap. 4. The function $D(x - x_i)$ is defined as a function where the Dirac function $\delta(x - \xi)$ exists continuously in the x-direction through the i,j-th void, namely the region from $x_i - b_{xi,j}/2$ to $x_i + b_{xi,j}/2$ in which ξ can take values from $x_i - b_{xi,j}/2$ to $x_i + b_{xi,j}/2$. Similarly, the function $D(y - y_j)$ is defined as a function where the Dirac function $\delta(y - \eta)$ exists continuously in the y-direction through the i,j-th void, in which η can take values from $y_j - b_{yi,j}/2$ to $y_j + b_{yi,j}/2$.

Meanwhile, the transverse shearing forces can be written as

$$Q_x = \kappa_x G_x h_0(w_{,x} + \theta_x) \quad (6.10)$$

$$Q_y = \kappa_y G_y h_0(w_{,y} + \theta_y) \quad (6.11)$$

The factors κ_x and κ_y are introduced to account for the nonuniformity of the shear strains over the cross section. The transverse shear stiffnesses $\kappa_x G_x h_0$ and $\kappa_y G_y h_0$ are determined from both the frame deformation and the transverse shear deformation of the plate elements, as given in the following section. Hence, Eq. (6.4) becomes

$$U = \frac{1}{2}\iint \Big\{ D_0 d\Big[(\theta_{x,x})^2 + 2\nu\theta_{x,x}\theta_{y,y} + (\theta_{y,y})^2 + \frac{1-\nu}{2}(\theta_{x,y} + \theta_{y,x})^2 \Big] + \kappa_x G_x h_0 (w_{,x} + \theta_x)^2 + \kappa_y G_y h_0 \big(w_{,y} + \theta_y\big)^2 \Big\} \mathrm{d}x\,\mathrm{d}y \tag{6.12}$$

Next, we consider the variation of the potential of external lateral loads and damping as follows:

$$\delta V = -\iint p(x, y, t)\delta w\,\mathrm{d}x\,\mathrm{d}y + \iint c\dot{w}\delta w\,\mathrm{d}x\,\mathrm{d}y - \int_c \big[m_x^*\delta\theta_x + m_{xy}^*\delta\theta_y\big]_0^{l_x}\mathrm{d}y - \int_c \big[m_y^*\delta\theta_y + m_{yx}^*\delta\theta_x\big]_0^{l_y}\mathrm{d}x - \int_c \big[v_x^*\delta w\big]_0^{l_x}\mathrm{d}y - \int_c \big[v_y^*\delta w\big]_0^{l_y}\mathrm{d}x \tag{6.13}$$

where p are lateral loads; v_x^*, m_x^*, and m_{xy}^* are external transverse force, external moment, and external twisting moment, respectively, prescribed on the mechanical boundary edges at $x = 0$ and $x = l_x$; v_y^*, m_y^*, and m_{yx}^* are external transverse forces, external moment, and external twisting moments, respectively, prescribed on the mechanical boundary edges at $y = 0$ and $y = l_y$.

Next, the kinetic energy T for current cellular plate can be given as

$$T = \frac{1}{2}\iint \big\{ I_p(x, y)[(\dot{\theta}_x)^2 + (\dot{\theta}_y)^2] + m_0\alpha_h(x, y)(\dot{w})^2 \big\}\mathrm{d}x\,\mathrm{d}y \tag{6.14}$$

in which the dot indicates differentiation with respect to time. The notations α_h, m_0, and I_p are defined as follows

$$\alpha_h(x, y) = 1 - \frac{h_{i,j}}{h_0} D(x - x_i)D(y - y_j) \tag{6.15}$$

$$m_0 = \rho h_0 \tag{6.16}$$

$$I_p = \frac{\rho h_0^3}{12}\mathrm{d}(x, y) \tag{6.17}$$

where ρ is the mass density of the plate.

Substituting Eqs. (6.12)–(6.14) into Hamilton's principle, we can obtain the differential equation of motion for current rectangular cellular plates as follows:

$$\delta w : m_0\alpha_h\ddot{w} - [\kappa_x G_x h_0(w_{,x} + \theta_x)]_{,x} - [\kappa_y G_y h_0(w_{,y} + \theta_y)]_{,y} - p + c\dot{w} = 0 \tag{6.18}$$

$$\delta\theta_x : I_p\ddot{\theta}_x + \kappa_x G_x h_0(w,_x + \theta_x) - [D_0 d(\theta_{x,x} + \nu\theta_{y,y})],_x$$
$$- \frac{1-\nu}{2}[D_0 d(\theta_{x,y} + \theta_{y,x})],_y = 0 \tag{6.19}$$
$$\delta\theta_y : I_p\ddot{\theta}_y + \kappa_y G_y h_0(w,_y + \theta_y) - [D_0 d(\theta_{y,y} + \nu\theta_{x,x})],_y$$
$$- \frac{1-\nu}{2}[D_0 d(\theta_{x,y} + \theta_{y,x})],_x = 0 \tag{6.20}$$

Also, the associated boundary conditions can be given simultaneously as follows:

$$w = w^* \quad \text{or } \kappa_x G_x h_0(w,_x + \theta_x) = v_x^* \tag{6.21}$$
$$\theta_x = \theta_x^* \quad \text{or } D_0 d(\theta_{x,x} + \nu\theta_{y,y}) = m_x^* \tag{6.22}$$
$$\theta_y = \theta_y^* \quad \text{or } \tfrac{1-\nu}{2} D_0 d(\theta_{x,y} + \theta_{y,x}) = m_{xy}^* \tag{6.23}$$

at $x=0$ and l_x; and

$$w = w^* \quad \text{or } \kappa_y G_y h_0(w,_y + \theta_y) = v_y^* \tag{6.24}$$
$$\theta_x = \theta_x^* \quad \text{or } \tfrac{1-\nu}{2} D_0 d(\theta_{x,y} + \theta_{y,x}) = m_{yx}^* \tag{6.25}$$
$$\theta_y = \theta_y^* \quad \text{or } D_0 d(\theta_{y,y} + \nu\theta_{x,x}) = m_y^* \tag{6.26}$$

at $y=0$ and l_y, where w^*, θ_x^*, and θ_y^* are boundary displacement components prescribed on the geometrical boundary. On the other hand, v_x^*, v_y^*, m_x^*, m_y^*, m_{xy}^*, and m_{yx}^* are boundary force and moments, respectively, prescribed on the mechanical boundary.

The above-mentioned governing equations are applicable to all rectangular plates including and excluding the variation of the stiffness and mass. The general governing equations contain the rectangular normal plates mentioned in Chap. 1 and the rectangular plates with voids mentioned in Chap. 4. For solid plates without voids, the coefficients α_h and d become all 1; the reduced governing equations for rectangular normal plates equals the theory proposed by Washizu [12] and Mindlin [9]. The governing equations of solid (normal) rectangular plates based on the Mindlin-Reissner hypothesis are given in Appendix 3. Furthermore, neglecting the transverse shear deformation, the governing equations proposed here reduce to the general equations for a rectangular plate with relative small voids excluding the transverse shear deformation, as already described in Chap. 4. Table 6.1 represents the relationships among the above-mentioned theories.

Table 6.1 Status of the proposed theory for rectangular plate

Rectangular plate	Deformation	Unknown displacement	Equations of motion	Note
Cellular plate	Transverse shear deformation Frame deformation (Mindlin-Reissner hypothesis)	w, θ_x, θ_y	Equations (6.18–6.20) Chap. 6	$d(x, y)$ $\alpha_h(x, y)$
Solid plate	Transverse shear deformation (Mindlin-Reissner hypothesis)	w, θ_x, θ_y	Equations (6.102–6.104) Chap. 6	$d(x, y) = 1$ $\alpha_h(x, y) = 1$
Void plate	Kirchhoff-Love hypothesis	w	Equation (4.25) Chap. 4	$d(x, y)$ $\alpha_h(x, y)$
Solid plate	Kirchhoff-Love hypothesis	w	Equation (1.87) Chap. 1	$d(x, y) = 1$ $\alpha_h(x, y) = 1$

6.3 Transverse Shear Stiffness of Cellular Plates

The bending and transverse shear stiffnesses of cellular plates decrease due to voids. The former reduction is already considered by the stiffness coefficient $d(x, y)$ given in Eq. (6.8). The latter reduction is determined from the frame deformation and the transverse shear deformation of the top and bottom platelets and of partitions surrounding a void, as shown in Fig. 6.1.

We consider the transverse shear stiffness of a cellular plate, as shown in Fig. 6.3. Assuming that the influence points of top and bottom platelets and of partitions surrounding a void are situated on the midspan of each member, the transverse shear stiffness $\kappa_x G_x h_0$ per unit width in x-direction of the cellular plate is obtained easily from Eq. (72) in Takabatake et al. [13, 14] by

$$\kappa_x G_x h_0 = \frac{1}{\frac{b^*_{xi,j}\left(\frac{b^*_{xi,j}}{I_{c1}+I_{c2}}+\frac{h_0}{I_b}\right)}{12E} + \frac{1}{\kappa G(h_0 - h_{i,j})} + \frac{b^*_{xi,j}}{\kappa G h_0 b_{x0}}} \tag{6.27}$$

in which G is the shear modulus and κ represents the distribution coefficient of transverse shear stress. The coefficient κ has been included to account for the nonuniformity of the shear stresses over the members. This equation is also obtained by adding the shear deformation of members, such as platelets and partition, to the equation given by Smith et al. [15]. In Eq. (6.27), the first term of denominator in the right side represents the frame deformation; the second and third terms represent the shear

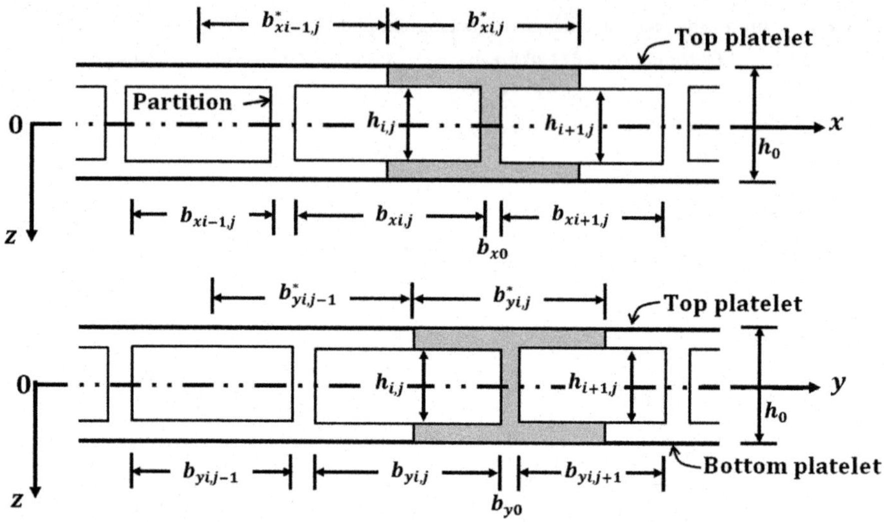

Fig. 6.3 Cross section of cellular plates [10]

deformations of only the top and bottom platelets and of partitions in current voided plate, respectively. The notations I_{c1}, I_{c2}, and I_b are defined as

$$I_{c1} = I_{c2} = \frac{1}{24}\left[\left(\frac{h_0 - h_{i,j}}{2}\right)^3 + \left(\frac{h_0 - h_{i+1,j}}{2}\right)^3\right] \tag{6.28}$$

$$I_b = \frac{b_{x0}^3}{12} \tag{6.29}$$

When the void is surrounded by partitions like shear wall, as shown in Fig. 6.4, the transverse shear stiffnesses for cellular plates must add the transverse shear stiffness of paralleled partitions (like shear walls) to the transverse shear stiffness $(\kappa_x G_x h_0)^*$ given by Eq. (6.27). Hence,

$$\kappa_x G_x h_0 = (\kappa_x G_x h_0)^* + \kappa G h_0 \frac{b_{y0}}{b^*_{yi,j}} \tag{6.30}$$

When a void passes through the x-direction, as shown in Fig. 6.5, the transverse shear stiffness in the x-direction of the cellular plate is based on only the shear stiffness of the paralleled partitions as the shear walls and is given as

$$\kappa_x G_x h_0 = \kappa G h_0 \frac{b_{y0}}{b^*_{yi,j}} \tag{6.31}$$

Similarly, the transverse shear stiffness $\kappa_y G_y h_0$ in the y-direction may be expressed by replacing b_{x0}, $b^*_{xi,j}$, and $b^*_{yi,j}$ with b_{y0}, b_{x0}, and $b^*_{yi,j}$, in the mentioned

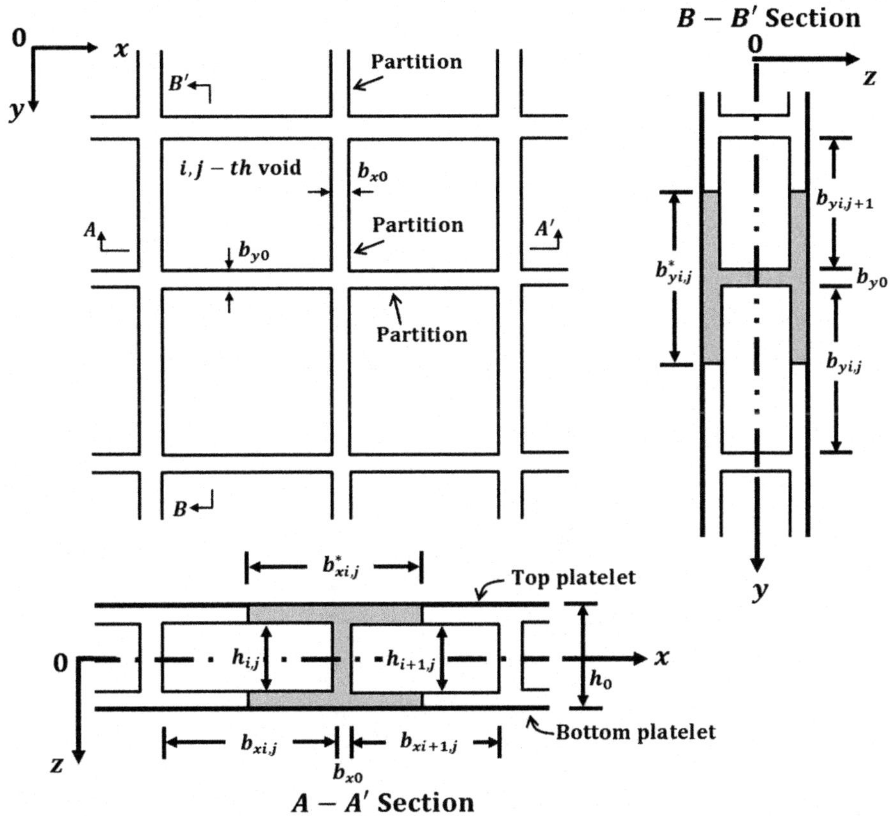

Fig. 6.4 Cellular plate with crossed partitions [10]

expressions of $\kappa_x G_x h_0$ and $b^*_{xi,j}$, respectively. Thus, the transverse shear stiffness of cellular plates has been given in simple form. It must be noticed that the method proposed here replaces the transverse shear stiffness of the cellular plate in only a field of the width $b^*_{xi,j}$ or $b^*_{yi,j}$ prescribed by the *i,j*-th void with an equivalent transverse shear stiffness.

Thus, the equivalent transverse shear stiffnesses per the *i,j*-th void are expressed in general as

$$\left.\begin{aligned} \kappa_x G_x h_0 &= \alpha_{Gxi,j} \kappa G h_0 \\ \kappa_y G_y h_0 &= \alpha_{Gyi,j} \kappa G h_0 \end{aligned}\right\} \tag{6.32}$$

in which the coefficients α_{Gx} and α_{Gy} indicating the variation of the transverse shear stiffness due to voids are defined as

$$\alpha_{Gxi,j} = \frac{\kappa_x G_x h_0}{\kappa G h_0} D(x - \hat{x}_i) D(y - \hat{y}_j) \tag{6.33}$$

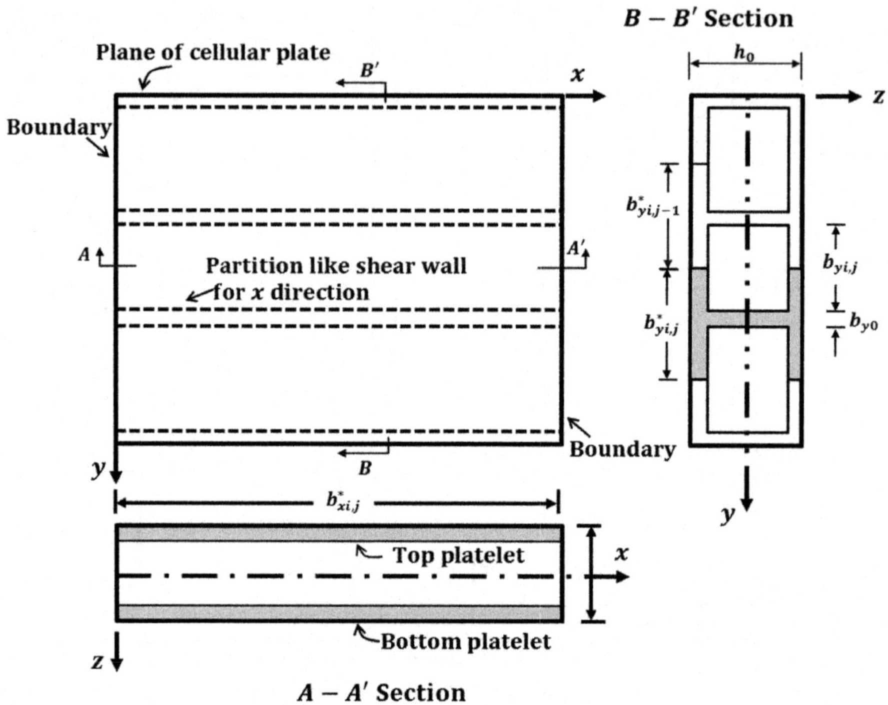

Fig. 6.5 Plate with voids passed through [10]

$$\alpha_{Gyi,j} = \frac{\kappa_y G_y h_0}{\kappa G h_0} D(x - \hat{x}_i) D(y - \hat{y}_j) \tag{6.34}$$

where $\hat{x}_i$ and $\hat{y}_j$ indicate the fields (from $x_i - b^*_{xi-1,j}/2$ to $x_i + b^*_{xi,j}/2$ and from $y_j - b^*_{yi,j-1,j}/2$ to $y_j + b^*_{yi,j}/2$, respectively) prescribed by the *i,j*-th void. It is noticed that for solid plates or a portion without voids, the coefficients α_{Gx} and α_{Gy} become all 1.

6.4 Stress Resultants and Stress Couples of Platelets and Partition

Once the transverse shearing forces Q_x and Q_y are obtained from Eqs. (6.10) and (6.11), the stress resultants and stress couples of the frame and partition parallel to the *x*-direction, as shown in Fig. 6.6, are calculated by the following process.

First, the transverse shearing forces $Q_{x\text{frame}}$ and $Q_{x\text{partition}}$ of the frame and partition located parallel to the *x*-direction are given by

$$Q_{x\text{frame}} = Q_x \frac{(\kappa_x G_x h_0)^*}{\kappa_x G_x h_0} \tag{6.35}$$

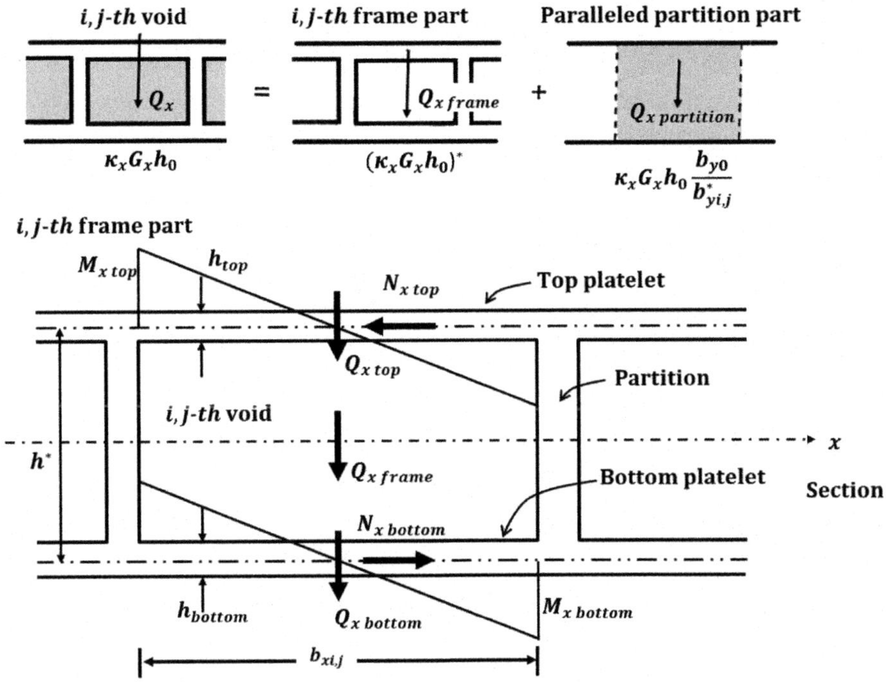

Fig. 6.6 Stress resultants and stress couples of top and bottom platelets [10]

$$Q_{x\text{partition}} = Q_x \frac{\kappa G h_0}{\kappa_x G_x h_0} \frac{b_{y0}}{b^*_{yi,j}} \tag{6.36}$$

where the frame is composed of the top and bottom platelets and of partition per void and excludes the partition parallel like shear wall to the direction.

Second, the transverse shearing force $Q_{x\text{frame}}$ of the frame located in the x-direction is divided by the transverse shearing forces $Q_{x\text{top}}$ and $Q_{x\text{bottom}}$ of the top and bottom platelets in proportion to the thickness as

$$Q_{x\text{top}} = Q_{x\text{frame}} \frac{h_{\text{top}}}{h_{\text{top}} + h_{\text{bottom}}} \tag{6.37}$$

$$Q_{x\text{bottom}} = Q_{x\text{frame}} \frac{h_{\text{bottom}}}{h_{\text{top}} + h_{\text{bottom}}} \tag{6.38}$$

where h_{top} and h_{bottom} are the slab's thickness of the top and bottom platelets, respectively.

Third, the flexural moments $M_{x\text{top}}$ and $M_{x\text{bottom}}$ of the top and bottom platelets at both ends of the i,j-th void are given by, as shown in Fig. 6.6,

$$M_{x\text{top}} = Q_{x\text{top}} \frac{b_{xi,j}}{2} \tag{6.39}$$

$$M_{x\text{bottom}} = Q_{x\text{bottom}} \frac{b_{xi,j}}{2} \tag{6.40}$$

Lastly, the axial forces $N_{x\text{top}}(= -N_{x\text{bottom}})$ of the top and bottom platelets, due to the flexural moments, are calculated by

$$N_{x\text{top}} = \frac{M_x + M_{x\text{top}} + M_{x\text{bottom}}}{h^*} \tag{6.41}$$

in which h^* is distance between middle surfaces of top and bottom platelets. Similar expressions for y-direction can be obtained.

6.5 Static Analysis

We consider the static solutions of general rectangular cellular plates by means of the Galerkin method. The deflections w and rotational angles θ_x and θ_y can be expressed by a series expansion as follows:

$$w(x, y) = \sum_{m=1} \sum_{n=1} w_{mn} f_{mn}(x, y) \tag{6.42}$$

$$\theta_x(x, y) = \sum_{m=1} \sum_{n=1} \theta_{xmn} g_{xmn}(x, y) \tag{6.43}$$

$$\theta_y(x, y) = \sum_{m=1} \sum_{n=1} \theta_{ymn} g_{ymn}(x, y) \tag{6.44}$$

in which w_{mn}, θ_x, and θ_y are unknown displacement coefficients, and f_{mn}, g_{xmn}, and g_{ymn} indicate shape functions satisfying the specified boundary conditions.

We also use the Galerkin method for the static problem of rectangular cellular plates. The Galerkin equations for the static problem can be given as

$$\delta w_{mn} : \sum_{\bar{m}=1} \sum_{\bar{n}=1} w_{\bar{m}\bar{n}} A_{1mn\bar{m}\bar{n}} + \sum_{\bar{m}=1} \sum_{\bar{n}=1} \theta_{x\bar{m}\bar{n}} A_{2mn\bar{m}\bar{n}} + \sum_{\bar{m}=1} \sum_{\bar{n}=1} \theta_{y\bar{m}\bar{n}} A_{3mn\bar{m}\bar{n}} = p_{mn} \tag{6.45}$$

$$\delta \theta_{xmn} : \sum_{\bar{m}=1} \sum_{\bar{n}=1} w_{\bar{m}\bar{n}} B_{1mn\bar{m}\bar{n}} + \sum_{\bar{m}=1} \sum_{\bar{n}=1} \theta_{x\bar{m}\bar{n}} B_{2mn\bar{m}\bar{n}} + \sum_{\bar{m}=1} \sum_{\bar{n}=1} \theta_{y\bar{m}\bar{n}} B_{3mn\bar{m}\bar{n}} = 0 \tag{6.46}$$

$$\delta \theta_{ymn} : \sum_{\bar{m}=1} \sum_{\bar{n}=1} w_{\bar{m}\bar{n}} C_{1mn\bar{m}\bar{n}} + \sum_{\bar{m}=1} \sum_{\bar{n}=1} \theta_{x\bar{m}\bar{n}} C_{2mn\bar{m}\bar{n}} + \sum_{\bar{m}=1} \sum_{\bar{n}=1} \theta_{y\bar{m}\bar{n}} C_{3mn\bar{m}\bar{n}} = 0 \tag{6.47}$$

where the coefficients $A_{1mn\bar{m}\bar{n}}, \ldots, C_{3mn\bar{m}\bar{n}}$ are given in Appendix 1 The unknown displacement coefficients are determined from solving a set of algebraic linear equations given by Eqs. (6.45)–(6.47).

Table 6.2 Lists of isotropic rectangular cellular plates [10]

Type	Plane	Section	$h_{i,j}$ (m)	$b_{xi,j}$ (m)	$b_{yi,j}$ (m)
P1	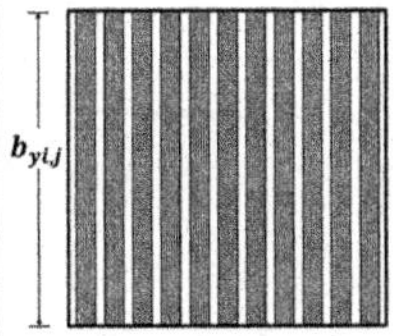	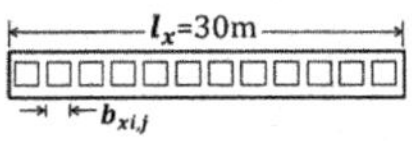	0.8	2.3	30.0
P2			0.8	2.4	30.0
P3	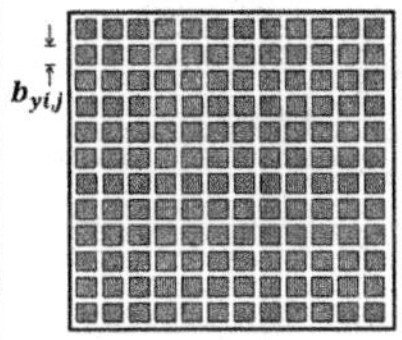	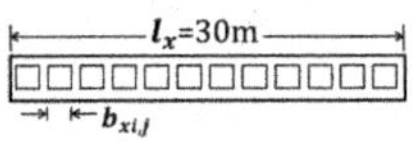	0.8	2.3	2.3
P4			0.8	2.4	2.4

6.6 Numerical Results for Static Calculation

In the preceding section, the static analysis for a rectangular cellular plate has been presented by means of the Galerkin method. Then, in order to examine the method proposed here, numerical calculations are performed for four types with simply supported and clamped cellular plates, as shown in Table 6.2. Data used are as follows: Young's modulus $E = 2.06 \times 10^{10}$ N/m^2; Poisson's ratio $\nu = 0.17$; slab's height $h_0 = 1$ m; span lengths $l_x = l_y = 30$ m; each void's height $h_{i,j} = 0.8$ m. For simplicity, the lateral load is assumed to be uniformly distributed load $p = 9.807 \times 10^6$ N/m^2; the mass density $\rho = 2440$ kg/m^3.

The shape functions used are

$$f_{mn}(x, y) = \sin \frac{m\pi x}{l_x} \sin \frac{n\pi y}{l_y} \tag{6.48}$$

$$g_{xmn}(x, y) = \cos \frac{m\pi x}{l_x} \sin \frac{n\pi y}{l_y} \tag{6.49}$$

$$g_{ymn}(x, y) = \sin \frac{m\pi x}{l_x} \cos \frac{n\pi y}{l_y} \tag{6.50}$$

for the simply supported plates and

$$f_{mn}(x, y) = \bar{x}^m(\bar{x}^2 - 1)\bar{y}^n(\bar{y}^2 - 1) \tag{6.51}$$

$$g_{xmn}(x, y) = \bar{x}^{m+1}(\bar{x}^2 - 1)\bar{y}^n(\bar{y}^2 - 1) \tag{6.52}$$

$$g_{ymn}(x, y) = \bar{x}^m(\bar{x}^2 - 1)\bar{y}^{n+1}(\bar{y}^2 - 1) \tag{6.53}$$

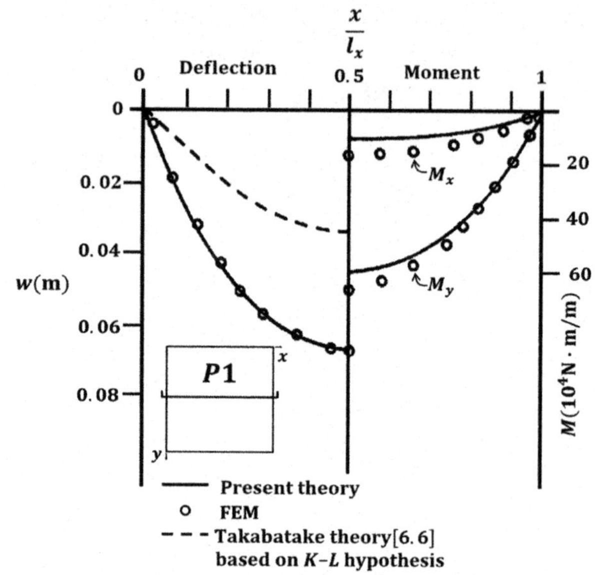

Fig. 6.7 w and M_y for a simply supported cellular plate [10]

for the clamped plate, in which $\bar{x}$ and $\bar{y}$ are defined as $\bar{x} = 2x/l_x$ and $\bar{y} = 2y/l_y$, respectively.

Figures 6.7 and 6.8 show the distributions of the deflections w and bending moments M_x and M_y for the simply supported and clamped plates, respectively. In these figures, the solid lines represent the numerical results obtained from the proposed theory; the circles represent the numerical results obtained by using FEM code NASTRAN; the broken lines indicate the numerical results obtained from theory [6] excluding the transverse shear deformation, which indicates Eq. (4.35) based on the Kirchhoff-Love hypothesis, as already stated in Chap. 4. It follows from these figures that the results obtained from the theory proposed here demonstrate relatively good agreement with the results obtained from the finite element method. The finite element method used here is based on isotropic and rectangular plate elements proposed on the top and bottom platelets and partitions of the voids, as shown in Fig. 6.9. The result obtained from the plate element used here in FEM is confirmed to show excellent agreement with the result using solid element to current plates.

Figure 6.10 represents the distributions of the transverse shearing forces, flexural moments, and axial forces of top and bottom platelets at adjacent midspan for clamped cellular plate P2. Figure 6.11 shows the distributions of the flexural moment of partition (at section B in Fig. 6.11) and the axial forces of top and bottom platelets (at section C in Fig. 6.11) for the same cellular plate. It is clarified from these figures that the simplified theory proposed here also shows agreement with the results using the finite element method. For the other plates, a similar good agreement is obtained.

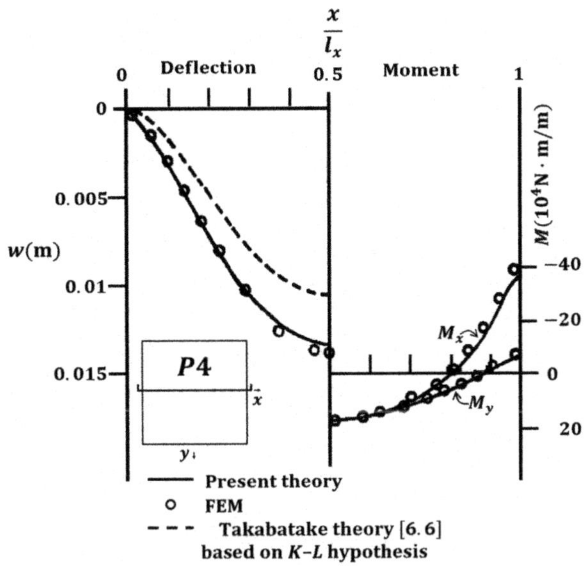

Fig. 6.8 w and M_y for a clamped cellular plate [10]

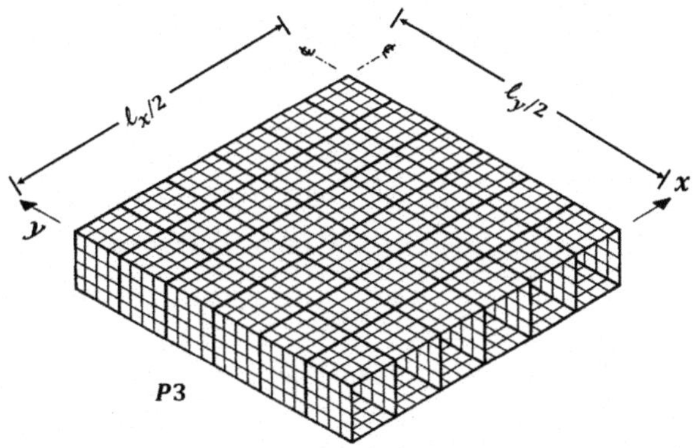

Fig. 6.9 Mesh of FEM [10]

The convergence of series expansion is very rapid for uniform loads. So, the consideration of 16 terms gives accuracy sufficient for all practical purposes.

6.7 Free Transverse Vibrations of Cellular Plates

We consider free transverse vibrations of a cellular plate. The method of separation of variables is also employed, assuming that

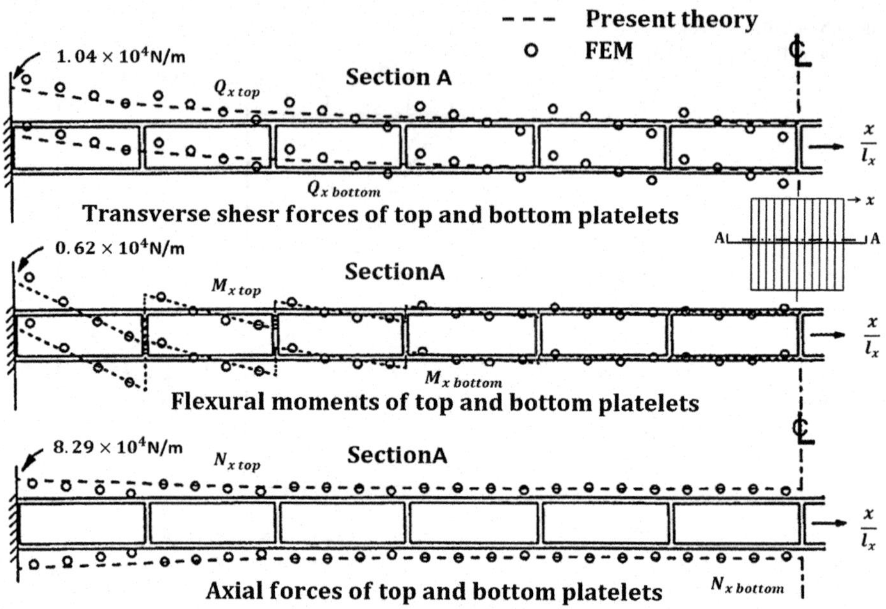

Fig. 6.10 Transverse shearing forces, flexural moments, and axial forces of top and bottom platelets for clamped P2 [10]

$$w(x, y, t) = \bar{w}(x, y)e^{i\omega t}(t) \tag{6.54}$$

$$\theta_x(x, y, t) = \bar{\theta}_x(x, y)e^{i\omega t}(t) \tag{6.55}$$

$$\theta_y(x, y, t) = \bar{\theta}_y(x, y)e^{i\omega t}(t) \tag{6.56}$$

in which $\bar{w}$, $\bar{\theta}_x$, and $\bar{\theta}_y$ are functions of x and y. The substitution of Eqs. (6.54)–(6.56) into the expressions for free transverse vibrations obtained from Eqs. (6.18)–(6.20) becomes

$$\delta w: \quad \omega^2 m_0 \alpha_h \bar{w} + [\kappa_x G_x h_0(\bar{w}_{,x} + \bar{\theta}_x)]_{,x} + [\kappa_y G_y h_0(\bar{w}_{,y} + \bar{\theta}_y)]_{,y} = 0 \tag{6.57}$$

$$\begin{aligned} \delta\theta_x: \quad & -\omega^2 I_p \bar{\theta}_x + \kappa_x G_x h_0(\bar{w}_{,x} + \bar{\theta}_x) - [D_0 d(\bar{\theta}_{x,x} + \nu\bar{\theta}_{y,y})]_{,x} \\ & - \frac{1-\nu}{2}[D_0 d(\bar{\theta}_{x,y} + \bar{\theta}_{y,x})]_{,y} = 0 \end{aligned} \tag{6.58}$$

$$\begin{aligned} \delta\theta_y: \quad & -\omega^2 I_p \bar{\theta}_y + \kappa_y G_y h_0(\bar{w}_{,y} + \bar{\theta}_y) - [D_0 d(\bar{\theta}_{y,y} + \nu\bar{\theta}_{x,x})]_{,y} \\ & - \frac{1-\nu}{2}[D_0 d(\bar{\theta}_{x,y} + \bar{\theta}_{y,x})]_{,x} = 0 \end{aligned} \tag{6.59}$$

in which ω is a constant.

The natural frequencies are also presented by means of the Galerkin method. The functions $\bar{w}$, $\bar{\theta}_x$, and $\bar{\theta}_y$ are expressed as

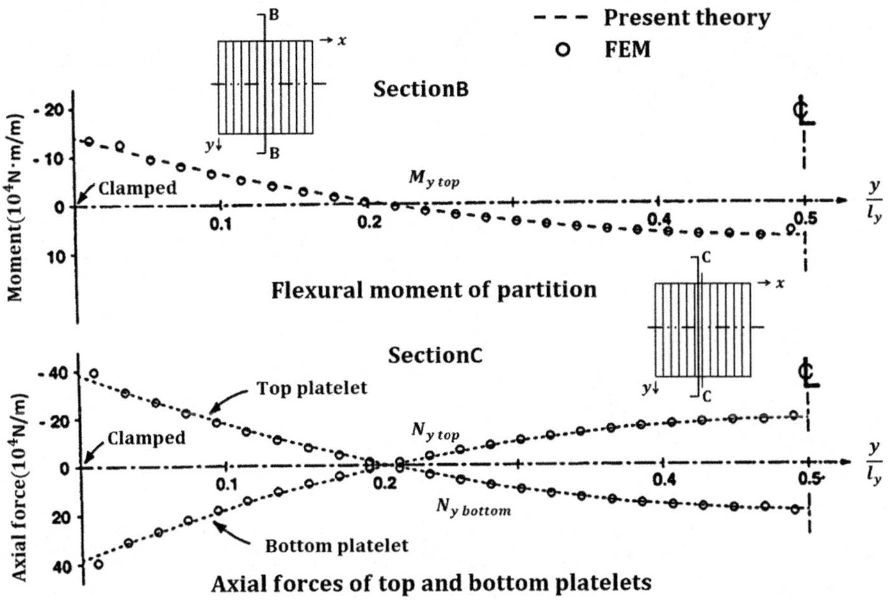

Fig. 6.11 Flexural moment of partition and axial forces of top and bottom plates for clamped P2 [10]

$$\bar{w}(x, y) = w_{mn}\, f_{mn}(x, y) \tag{6.60}$$

$$\bar{\theta}_x(x, y) = \theta_{xmn}\, g_{xmn}(x, y) \tag{6.61}$$

$$\bar{\theta}_y(x, y) = \theta_{ymn}\, g_{ymn}(x, y) \tag{6.62}$$

in which w_{mn}, θ_{xmn}, and θ_{ymn} are unknown displacement coefficients; and f_{mn}, g_{xmn}, and g_{ymn} are shape functions satisfying the specified boundary conditions. Hence, the Galerkin equations of Eqs. (6.57)–(6.59) become

$$\delta w_{mn}:\quad w_{\bar{m}\bar{n}}(A_{1mn\bar{m}\bar{n}} + \omega^2 m_0\, F_{1mn\bar{m}\bar{n}}) + \theta_{x\bar{m}\bar{n}} A_{2mn\bar{m}\bar{n}} + \theta_{y\bar{m}\bar{n}} A_{3mn\bar{m}\bar{n}} = 0 \tag{6.63}$$

$$\delta\theta_{xmn}:\quad w_{\bar{m}\bar{n}} B_{1mn\bar{m}\bar{n}} + \theta_{x\bar{m}\bar{n}}(B_{2mn\bar{m}\bar{n}} - \omega^2 F_{2mn\bar{m}\bar{n}}) + \theta_{y\bar{m}\bar{n}} B_{3mn\bar{m}\bar{n}} = 0 \tag{6.64}$$

$$\delta\theta_{ymn}:\quad w_{\bar{m}\bar{n}} C_{1mn\bar{m}\bar{n}} + \theta_{x\bar{m}\bar{n}} C_{2mn\bar{m}\bar{n}} + \theta_{y\bar{m}\bar{n}}(C_{3mn\bar{m}\bar{n}} - \omega^2 F_{3mn\bar{m}\bar{n}}) = 0 \tag{6.65}$$

in which $F_{1mn\bar{m}\bar{n}}$, $F_{2mn\bar{m}\bar{n}}$, and $F_{3mn\bar{m}\bar{n}}$ are defined as

$$F_{1mn\bar{m}\bar{n}} = \int_0^{l_x}\int_0^{l_y} f_{\bar{m}\bar{n}} f_{mn}\,\mathrm{d}x\,\mathrm{d}y - \sum_{i=1}\sum_{j=1}\int_0^{l_x}\int_0^{l_y}\left(\frac{h_{i,j}}{h_0}\right) D(x - x_i)\, D(y - y_j) f_{\bar{m}\bar{n}} f_{mn}\,\mathrm{d}x\,\mathrm{d}y \tag{6.66}$$

$$F_{2mn\bar{m}\bar{n}} = \frac{\rho h_0^3}{12}\left[\int_0^{l_x}\int_0^{l_y} g_{x\bar{m}\bar{n}}\, g_{xmn} \mathrm{d}x\, \mathrm{d}y \right.$$

$$\left. - \sum_{i=1}\sum_{j=1}\int_0^{l_x}\int_0^{l_y} \alpha_{i,j}\, D(x-x_i)\, D(y-y_j) g_{x\bar{m}\bar{n}} g_{xmn} \mathrm{d}x\, \mathrm{d}y \right] \tag{6.67}$$

$$F_{3mn\bar{m}\bar{n}} = \frac{\rho h_0^3}{12}\left[\int_0^{l_x}\int_0^{l_y} g_{y\bar{m}\bar{n}}\, g_{ymn} \mathrm{d}x\, \mathrm{d}y \right.$$

$$\left. - \sum_{i=1}\sum_{j=1}\int_0^{l_x}\int_0^{l_y} \alpha_{i,j}\, D(x-x_i)\, D(y-y_j) g_{y\bar{m}\bar{n}} g_{ymn} \mathrm{d}x\, \mathrm{d}y \right] \tag{6.68}$$

The natural frequencies ω_{mn} are obtained by solving the coupled equations composed of Eqs. (6.63)–(6.65) as eigenvalue problems for a prescribed value of $m = \bar{m}$ and $n = \bar{n}$.

Then, for practical uses, we consider approximate expressions for the natural frequencies. Equations (6.63–6.65) are coupled equations concerning w_{mn}, θ_{xmn}, θ_{ymn}, and ω_{mn}. Expressing θ_{xmn} and θ_{ymn} in terms of both w_{mn} and ω^2 from Eqs. (6.63)–(6.65) and neglecting the terms of $\left(\omega^2\right)^2$ and $\left(\omega^2\right)^3$ in a cubic equation with respect to ω^2, the m,n-th natural frequency ω_{mn} is obtained as

$$\omega_{mn} \approx \sqrt{\frac{K_{mn}}{F_{mn}}} \tag{6.69}$$

in which K_{mn} and F_{mn} are defined as

$$\begin{aligned} K_{mn} = {} & A_{1mnmn}(-B_{2mnmn}C_{3mnmn} + B_{3mnmn}C_{2mnmn}) \\ & + A_{2mnmn}(B_{1mnmn}C_{3mnmn} - C_{1mnmn}B_{3mnmn}) \\ & + A_{3mnmn}(C_{1mnmn}B_{2mnmn} - B_{1mnmn}C_{2mnmn}) \end{aligned} \tag{6.70}$$

$$\begin{aligned} F_{mn} = {} & A_{1mnmn}(-B_{2mnmn}F_{3mnmn} - C_{3mnmn}F_{2mnmn}) \\ & + m_0 F_{1mnmn}(B_{2mnmn}C_{3mnmn} - B_{3mnmn}C_{2mnmn}) \\ & + A_{2mnmn}B_{1mnmn}F_{3mnmn} + A_{3mnmn}C_{1mnmn}F_{2mnmn} \end{aligned} \tag{6.71}$$

Equation (6.69) proposed here for the natural frequency is uncoupled equation and can directly determine ω_{mn}.

6.8 Numerical Results for Natural Frequencies

In the preceding section, the natural frequencies for a rectangular cellular plate have been presented in an approximate form by means of the Galerkin method. Then, in order to examine the exactness of the proposed method, numerical computations are performed for the previous rectangular cellular plates, as shown in Table 6.2. The shape function is made up of the well-known natural functions of beams, for instance, as given in Szilard [16] and Volterra and Zachmanoglou [17]. The current shape functions used are Eqs. (6.48)–(6.50) for simply supported cellular plates and the following natural functions of clamped beams for clamped cellular plate:

$$\begin{aligned} f_{mn}(x, y) = &\left\{\cosh\left(\frac{\lambda_m x}{l_x}\right) - \cos\left(\frac{\lambda_m x}{l_x}\right) - \frac{\cosh\lambda_m - \cos\lambda_m}{\sinh\lambda_m - \sin\lambda_m}\left[\sinh\left(\frac{\lambda_m x}{l_x}\right) - \sin\left(\frac{\lambda_m x}{l_x}\right)\right]\right\} \\ &\times\left\{\cosh\left(\frac{\lambda_n y}{l_y}\right) - \cos\left(\frac{\lambda_n y}{l_y}\right) - \frac{\cosh\lambda_n - \cos\lambda_n}{\sinh\lambda_n - \sin\lambda_n}\left[\sinh\left(\frac{\lambda_n y}{l_y}\right) - \sin\left(\frac{\lambda_n y}{l_y}\right)\right]\right\} \end{aligned} \tag{6.72}$$

$$\begin{aligned} g_{xmn}(x, y) = &\left\{\sinh\left(\frac{\lambda_m x}{l_x}\right) + \sin\left(\frac{\lambda_m x}{l_x}\right) - \frac{\cosh\lambda_m - \cos\lambda_m}{\sinh\lambda_m - \sin\lambda_m}\left[\cosh\left(\frac{\lambda_m x}{l_x}\right) - \cos\left(\frac{\lambda_m x}{l_x}\right)\right]\right\} \\ &\times\left\{\cosh\left(\frac{\lambda_n y}{l_y}\right) - \cos\left(\frac{\lambda_n y}{l_y}\right) - \frac{\cosh\lambda_n - \cos\lambda_n}{\sinh\lambda_n - \sin\lambda_n}\left[\sinh\left(\frac{\lambda_n y}{l_y}\right) - \sin\left(\frac{\lambda_n y}{l_y}\right)\right]\right\} \end{aligned} \tag{6.73}$$

$$\begin{aligned} g_{ymn}(x, y) = &\left\{\cosh\left(\frac{\lambda_m x}{l_x}\right) - \cos\left(\frac{\lambda_m x}{l_x}\right) - \frac{\cosh\lambda_m - \cos\lambda_m}{\sinh\lambda_m - \sin\lambda_m}\left[\sinh\left(\frac{\lambda_m x}{l_x}\right) - \sin\left(\frac{\lambda_m x}{l_x}\right)\right]\right\} \\ &\times\left\{\sinh\left(\frac{\lambda_n y}{l_y}\right) + \sin\left(\frac{\lambda_n y}{l_y}\right) - \frac{\cosh\lambda_n - \cos\lambda_n}{\sinh\lambda_n - \sin\lambda_n}\left[\cosh\left(\frac{\lambda_n y}{l_y}\right) - \cos\left(\frac{\lambda_n y}{l_y}\right)\right]\right\} \end{aligned} \tag{6.74}$$

where λ_m and λ_n are well-known constants, as already given in Sect. 1.8, and take the following values: $\lambda_1 = 4.7300$; $\lambda_2 = 7.8532$; $\lambda_3 = 10.9956$; $\lambda_4 = 14.1372$; $\lambda_m \approx (2m + 1)\pi/2$ for m large.

Table 6.3 represents the natural frequencies for Type P1. It shows that the differences between the results obtained using Eqs. (6.63)–(6.65), indicated with "Rigorous", and the approximate results obtained using Eq. (6.69), indicated with "Approximate", are negligible in practical use. The results obtained from the proposed theory also demonstrate good agreement with results obtained from FEM. For the other plates, a similar agreement is obtained.

6.9 Forced Vibration of Cellular Plates

We consider forced vibrations of the current rectangular cellular plate. Since the equation of motion for cellular plates is already given in Eqs. (6.18)–(6.20), the general solutions are also assumed to be the form

Table 6.3 Natural frequencies of simply supported and clamped cellular plates P1 [10]

Mode	Analytical method	Simply supported (rad/s)			Clamped (rad/s)		
		$m=1$	$m=2$	$m=3$	$m=1$	$m=2$	$m=3$
$n=1$	Rigorous	18.66	29.42	41.34	32.80	41.14	51.34
	Approximate	18.64	29.39	41.29	32.76	41.10	51.29
	FEM	18.49	28.95	40.92	33.01	40.04	49.63
	Rigorous	53.42	60.40	73.39	79.66	86.20	95.96
$n=2$	Approximate	53.16	61.19	73.42	79.23	85.87	95.64
	FEM	55.69	63.24	74.52	79.39	84.59	92.90
	Rigorous	111.00	112.70	124.22	145.37	149.64	157.81
$n=3$	Approximate	109.58	114.16	124.17	143.40	148.17	156.51
	FEM	110.72	113.35	126.77	138.80	142.67	158.90

In which m = mode number in x-direction, n = mode number in y-direction

$$w(x, y, t) = \sum_{m=1} \sum_{n=1} f_{mn}(x, y)\, w_{mn}(t) \tag{6.75}$$

$$\theta_x(x, y, t) = \sum_{m=1} \sum_{n=1} g_{xmn}(x, y)\, \theta_{xmn}(t) \tag{6.76}$$

$$\theta_y(x, y, t) = \sum_{m=1} \sum_{n=1} g_{ymn}(x, y)\, \theta_{ymn}(t) \tag{6.77}$$

in which w_{mn}, θ_{xmn}, and θ_{ymn} are unknown functions with respect to time t.

Then, applying Eqs. (6.75)–(6.77) into (6.18)–(6.20), we have

$$\begin{aligned} \delta w_{mn} : \quad & \sum_{\bar{m}=1} \sum_{\bar{n}=1} (\ddot{w}_{\bar{m}\bar{n}} m_0 F_{1mn\bar{m}\bar{n}} + 2h_{\bar{m}\bar{n}} \omega_{\bar{m}\bar{n}} m_0 \dot{w}_{\bar{m}\bar{n}} F_{1mn\bar{m}\bar{n}} \\ & - w_{\bar{m}\bar{n}} A_{1mn\bar{m}\bar{n}} - \theta_{x\bar{m}\bar{n}} A_{2mn\bar{m}\bar{n}} - \theta_{y\bar{m}\bar{n}} A_{3mn\bar{m}\bar{n}}) - P_{mn} = 0 \end{aligned} \tag{6.78}$$

$$\delta\theta_{xmn} : \quad \sum_{\bar{m}=1} \sum_{\bar{n}=1} (\ddot{\theta}_{x\bar{m}\bar{n}} F_{2mn\bar{m}\bar{n}} + w_{\bar{m}\bar{n}} B_{1mn\bar{m}\bar{n}} + \theta_{x\bar{m}\bar{n}} B_{2mn\bar{m}\bar{n}} + \theta_{y\bar{m}\bar{n}} B_{3mn\bar{m}\bar{n}}) = 0 \tag{6.79}$$

$$\delta\theta_{ymn} : \quad \sum_{\bar{m}=1} \sum_{\bar{n}=1} (\ddot{\theta}_{y\bar{m}\bar{n}} F_{3mn\bar{m}\bar{n}} + w_{\bar{m}\bar{n}} C_{1mn\bar{m}\bar{n}} + \theta_{x\bar{m}\bar{n}} C_{2mn\bar{m}\bar{n}} + \theta_{y\bar{m}\bar{n}} C_{3mn\bar{m}\bar{n}}) = 0 \tag{6.80}$$

in which h_{mn} are damping constant for the m,n-th mode shape of the cellular plate. The dynamic response is obtained by solving Eqs. (6.78)–(6.80). The coefficients in these equations have nondiagonal terms $m \neq \bar{m}$ and/or $n \neq \bar{n}$ due to the effect of the voids. Therefore, the solution must be based on either numerical computation, such as step-by-step integration methods, or the approximate solution proposed in the following section.

6.10 Approximate Solutions for Forced Vibrations

For practical use, we propose an approximate solution in the closed form. Assuming that the behavior of a cellular plate is now dominated by only the diagonal terms in the coefficients of Eqs. (6.78)–(6.80) with coupled form, the reduced equations become of uncoupled form with respect to m and $\bar{m}$, and, n and $\bar{n}$, respectively. Furthermore, neglecting the rotatory inertia's terms in the uncoupled equations, we have the following differential equation in terms of only w_{mn}:

$$\ddot{w}_{mn} + 2a_{mn}\dot{w}_{mn} + b_{mn}w_{mn} = Q_{mn}(t) \tag{6.81}$$

in which the coefficients are defined as

$$a_{mn} \equiv h_{mn}\omega_{mn} \tag{6.82}$$

$$b_{mn} \equiv -\frac{A^*_{mn}}{m_0 F_{1mnmn}} \tag{6.83}$$

$$Q_{mn}(t) \equiv \frac{P_{mn}}{m_0 F_{1mnmn}} \tag{6.84}$$

in which

$$\begin{aligned} A^*_{mn} = A_{1mnmn} - \frac{1}{\Delta_{mn}}[&A_{2mnmn}(C_{3mnmn}B_{1mnmn} - B_{3mnmn}C_{1mnmn}) \\ &+ A_{3mnmn}(B_{2mnmn}C_{1mnmn} - B_{1mnmn}C_{2mnmn})] \end{aligned} \tag{6.85}$$

where

$$\Delta_{mn} = B_{2mnmn}C_{3mnmn} - B_{3mnmn}C_{2mnmn} \tag{6.86}$$

The general solution of Eq. (6.81) can be expressed as

$$\begin{aligned} w_{mn} &= \mathrm{e}^{-a_{mn}t}(C_1 \sin\alpha_{0mn}t + C_2\cos\alpha_{0mn}t) \\ &+ \frac{1}{\alpha_{0mn}}\int_0^t \mathrm{e}^{-a_{mn}(t-\tau)}\sin\alpha_{0mn}(t-\tau)Q_{mn}(\tau)\,\mathrm{d}\tau \end{aligned} \tag{6.87}$$

in which C_1 and C_2 are constants, and α_{0mn} are defined as

$$\alpha_{0mn} \equiv \sqrt{b_{mn} - a^2_{mn}} \tag{6.88}$$

Thus, the dynamic deflections of a rectangular cellular plate are easily determined by substituting Eq. (6.87) into (6.75).

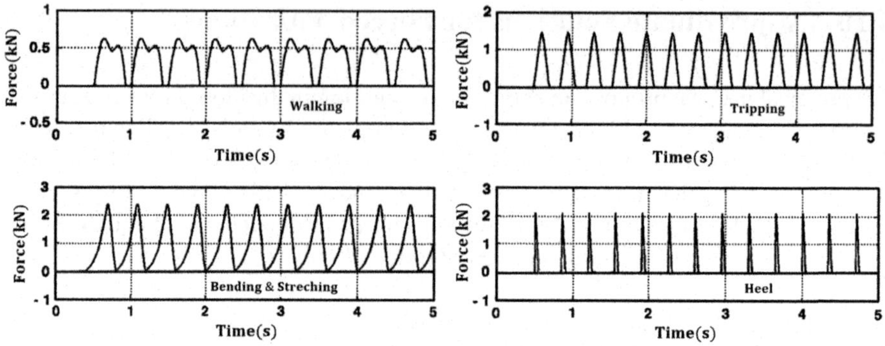

Fig. 6.12 Test loads for floors [10]

6.11 Numerical Results for Dynamic Responses

In order to examine the exactness of the closed-form approximate solution proposed here, numerical computations are carried out for the previous cellular plates, as shown in Table 6.2. The damping constants h_{mn} are 0.03 for all modes. The following two types of external lateral loads without the additional mass are assumed: a harmonic and uniformly distributed force $p = p^* \sin(\omega_p t)$, in which $p^* = 9.807 \times 10^6$ N/m^2 and $\omega_p = 62.83$ rad/s (10 cps), and test loads checking the disturbing effect of the vibration of floors caused by people walking and other every usage, as shown in Fig. 6.12. The test loads consist of four types of walking, tripping, bending and stretching, and heel, as already given in the preceding chapters. They act as a concentrated load at the midspan. The current natural functions used are the same ones as used in the free vibrations.

Figure 6.13 represents the time history of the dynamic deflections at the midpoint of simply supported cellular plates of Type P1 subjected to the harmonic load. In this figure, the solid lines represent values obtained from Eqs. (6.78)–(6.80) using the linear acceleration method; the broken lines represent values obtained from the approximate solution Eq. (6.87); and the solid lines with circles indicate values obtained from FEM. The difference between solid lines and broken lines is too small to plot. Table 6.4 compares the results obtained from various analytical methods for the maximum dynamic deflections at the midspan of the cellular plates subjected to the harmonic load. Figure 6.14 shows the time history of the dynamic deflections at the midpoint of clamped cellular plates of P4, subjected to the test loads. The numerical results demonstrate that the approximate solution proposed here is also applicable to the dynamic analyses of cellular plates, in practical use.

For these numerical models, the calculations using the present theory are remarkably fast as compared to those using FEM and the input data are very simple. Since the series in the theory proposed here converges very rapidly, the consideration of nine terms gives accuracy sufficient for all practical purpose.

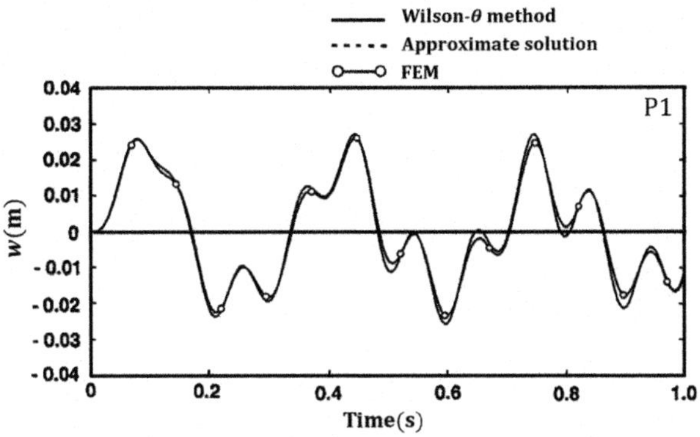

Fig. 6.13 Dynamic deflections of a simply supported plate P1 subjected to uniform load [10]

Table 6.4 Maximum dynamic deflection of cellular plates [10]

Type	Analytical method	Maximum dynamic deflection	
		Simply supported (m)	Clamped (m)
P1	Wilson-θ method	0.02723	0.01875
	Approximate solution	0.02715	0.02223
	FEM	0.02624	0.01874
P2	Wilson-θ method	0.03964	0.02608
	Approximate solution	0.03971	0.02923
	FEM		0.02455
P3	Wilson-θ method	0.02247	0.02279
	Approximate solution	0.02291	0.02281
	FEM	0.02111	0.02156
P4	Wilson-θ method	0.02730	0.02731
	Approximate solution	0.02733	0.02905
	FEM		0.02638

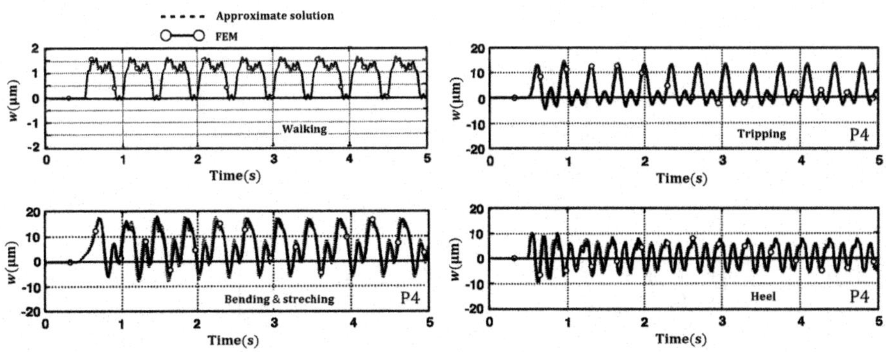

Fig. 6.14 Dynamic deflections of a clamped plate P4 subjected to test loads [10]

6.12 Conclusions

The general and simple analysis methods for the static and dynamic problems of an isotropic rectangular cellular plate with arbitrarily disposed voids have been proposed by considering the transverse shear deformation along with frame deformation. It is clarified from numerical computations that the simplified theory proposed here is usable in the preliminary and final stages of designs of such a cellular plate. Also, the governing equations proposed in this chapter include the equations presented in Chaps. 1 and 4, as already shown in Table 6.1.

Appendix 1: Coefficients and Load Terms Used in Eqs. (6.45)–(6.47)

The coefficients used in Eqs. (6.45)–(6.47) are defined as

$$A_{1mn\bar{m}\bar{n}} = \kappa G h_0 \int_0^{l_x}\int_0^{l_y} \left[(\alpha_{Gx} f_{\bar{m}\bar{n},x})_{,x} + (\alpha_{Gy} f_{\bar{m}\bar{n},y})_{,y}\right] f_{mn} \mathrm{d}x\,\mathrm{d}y \tag{6.89}$$

$$A_{2mn\bar{m}\bar{n}} = \kappa G h_0 \int_0^{l_x}\int_0^{l_y} (\alpha_{Gx} g_{x\bar{m}\bar{n}})_{,x} f_{mn} \mathrm{d}x\,\mathrm{d}y \tag{6.90}$$

$$A_{3mn\bar{m}\bar{n}} = \kappa G h_0 \int_0^{l_x}\int_0^{l_y} (\alpha_{Gy} g_{y\bar{m}\bar{n}})_{,y} f_{mn} \mathrm{d}x\,\mathrm{d}y \tag{6.91}$$

$$B_{1mn\bar{m}\bar{n}} = \kappa G h_0 \int_0^{l_x}\int_0^{l_y} \alpha_{Gx,x} f_{\bar{m}\bar{n},x} g_{xmn} \mathrm{d}x\,\mathrm{d}y \tag{6.92}$$

$$B_{2mn\bar{m}\bar{n}} = \int_0^{l_x}\int_0^{l_y} \Big[\kappa G h_0\, \alpha_{Gx} g_{x\bar{m}\bar{n}} - D_0(\mathrm{d}g_{x\bar{m}\bar{n},x})_{,x} - \frac{1-\upsilon}{2} D_0(\mathrm{d}g_{x\bar{m}\bar{n},y})_{,y}\Big] g_{xmn}\, \mathrm{d}x\,\mathrm{d}y \tag{6.93}$$

$$B_{3mn\bar{m}\bar{n}} = D_0 \int_0^{l_x}\int_0^{l_y} \left[-\upsilon(\mathrm{d}g_{y\bar{m}\bar{n},y})_{,x} - \frac{1-\upsilon}{2}(\mathrm{d}g_{y\bar{m}\bar{n},x})_{,y}\right] g_{xmn}\, \mathrm{d}x\,\mathrm{d}y \tag{6.94}$$

$$C_{1mn\bar{m}\bar{n}} = \kappa G h_0 \int_0^{l_x}\int_0^{l_y} \alpha_{Gx} f_{\bar{m}\bar{n},y}\, g_{ymn}\, \mathrm{d}x\,\mathrm{d}y \tag{6.95}$$

$$C_{2mn\bar{m}\bar{n}} = D_0 \int_0^{l_x}\int_0^{l_y} \left[-\upsilon (\mathrm{d}g_{x\bar{m}\bar{n},x})_{,y} - \frac{1-\upsilon}{2} (\mathrm{d}g_{x\bar{m}\bar{n},y})_{,x} \right] g_{ymn}\, \mathrm{d}x\, \mathrm{d}y \tag{6.96}$$

$$C_{3mn\bar{m}\bar{n}} = \int_0^{l_x}\int_0^{l_y} \left[\kappa G h_0\, \alpha_{Gy} g_{y\bar{m}\bar{n}} - D_0 (\mathrm{d}g_{y\bar{m}\bar{n},y})_{,y} - \frac{1-\upsilon}{2} D_0 (\mathrm{d}g_{y\bar{m}\bar{n},x})_{,x} \right] g_{ymn}\, \mathrm{d}x\, \mathrm{d}y \tag{6.97}$$

and the terms of external loads are

$$P_{mn} = \int_0^{l_x}\int_0^{l_y} p\, f_{mn}\, \mathrm{d}x\, \mathrm{d}y \tag{6.98}$$

They may be calculated easily by the method shown in Appendix 2.

Appendix 2: Calculation Including an Extended Dirac Function *D(x − xi)*

The integral calculation including the extended Dirac function $D\,(x - x_i)$ can be written as stated in Eqs. (4.8)–(4.11) in Chap. 4,

$$\int_0^{l_x} D(x - x_i)\, f(x)\, \mathrm{d}x = \int_{x_i-(b_{xi,j}/2)}^{x_i+(b_{xi,j}/2)} f(\xi)\mathrm{d}\xi \tag{6.99}$$

in which ξ is a supplementary variable of x. The n-th derivatives of the extended Dirac functions can therefore be expressed as

$$\int_0^{l_x} D^{(n)}(x - x_i)\, f(x)\, \mathrm{d}x = \int_{x_i-(b_{xi,j}/2)}^{x_i+(b_{xi,j}/2)} (-1)^n f^{(n)}(\xi)\, \mathrm{d}\xi \tag{6.100}$$

in which the superscripts enclosed within parentheses indicate the differential order. For calculations including the extended Dirac function $D(y - y_j)$, similar expressions may be obtained.

When the conditions $b_{xi,j} \ll l_x$ and $b_{yi,j} \ll l_y$ are satisfied, the extended Dirac functions $D(x - x_i)$ and $D(y - y_j)$ are approximately related to the Dirac delta functions, $\delta(x - x_i)$ and $\delta(y - y_j)$, by:

$$D(x - x_i) \approx b_{xi,j}\,\delta\,(x - x_i)$$
$$D(y - y_j) \approx b_{yi,j}\,\delta\,(y - y_j) \tag{6.101}$$

Appendix 3: The Governing Equations for Solid Plates Without Voids Based on Mindlin-Reissner Hypothesis

In Sect. 6.2, the governing equations of rectangular cellular plates based on Mindlin-Reissner hypothesis had been stated to easily transform into rectangular solid (normal) plates without voids based on the same Mindlin-Reissner hypothesis by changing α_h and d to 1 and $\kappa_x G_x$ and $\kappa_y G_y$ to κG. We introduce the governing equations of rectangular solid plates based on Mindlin-Reissner hypothesis for practical uses. From Eqs. (6.18), (6.19), (6.20), (6.21), (6.22), (6.23), (6.24), (6.25), (6.26), we have

$$\delta w : m_0\ddot{w} - [\kappa G h_0(w,_x + \theta_x)],_x - [\kappa G h_0(w,_y + \theta_y)],_y - p + c\dot{w} = 0 \tag{6.102}$$

$$\delta\theta_x : I_p\ddot{\theta}_x + \kappa G h_0(w,_x + \theta_x) - [D_0\,(\theta_{x,x} + \nu\theta_{y,y})],_x - \frac{1-\nu}{2}[D_0(\theta_{x,y} + \theta_{y,x})],_y = 0 \tag{6.103}$$

$$\delta\theta_y : I_p\ddot{\theta}_y + \kappa G h_0(w,_y + \theta_y) - [D_0\,(\theta_{y,y} + \nu\theta_{x,x})],_y - \frac{1-\nu}{2}[D_0\,(\theta_{x,y} + \theta_{y,x})],_x = 0 \tag{6.104}$$

together with the associated boundary conditions

$$w = w^* \quad \text{or}\ \kappa G h_0(w,_x + \theta_x) = v_x^* \tag{6.105}$$

$$\theta_x = \theta_x^* \quad \text{or}\ D_0\,(\theta_{x,x} + \nu\theta_{y,y}) = m_x^* \tag{6.106}$$

$$\theta_y = \theta_y^* \quad \text{or}\ \tfrac{1-\nu}{2} D_0\,(\theta_{x,y} + \theta_{y,x}) = m_{xy}^* \tag{6.107}$$

at $x = 0$ and l_x; and

$$w = w^* \quad \text{or}\ \kappa G h_0(w,_y + \theta_y) = v_y^* \tag{6.108}$$

$$\theta_x = \theta_x^* \quad \text{or}\ \tfrac{1-\nu}{2} D_0(\theta_{x,y} + \theta_{y,x}) = m_{yx}^* \tag{6.109}$$

$$\theta_y = \theta_y^* \quad \text{or}\ D_0\,(\theta_{y,y} + \nu\theta_{x,x}) = m_y^* \tag{6.110}$$

at $y = 0$ and l_y, in which w^*, θ_x^*, and θ_y^* are displacement components prescribed on the geometrical boundary; v_x^*, v_y^*, m_x^*, m_y^*, m_{xy}^*, and m_{yx}^* are vertical forces, moments, respectively, prescribed on the mechanical boundary.

References

1. Crisfield, M.A., Twemlow, R.P.: The equivalent plate approach for the analysis of cellular structures. Civil Eng. Public Works, 259–263 (1971) (March)
2. Cope, R.J., Harris, G., Sawko, F.: A new approach to the analysis of cellular bridge decks. Anal. Struct. Syst. Torsion ACI **35**, 185–210 (1973)
3. Holmberg, A.: Shear-weak beams an elastic foundation. Int. Assoc. Bridge Struct. Eng. (IABSE) Zurich, **10** (1960)
4. Sawko, F., Cope, R.J.: Analysis of multi-cell bridges without transverse diaphragms—a finite element approach. Struct. Eng. **47**(11), 455–460 (1969)
5. Elliott, G.: Partial loading on orthotropic plates. Cement Concr. Assoc. Tech. Rep. Lond. **42**, 519 (1978)
6. Takabatake, H.: Static analyses of elastic plates with voids. Int. J. Solids Struct. **28**(2), 179–196 (1991)
7. Takabatake, H.: Dynamic analyses of elastic plates with voids. Int. J. Solids Struct. **28**(7), 879–895 (1991)
8. Reissner, E.: The effect of transverse shear deformation on the bending of elastic plates. J. Appl. Mech. ASME **12**, A69–A77 (1945)
9. Mindlin, R.D.: Influence of rotatory inertia and shear on flexural motions of isotropic, elastic plates. J. Appl. Mech. ASME **18**, 31–38 (1951)
10. Takabatake, H., Yanagisawa, N., Kawano, T.: A simplified analysis of rectangular cellular plates. Int. J. Solids Struct. **33**(14), 2055–2074 (1996)
11. Takabatake, H.: Lateral buckling of I beams with web stiffeners and batten plates. Int. J. Solids Struct. **24**(10), 1003–1019 (1988)
12. Washizu, K.: Variational Methods in Elasticity and Plasticity, 3rd edn. Pergamon Press, New York (1982)
13. Takabatake, H., Mukai, H., Hirano, T.: Doubly symmetric tube structures. I: Static analysis. J. Struct. Eng. ASCE, **119**, 1981–2001 (1993)
14. Takabatake, H., Mukai, H., Hirano, T.: Doubly symmetric tube structures. II: Dynamic analysis. J. Struct. Eng. ASCE, **119**, 2002–2016 (1993)
15. Smith, B.S., Kuster, M., Hoenderkamp, J.C.D.: Generalized method for estimating drift in high-rise structures. J. Struct. Eng. ASCE **110**(7), 1549–2072 (1984)
16. Szilard, R.: Theory and Analysis of Plates-Classical and Numerical Methods. Prentice-Hall, Englewood Cliffs (1974)
17. Volterra, E., Zachmanoglou, E.C.: Dynamics of Vibrations. Charles E. Merrill Publishing Company, Ohio (1965)

Chapter 7
Static and Dynamic Analyses of Circular Cellular Plates

7.1 Introduction

In the preceding chapter, we have considered static and dynamic analyses of rectangular cellular plates based on the Mindlin-Reissner hypothesis. This chapter presents the same problem for circular cellular plates which have often been used as building slabs with long spans, and it corresponds to the extension of Chap. 5 based on Kirchhoff-Love hypothesis.

This chapter presents a general and simplified method for a circular cellular plate with arbitrarily positioned voids, in which bending and transverse shear deformations along with the frame deformation are considered, as shown in Fig. 7.1. The discontinuous variations of the bending and transverse shear stiffnesses and of mass are also treated as a continuous function by means of an extended Dirac function. This treatment is independent of the well-known equivalent plate analogy for circular plates.

First, the general governing equations for a circular cellular plate with the transverse shear deformation along with the frame deformation are presented by changing the theory for a rectangular cellular plate, proposed in Chap. 6, from Cartesian coordinate system to polar coordinate one. Second, the static solutions are presented by the use of the Galerkin method. Third, the natural frequencies are presented by means of the Galerkin method, and an approximate expression for the natural frequency is proposed. Last, the approximate solutions for the forced vibrations are proposed to general external loads. The theory proposed here is examined from numerical results by comparing the results obtained from the proposed theory for simply supported and clamped cellular plates with the results obtained from the FEM code NASTRAN. The methodology proposed in this chapter is mainly based on Ref. [1]. Theoretical development in this chapter is basically the same as the method developed in Chap. 6 except for the coordinate transformation.

H. Takabatake, *Simplified Analytical Methods of Elastic Plates*,
https://doi.org/10.1007/978-981-13-0086-8_7

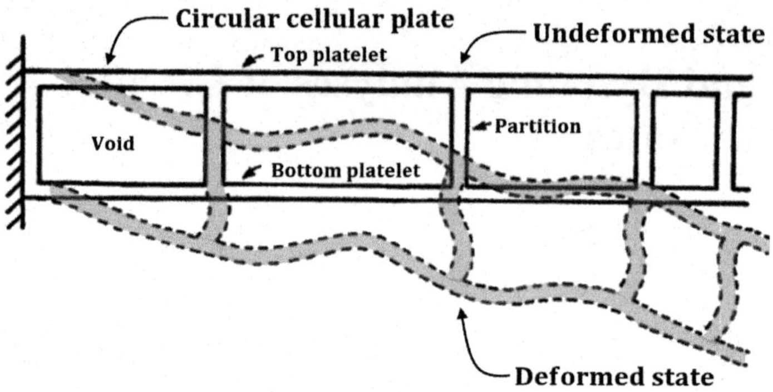

Fig. 7.1 Frame deformation of circular cellular plate [1]

7.2 Governing Equations of a Circular Cellular Plate with Transverse Shear Deformations Along with Frame Deformation

We consider a plate with arbitrarily positioned voids, as shown in Fig. 7.2. A polar coordinate system r, ϕ, z is employed. Each void is assumed to be a fan parallelepiped whose ridgelines are parallel to the axis r or the axis ϕ and which is symmetrically positioned with respect to the middle plane of the plate. The position of the i, j-th void is indicated by the coordinate value $\left(r_i, \phi_j\right)$ of the midpoint of the void; the widths in the r and ϕ directions of the void are $b_{ri,j}$ and $b_{\phi i,j}$, respectively, and its height is $h_{i,j}$. The size and position of each void are arbitrary except for the assumptions mentioned above.

This chapter considers the bending of an isotropic circular cellular plate subject to small deformations, including the transverse shear deformation along with the frame deformation. The deformation of cellular plates is assumed to be adequately defined by describing the geometry of its middle surface, which is a surface that bisects the plate thickness, h_0, at each point. Then, considering the in-plane mean transverse deformation, the displacement components, $\bar{U}$, $\bar{V}$, and $\bar{W}$, on a general point of the cellular plate are given as

$$\bar{U}(r, \phi, t) = z\,\theta_r(r, \phi, t) \tag{7.1}$$

$$\bar{V}(r, \phi, t) = z\,\theta_\phi(r, \phi, t) \tag{7.2}$$

$$\bar{W}(r, \phi, t) = w(r, \phi, t) \tag{7.3}$$

in which w is the lateral displacement on the middle surface; θ_r and θ_ϕ are the rotational angles about the axes ϕ and r, respectively, on the middle surface; and z is the distance measured from the middle surface of the plate. These displacement and rotational angles are defined as positive when they point toward the positive direction

Fig. 7.2 Details of void [1]

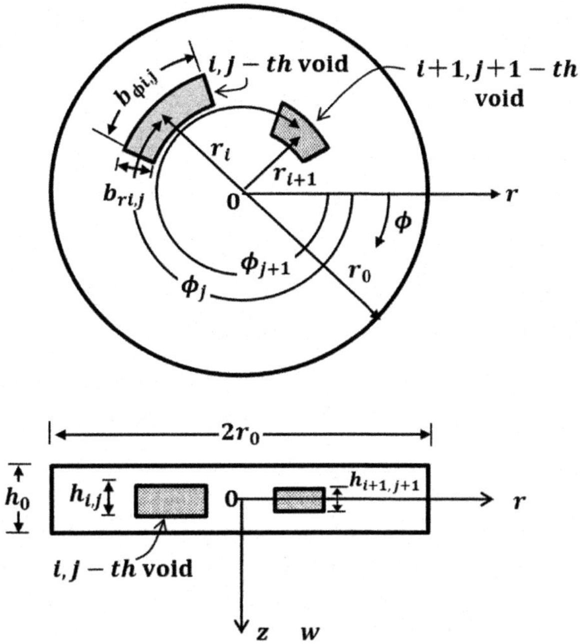

of the coordinate axis z and the clockwise with respect to the ϕ- and r-coordinate axes, respectively.

In Chap. 6, the governing equations for rectangular cellular plate with the transverse shear deformations along with frame deformation are given from Eqs. (6.18)–(6.26).

The current governing equations are proposed by translating the governing equations for rectangular cellular plates into circular cellular plates. It must be noticed that the translation from the rotational angles θ_x and θ_y in Cartesian coordinate to the rotational angles θ_r and θ_ϕ in polar coordinates for rotational angles is a one-order higher translation than the one for the lateral deflection w, as shown in Appendix 1. Thus, the governing equations for current circular cellular plates subject to arbitral general external loads become

$$\delta w: \quad m_0\alpha_m\ddot{w} - [\kappa_r G_r h_0(w_{,r} + \theta_r)]_{,r} - (\kappa_\phi G_\phi h_0)_{,\phi}\left(\frac{1}{r^2}w_{,\phi} + \frac{1}{r^2}\theta_\phi\right)$$

$$-\kappa_\phi G_\phi h_0\left(\frac{1}{r}w_{,r} + \frac{1}{r^2}w_{,\phi\phi} + \frac{1}{r}\theta_r + \frac{1}{r^2}\theta_{\phi,\phi}\right) - p + c\dot{w} = 0$$

$$\delta\theta_r: \quad I_p\ddot{\theta}_r + \kappa_r G_r h_0(w_{,r} + \theta_r) - D_0 d_{,r}\left[\theta_{r,r} + \nu\left(\frac{1}{r}\theta_r + \frac{1}{r^2}\theta_{\phi,\phi}\right)\right] \tag{7.4}$$

$$
\begin{aligned}
&- D_0\mathrm{d}\left[\theta_{r,rr} + \nu\left(\frac{1}{r}\theta_{r,r} - \frac{2}{r^3}\theta_{\phi,\phi} + \frac{1}{r^2}\theta_{\phi,r\phi} - \frac{1}{r^2}\theta_r\right)\right] \\
&- \frac{(1-\nu)}{2}D_0\mathrm{d}_{,\phi}\left(\frac{1}{r^2}\theta_{r,\phi} - \frac{2}{r^3}\theta_\phi + \frac{1}{r^2}\theta_{\phi,r}\right) \\
&- \frac{(1-\nu)}{2}D_0\mathrm{d}\left(\frac{2}{r}\theta_{r,r} + \frac{1}{r^2}\theta_{r,\phi\phi} + \frac{1}{r^2}\theta_{\phi,r\phi} - \frac{1}{r^2}\theta_r - \frac{4}{r^3}\theta_{\phi,\phi}\right) = 0 \qquad (7.5)
\end{aligned}
$$

$$
\begin{aligned}
\delta\theta_\phi:\ &\frac{1}{r}I_p\ddot{\theta}_\phi + \kappa_\phi G_\phi h_0\left(\frac{1}{r}w_{,\phi} + \frac{1}{r}\theta_\phi\right) - D_0\mathrm{d}_{,\phi}\left(\frac{1}{r^2}\theta_r + \frac{1}{r^3}\theta_{\phi,\phi} + \nu\frac{1}{r}\theta_{r,r}\right) \\
&- D_0\mathrm{d}\left[-\frac{2}{r^3}\theta_\phi + \frac{1}{r^2}\theta_{\phi,r} + \frac{2}{r^2}\theta_{r,\phi} + \frac{1}{r^3}\theta_{\phi,\phi\phi}\right. \\
&\left. + \nu\left(\frac{1}{r}\theta_{r,r\phi} + \frac{2}{r^3}\theta_\phi - \frac{1}{r^2}\theta_{\phi,r} - \frac{1}{r^2}\theta_{r,\phi}\right)\right] \\
&- \frac{(1-\nu)}{2}D_0\mathrm{d}_{,r}\left(\frac{1}{r}\theta_{r,\phi} - \frac{2}{r^2}\theta_\phi + \frac{1}{r}\theta_{\phi,r}\right) \\
&- \frac{(1-\nu)}{2}D_0\mathrm{d}\left(\frac{1}{r}\theta_{r,r\phi} + \frac{4}{r^3}\theta_\phi - \frac{3}{r^2}\theta_{\phi,r} - \frac{1}{r^2}\theta_{r,\phi} + \frac{1}{r}\theta_{\phi,rr}\right) = 0 \qquad (7.6)
\end{aligned}
$$

together with the associated boundary conditions:

$$
w = w^* \quad \text{or} \quad \kappa_r G_r h_0(w_{,r} + \theta_r) = v_r^* \qquad (7.7)
$$

$$
\theta_r = \theta_r^* \quad \text{or} \quad D_0\mathrm{d}\left[\theta_{r,r} + \nu\left(\frac{1}{r}\theta_r + \frac{1}{r^2}\theta_{\phi,\phi}\right)\right] = m_r^* \qquad (7.8)
$$

$$
\theta_\phi = \theta_\phi^* \quad \text{or} \quad \frac{1-\nu}{2}D_0\mathrm{d}\left(\frac{1}{r}\theta_{r,\phi} - \frac{2}{r^2}\theta_\phi + \frac{1}{r}\theta_{\phi,r}\right) = m_{r\phi}^* \qquad (7.9)
$$

at $r = r_0$, in which p are lateral loads; w^*, θ_r^*, and θ_ϕ^* are prescribed on the geometrical boundary conditions; and v_r^*, m_r^*, and $m_{r\phi}^*$ are prescribed on the mechanical boundary conditions at $r = r_0$. The flexural rigidity and the mass per unit area of solid plates are $D_0 = Eh_0^3/[12-(1-\upsilon)^2]$ and $m_0 = \rho h_0$, respectively. The stiffness coefficient $\mathrm{d}(r,\phi)$, mass coefficient $\alpha_h(r,\phi)$, and rotatory inertia $I_p(r,\phi)$ are defined as

$$
\mathrm{d}(r,\phi) = 1 - \left(\frac{h_{i,j}}{h_0}\right)^3 D(r-r_i)D(\phi-\phi_j) \qquad (7.10)
$$

$$
\alpha_h(r,\phi) = 1 - \frac{h_{i,j}}{h_0}D(r-r_i)D(\phi-\phi_j) \qquad (7.11)
$$

$$
I_p(r,\phi) = \frac{\rho h_0^3}{12}\left[1 - \left(\frac{h_{i,j}}{h_0}\right)^3 D(r-r_i)D(\phi-\phi_j)\right] \qquad (7.12)
$$

Here, $D(r-r_i)$ and $D(\phi-\phi_j)$ are the extended Dirac functions in the polar coordinate system, as explained in Chap. 5.

Table 7.1 Status of the proposed theory for circular plate

Circular plate	Deformation	Unknown displacement	Equations of motion	Note
Cellular plate	Transverse shear deformation Frame deformation (Mindlin-Reissner hypothesis)	w, θ_r, θ_ϕ	Equations (7.4)–(7.6) In this chapter	$\mathrm{d}(r, \phi)$ $\alpha_h(r, \phi)$
Solid plate	Transverse shear deformation (Mindlin-Reissner hypothesis)	w, θ_r, θ_ϕ	Equations (7.89)–(7.91) In this chapter	$\mathrm{d}(r, \phi) = 1$ $\alpha_h(r, \phi) = 1$
Void plate	Kirchhoff-Love hypothesis	w	Equation (5.21) Chapter 5	$\mathrm{d}(r, \phi)$ $\alpha_h(r, \phi)$
Solid plate	Kirchhoff-Love hypothesis	w	Chapter 2	$\mathrm{d}(r, \phi) = 1$ $\alpha_h(r, \phi) = 1$

The bending and twisting moments and transverse shear forces per unit width, M_r, M_φ, $M_{r\phi}$, Q_r, and Q_ϕ, respectively, are given by

$$M_r = D_0 \mathrm{d}(r, \phi)\left[\theta_{r,r} + \nu\left(\frac{1}{r}\theta_r + \frac{1}{r^2}\theta_{\phi,\phi}\right)\right] \tag{7.13}$$

$$M_\phi = D_0 \mathrm{d}(r, \phi)\left(\frac{1}{r}\theta_r + \frac{1}{r^2}\theta_{\phi,\phi} + \nu\theta_{r,r}\right) \tag{7.14}$$

$$M_{r\phi} = \frac{1-\nu}{2} D_0 \mathrm{d}(r, \phi)\left(\frac{1}{r}\theta_{r,\phi} - \frac{1}{r^2}\theta_\phi + \frac{1}{r}\theta_{\phi,r}\right) \tag{7.15}$$

$$Q_r = \kappa_r G_r h_0 (w_{,r} + \theta_r) \tag{7.16}$$

$$Q_\phi = \kappa_\phi G_\phi h_0\left(\frac{1}{r} w_{,\phi} + \frac{1}{r}\theta_\phi\right) \tag{7.17}$$

in which E is Young's modulus and ν is Poisson's ratio. The values at $r = 0$ can be calculated by means of L'Hopital's rule.

The above-mentioned governing equations for circular cellular plates are applicable to general problems. When the transverse shear deformation along with local deformation is neglected, the governing equations given in Eqs. (7.4)–(7.6) reduce to the theory proposed under the Kirchhoff-Love hypothesis prescribed in Chap. 5, as given in Ref. [2]. Table 7.1 demonstrates the relations among the theory proposed here and theories for the various circular plates.

The transverse shear stiffnesses $\kappa_r G_r h_0$ and, $\kappa_\phi G_\phi h_0$ are determined from both the frame deformation and the transverse shear deformation of the plate elements, as shown in the following section.

7.3 Transverse Shear Stiffness of Cellular Plates

The bending and transverse shear stiffnesses of cellular plates decrease due to voids. The reduction of the bending stiffness is already considered by the stiffness coefficient $d(r, \phi)$ given in Eq. (7.10). The reduction of the transverse shear stiffness is determined from the frame deformation and the transverse shear deformation of the top and bottom platelets and of partitions surrounding a void.

We consider the transverse shear stiffness of a circular cellular plate, as shown in Fig. 7.3. Assuming that the influence points of top and bottom platelets and of partitions surrounding a void are situated on the midspan of each member, the transverse shear stiffness $\kappa_r G_r h_0$ per unit width in the radial direction of the cellular plate is obtained easily like Eq. (6.27).

$$\kappa_r G_r h_0 = \frac{1}{\dfrac{b^*_{ri,j}\left(\frac{b^*_{ri,j}}{I_{c1}+I_{c2}}+\frac{h_0}{I_b}\right)}{12E} + \dfrac{1}{\kappa G h_0 (1-\frac{h_{i,j}}{h_0})} + \dfrac{b^*_{ri,j}}{\kappa G h_0 \frac{b_{r1}+b_{r2}}{2}}} \tag{7.18}$$

In Eq. (7.18), the first term of denominator in the right-hand side represents the frame deformation; the second and third terms represent the shear deformations of only the top and bottom platelets and of partitions in current cellular plate, respec-

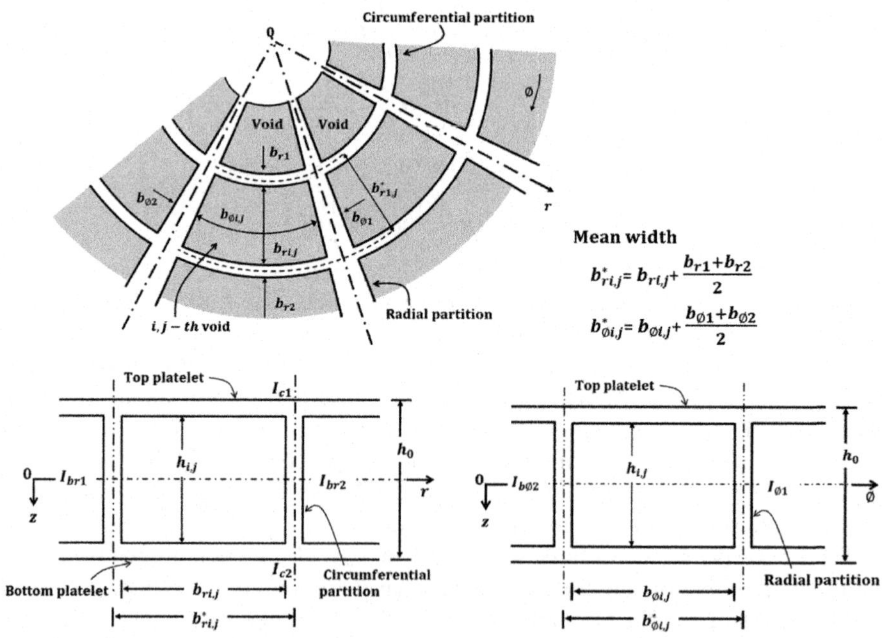

Fig. 7.3 Cross section of cellular plates [1]

tively. The inertia moments of the top and bottom platelets and circumferential partition, I_{c1},I_{c2} and I_b, are defined as

$$I_{c1} = I_{c2} = \frac{1}{12}\left(\frac{h_0 - h_{i,j}}{2}\right)^3 \tag{7.19}$$

$$I_b = \frac{b_{r1}^3 + b_{r2}^3}{24} \tag{7.20}$$

When the void is surrounded by radial partitions like shear wall, as shown in Fig. 7.3, the transverse shear stiffnesses for cellular plates must be added with the transverse shear stiffness of radial partitions (like shear wall) to the transverse shear stiffness given by Eq. (7.18). Hence, we have

$$\kappa_r G_r h_0 = (\kappa_r G_r h_0)^* + \kappa G h_0 \frac{b_{\phi 1} + b_{\phi 2}}{k_1 b_{\phi i,j}^*} \tag{7.21}$$

in which G is the shear modulus, and κ denotes the distribution coefficient of the transverse shear stress. $(\kappa_r G_r h_0)^*$ indicates the value obtained from Eq. (7.18). The value of k_1 is 2 as $b_{\phi 1}$ and $b_{\phi 2}$ exist, and 1 as $b_{\phi 1}$ or $b_{\phi 2}$ vanishes. The transverse shear stiffness $\kappa_\phi G_\phi h_0$ may be written in similar fashion.

Thus, the transverse shear stiffness of circular cellular plates has been presented. It must be noticed that the transverse shear stiffness of the cellular plate in a field of the mean widths $b_{ri,j}^*$ and $b_{\phi i,j}^*$ prescribed by the i, j-th void is replaced with an equivalent transverse shear stiffness. Therefore, the transverse shear stiffness of the solid part, such as the center core, has $\kappa G h_0$, being the stiffness of solid plates. Thus, the equivalent transverse shear stiffnesses per the i, j-th void are expressed, in general, as

$$\kappa_r G_r h_0 = \alpha_{Gri,j}\ \kappa G h_0 \tag{7.22}$$

$$\kappa_\phi G_\phi h_0 = \alpha_{G\phi i,j}\ \kappa G h_0 \tag{7.23}$$

in which the transverse shear coefficients $\alpha_{Gri,j}$ and $\alpha_{G\phi i,j}$ are defined as

$$\alpha_{Gri,j} = \frac{\kappa_r G_r h_0}{\kappa G h_0} D(r - \hat{r}_i)\, D(\phi - \hat{\phi}_j) \tag{7.24}$$

$$\alpha_{G\phi i,j} = \frac{\kappa_\phi G_\phi h_0}{\kappa G h_0} D(r - \hat{r}_i) D(\phi - \hat{\phi}_j) \tag{7.25}$$

where $\hat{r}_i$ and $\hat{\phi}_j$ indicate the fields (from $r_i - b_{ri-1,j}^*/2$ to $r_i + b_{ri,j}^*/2$ and from $\phi_i - b_{\phi i-1,j}^*/2$ to $\phi_i + b_{\phi i,j}^*/2$, respectively) prescribed by the i, j-th void. For solid plates or a portion without voids, these coefficients α_{Gr} and $\alpha_{G\phi}$ become 1.

7.4 Stress Resultants and Stress Couples of Platelets and Partition

Once the transverse shear forces Q_r and Q_ϕ are obtained, the stress resultants and stress couples of the frame and partition parallel to the radial direction, as shown in Fig. 7.4, are given as follows: The transverse shear forces $Q_{r\text{frame}}$ and $Q_{r\text{partition}}$ of the frame and partition located parallel to the radial direction are given as

$$Q_{r\text{frame}} = Q_r \frac{(\kappa_r G_r h_0)^*}{\kappa_r G_r h_0} \tag{7.26}$$

$$Q_{r\text{partition}} = Q_r \frac{\kappa G h_0}{\kappa_r G_r h_0} \frac{b_{\phi 1} + b_{\phi 2}}{k_1 b^*_{\phi i,j}} \tag{7.27}$$

in which the frame is composed of the top and bottom platelets and of partition per void and excludes the partition parallel like shear wall to the direction. The transverse shear force $Q_{r\text{frame}}$ is divided by the transverse shear forces, $Q_{r\text{top}}$ and $Q_{r\text{bottom}}$, of the top and bottom platelets in proportion to the thickness as

$$Q_{r\text{top}} = Q_{r\text{frame}} \frac{h_{\text{top}}}{h_{\text{top}} + h_{\text{bottom}}} \tag{7.28}$$

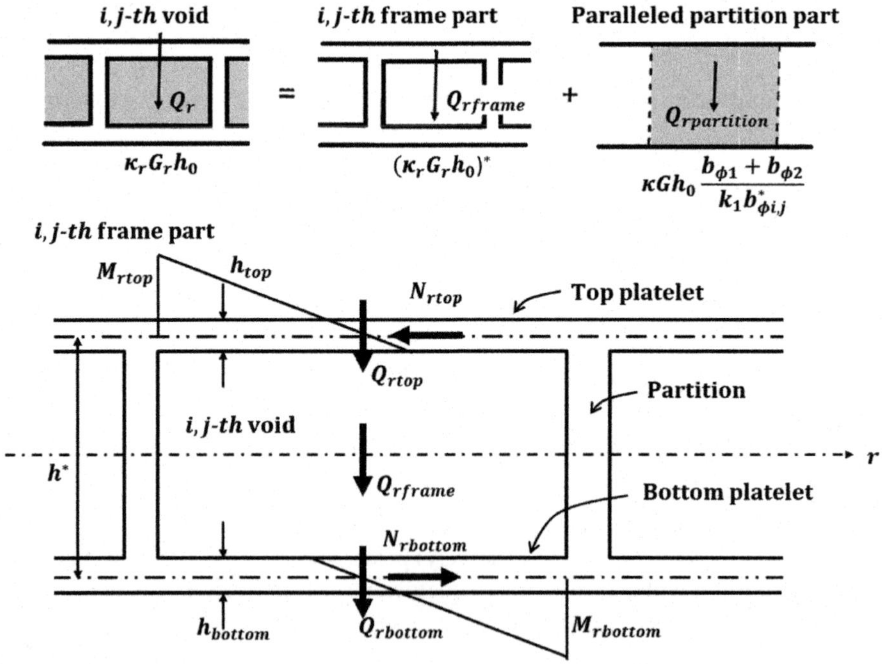

Fig. 7.4 Stress resultants and stress couples of top and bottom platelets [1]

$$Q_{r\text{bottom}} = Q_{r\text{frame}} \frac{h_{\text{bottom}}}{h_{\text{top}} + h_{\text{bottom}}} \tag{7.29}$$

in which h_{top} and h_{bottom} are the slab's thickness of the top and bottom platelets, respectively. Hence, the flexural moments $M_{r\text{top}}$ and $M_{r\text{bottom}}$ of the top and bottom platelets at both the ends of the i, j-th void are given by

$$M_{r\text{top}} = Q_{r\text{top}} \frac{b_{ri,j}}{2} \tag{7.30}$$

$$M_{r\text{bottom}} = Q_{r\text{bottom}} \frac{b_{ri,j}}{2} \tag{7.31}$$

Lastly, the axial forces $N_{r\text{top}}$ $(= -\ N_{r\text{bottom}})$ of the top and bottom platelets, due to the flexural moments, are calculated by

$$N_{r\text{top}} = \frac{M_r + M_{r\text{top}} + M_{r\text{bottom}}}{h^*} \tag{7.32}$$

in which h^* is distance between the middle surfaces of top and bottom platelets. Similar expressions for circumferential direction may be obtained.

7.5 Static Analysis

The static solutions of circular cellular plate are presented by means of the Galerkin method. The deflections and rotational angles can be expressed by a series expansion as follows

$$w(r, \phi) = \sum_{m=1} \sum_{n=0} w_{mn}\ f_{mn}(r, \phi) \tag{7.33}$$

$$\theta_r(r, \phi) = \sum_{m=1} \sum_{n=0} \theta_{rmn}\ g_{rmn}(r, \phi) \tag{7.34}$$

$$\theta_\phi(r, \phi) = \sum_{m=1} \sum_{n=0} \theta_{\phi\, mn}\ g_{\phi\, mn}(r, \phi) \tag{7.35}$$

in which w_{mn}, θ_{rmn}, and $\theta_{\phi\, mn}$ are unknown displacement coefficients; f_{mn}, g_{rmn}, and $g_{\phi mn}$ represent shape functions satisfying the specified boundary conditions.

The Galerkin equations for the static problem may be written as

$$\delta\, w_{mn} : \sum_{\bar{m}=1} \sum_{\bar{n}=0} w_{\bar{m}\bar{n}}\ A_{1mn\bar{m}\bar{n}} + \sum_{\bar{m}=1} \sum_{\bar{n}=0} \theta_{r\bar{m}\bar{n}}\ A_{2mn\bar{m}\bar{n}} + \sum_{\bar{m}=1} \sum_{\bar{n}=0} \theta_{\phi\, \bar{m}\bar{n}}\ A_{3mn\bar{m}\bar{n}} = P_{mn} \tag{7.36}$$

$$\delta\,\theta_{rmn}:\quad \sum_{\bar{m}=1}\sum_{\bar{n}=0} w_{\bar{m}\bar{n}}\,B_{1mn\bar{m}\bar{n}} + \sum_{\bar{m}=1}\sum_{\bar{n}=0} \theta_{r\bar{m}\bar{n}}\,B_{2mn\bar{m}\bar{n}} + \sum_{\bar{m}=1}\sum_{\bar{n}=0} \theta_{\phi\,\bar{m}\bar{n}}\,B_{3mn\bar{m}\bar{n}} = 0 \tag{7.37}$$

$$\delta\,\theta_{\phi\,mn}:\quad \sum_{\bar{m}=1}\sum_{\bar{n}=0} w_{\bar{m}\bar{n}}\,C_{1mn\bar{m}\bar{n}} + \sum_{\bar{m}=1}\sum_{\bar{n}=0} \theta_{r\bar{m}\bar{n}}\,C_{2mn\bar{m}\bar{n}} + \sum_{\bar{m}=1}\sum_{\bar{n}=0} \theta_{\phi\,\bar{m}\bar{n}}\,C_{3mn\bar{m}\bar{n}} = 0 \tag{7.38}$$

in which the coefficients $A_{1mn\bar{m}\bar{n}}, \ldots, C_{3mn\bar{m}\bar{n}}$ are given in Appendix 2. The unknown displacement coefficients are obtained from solving a set of algebraic linear equations given by Eqs. (7.36)–(7.38).

7.6 Numerical Results for Static Problem

A static analysis for a circular cellular plate has been presented by means of the Galerkin method. In order to examine the proposed method, numerical calculations are performed for two types of simply supported and clamped circular cellular plates, as shown in Fig. 7.5. Data used are as follows: Young's modulus $E = 2.06 \times 10^{10}$ N/m^2; Poisson's ratio $\nu = 0.17$; slab's height $h_0 = 1$ m; span radius $r_0 = 8$ m; each void's height $h_{i,j} = 0.8$ m. For simplicity, the two cases of the lateral loads are considered as a uniformly distributed load $p = 9.807 \times 10^6$ N/m^2 and a unit concentrated load at the midspan and the mass density $\rho = 2440$ kg/m^3.

The shape functions used are given by modifying the well-known solution for circular solid (normal) plates subjected to the same load. Hence, the shape functions for the simply supported plate are

$$f_{mn}(r,\phi) = \left[\left(\frac{r}{r_0}\right)^4 - 2\frac{3+2\nu}{1+\nu}\left(\frac{r}{r_0}\right)^2 + \frac{5+\nu}{1+\nu}\right]^m \cos n\phi \tag{7.39}$$

for the uniform load and

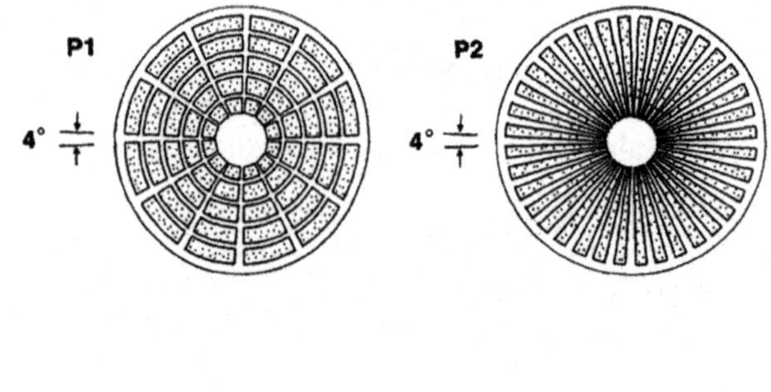

Fig. 7.5 Isotropic circular cellular plates [1]

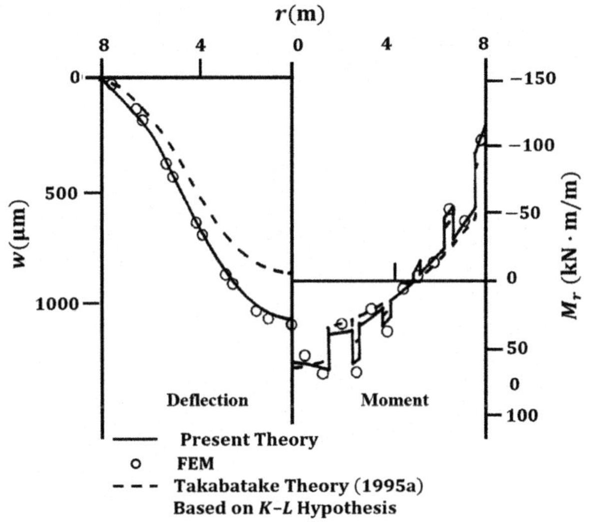

Fig. 7.6 w and M_r for a clamped cellular plate subject to uniform load [1]

$$f_{mn}(r,\phi)=\left[\left\{\frac{3+\nu}{1+\nu}\left[1-\left(\frac{r}{r_0}\right)^2\right]+2\left(\frac{r}{r_0}\right)^2\ln\left(\frac{r}{r_0}\right)\right\}(1-\nu^2)\right]^m\cos n\phi \tag{7.40}$$

for the concentrated load, as given in Ref. [3]. The other shape functions are given by

$$g_{rmn}(r,\phi)=f_{mn}(r,\phi)_{,r} \tag{7.41}$$

$$g_{\phi mn}(r,\phi)=f_{mn}(r,\phi)_{,\phi} \tag{7.42}$$

Similarly, the shape functions for clamped circular plates may be obtained. The values of m take odd numbers for $n=0$ and even numbers for $n\geq 1$.

Figures 7.6 and 7.7 show the distributions of the deflections w and bending moment M_r for the clamped plates P1, subjected to the uniform load and concentrated load, respectively. In these figures, the solid lines indicate the numerical results obtained from the proposed theory using Eqs. (7.36)–(7.38); the circles indicate the numerical results obtained by using the FEM code NASTRAN; the broken lines indicate the numerical results obtained from Eq. (5.25) presented in Chap. 5 excluding the transverse shear deformation. The results obtained from the present theory demonstrated relatively good agreement with the results obtained from the finite element method. The finite element method used here is based on isotropic and rectangular plate elements proposed on the top and bottom platelets and partitions of the voids. The result obtained from the plate element used here in the FEM is confirmed to show excellent agreement with the result using a solid element to current plates.

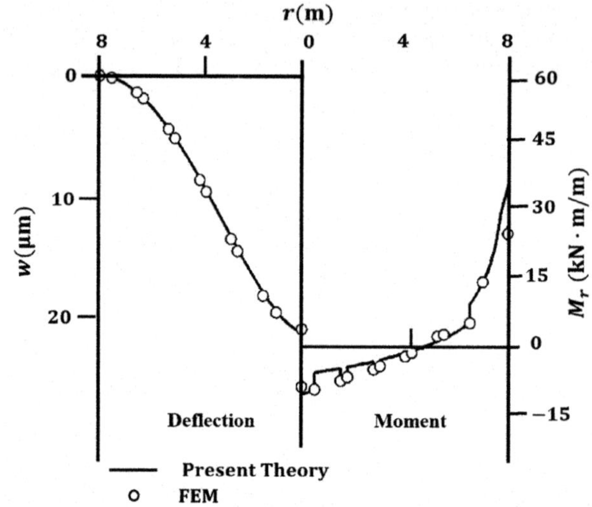

Fig. 7.7 w and M_r for a clamped cellular plate subject to concentrated load [1]

The convergence of series expansion is very rapid for uniform loads. So, the consideration of nine terms gives accuracy sufficient for all practical purposes.

7.7 Free Transverse Vibrations of Cellular Plates

We consider free transverse vibrations of a circular cellular plate. The method of separation of variables is also employed, assuming that

$$w(r, \phi, t) = \bar{w}(r, \phi)e^{i\omega t}(t) \tag{7.43}$$

$$\theta_r(r, \phi, t) = \bar{\theta}_r(r, \phi)e^{i\omega t}(t) \tag{7.44}$$

$$\theta_\phi(r, \phi, t) = \bar{\theta}_\phi(r, \phi)e^{i\omega t}(t) \tag{7.45}$$

in which $\bar{w}$, $\bar{\theta}_r$, and $\bar{\theta}_\phi$ are functions of r and ϕ.

The natural frequencies are presented by means of the Galerkin method. The functions $\bar{w}$, $\bar{\theta}_r$, and $\bar{\theta}_\phi$ are expressed as

$$\bar{w}(r, \phi) = \sum_{m=1}\sum_{n=0} w_{mn}\ f_{mn}(r, \phi) \tag{7.46}$$

$$\bar{\theta}_r(r, \phi) = \sum_{m=1}\sum_{n=0} \theta_{rmn}\ g_{rmn}(r, \phi) \tag{7.47}$$

$$\bar{\theta}_\phi(r, \phi) = \sum_{m=1}\sum_{n=0} \theta_{\phi mn}\ g_{\phi mn}(r, \phi) \tag{7.48}$$

in which w_{mn}, θ_{rmn}, and $\theta_{\phi\,mn}$ are unknown displacement coefficients, and f_{mn}, g_{rmn}, and $g_{\phi\,mn}$ indicate shape functions satisfying the specified boundary conditions. Hence, the Galerkin equations for free transverse vibration become

$$\delta\, w_{mn}: \quad \sum_{\bar{m}=1}\sum_{\bar{n}=0}[w_{\bar{m}\bar{n}}\,(A_{1mn\bar{m}\bar{n}} - \omega^2 F_{1mn\bar{m}\bar{n}}) + \theta_{r\bar{m}\bar{n}} A_{2mn\bar{m}\bar{n}} + \theta_{\phi\,\bar{m}\bar{n}} A_{3mn\bar{m}\bar{n}}] = 0 \tag{7.49}$$

$$\delta\, \theta_{rmn}: \quad \sum_{\bar{m}=1}\sum_{\bar{n}=0}[w_{\bar{m}\bar{n}}\, B_{1mn\bar{m}\bar{n}} + \theta_{r\bar{m}\bar{n}}\,(B_{2mn\bar{m}\bar{n}} - \omega^2 F_{2mn\bar{m}\bar{n}}) + \theta_{\phi\,\bar{m}\bar{n}} B_{3mn\bar{m}\bar{n}}] = 0 \tag{7.50}$$

$$\delta\, \theta_{\phi mn}: \quad \sum_{\bar{m}=1}\sum_{\bar{n}=0}[w_{\bar{m}\bar{n}}\, C_{1mn\bar{m}\bar{n}} + \theta_{r\bar{m}\bar{n}}\, C_{2mn\bar{m}\bar{n}} + \theta_{\phi\,\bar{m}\bar{n}}(C_{3mn\bar{m}\bar{n}} - \omega^2 F_{3mn\bar{m}\bar{n}})] = 0 \tag{7.51}$$

in which $A_{1mn\bar{m}\bar{n}}, \ldots, B_{1mn\bar{m}\bar{n}}, \ldots, C_{3mn\bar{m}\bar{n}}$ are the same as the coefficients in Eqs. (7.36) to (7.38) for the static problem. On the other hand, $F_{1mn\bar{m}\bar{n}}$, $F_{2mn\bar{m}\bar{n}}$, and $F_{3mn\bar{m}\bar{n}}$ are defined as

$$F_{1mn\bar{m}\bar{n}} = \int_0^{2\pi}\int_0^{r_0} m_0 \alpha_h\, f_{\bar{m}\bar{n}}\, f_{mn}\, \mathrm{d}r r\, \mathrm{d}\phi \tag{7.52}$$

$$F_{2mn\bar{m}\bar{n}} = \int_0^{2\pi}\int_0^{r_0} I_p\, g_{r\bar{m}\bar{n}}\, g_{rmn}\, \mathrm{d}r r\, \mathrm{d}\phi \tag{7.53}$$

$$F_{3mn\bar{m}\bar{n}} = \int_0^{2\pi}\int_0^{r_0} \frac{1}{r} I_p\, g_{\phi\bar{m}\bar{n}}\, g_{\phi mn}\, \mathrm{d}r r\, \mathrm{d}\phi \tag{7.54}$$

The natural frequencies ω_{mn} are obtained by solving Eqs. (7.49)–(7.51) as eigenvalue problems. Then, for practical use, the approximate expressions for the natural frequencies are considered, as mentioned in the preceding chapter. Considering only the diagonal terms to transform the coupled Eqs. (7.49)–(7.51) into uncoupled forms, the approximate solution for the natural frequencies is obtained from the following square (for $n = 0$) and cubic (for $n > 0$) equations with respect to ω^2:

$$\begin{bmatrix} A_{1m0m0} - \omega^2 F_{1m0m0} & F_{2m0m0} \\ B_{1m0m0} & B_{2m0m0} - \omega^2 F_{2m0m0} \end{bmatrix} \begin{bmatrix} \omega_{m0} \\ \theta_{rm0} \end{bmatrix} = 0 \quad (\text{for}\, n = 0) \tag{7.55}$$

$$\begin{bmatrix} A_{1mnmn} - \omega^2 F_{1mnmn} & A_{2mnmn} & A_{3mnmn} \\ B_{1mnmn} & B_{2mnmn} - \omega^2 F_{2mnmn} & B_{3mnmn} \\ C_{1mnmn} & C_{2mnmn} & C_{3mnmn} - \omega^2 F_{3mnmn} \end{bmatrix} \begin{bmatrix} w_{mn} \\ \theta_{rmn} \\ \theta_{\phi mn} \end{bmatrix} = 0$$

(for $n > 0$) (7.56)

7.8 Numerical Results for Natural Frequencies

The natural frequencies for a circular cellular plate have been presented in an approximate form by means of the Galerkin method. Then, to examine the proposed method, numerical computations are performed for the same models used in the static problem, as shown in Fig. 7.5. The following shape functions are used:

$$f_{mn}(r,\phi) = \sin\left[\frac{m\pi}{2}\left(1 - \frac{r}{r_0}\right)\right]\cos n\phi \quad \text{(for simply supported plate)} \tag{7.57}$$

$$f_{mn}(r,\phi) = \sin\left[\frac{m\pi}{2}\left(1 - \frac{r}{r_0}\right)\right]\sin\left[\frac{\pi}{2}\left(1 - \frac{r}{r_0}\right)\right]\cos n\phi \quad \text{(for clamped plate)} \tag{7.58}$$

in which the values of m take odd numbers for $n=0$ and even numbers for $n>1$. The other shape functions $g_{rmn}(r,\phi)$ and $g_{\phi mn}(r,\phi)$ are obtained from the relation of Eqs. (7.41) and (7.42).

Tables 7.2 and 7.3 show the natural frequencies for Pl and P2, respectively. The approximate solution can be used for lower natural frequencies. The results obtained

Table 7.2 Natural frequencies of simply supported and clamped cellular plates P1, in which m = mode number in radial direction and n = mode number in circumferential direction [1]

Mode	Analytical method	Simply supported (rad/s)			Clamped (rad/s)		
		$m=1$	$m=2$	$m=3$	$m=1$	$m=2$	$m=3$
$n=0$	Rigorous	53.25	290.29	601.91	113.53	360.49	663.38
	Approximate	53.29	268.65	580.00	114.58	361.90	647.36
	FEM	56.31	292.44	603.86	112.70	370.90	724.61
$n=1$	Rigorous	178.88	452.12	770.01	236.03	534.71	816.42
	Approximate	141.23	421.47	748.49	198.84	501.36	799.65
	FEM	158.00	480.36	815.82	233.58	563.37	911.70
$n=2$	Rigorous	281.59	617.96	996.16	347.49	689.49	1087.00
	Approximate	246.81	528.80	841.94	375.74	755.74	1031.62
	FEM	282.32	655.49	1192.44	363.92	733.09	1236.17

Table 7.3 Natural frequencies of simply supported and clamped cellular plates P2, in which m = mode number in radial direction and n = mode number in circumferential direction [1]

Mode	Analytical method	Simply supported (rad/s)			Clamped (rad/s)		
		$m=1$	$m=2$	$m=3$	$m=1$	$m=2$	$m=3$
$n=0$	Rigorous	52.51	317.24	724.83	118.50	418.05	842.84
	Approximate	52.54	287.85	681.59	119.02	419.11	809.85
	FEM	53.44	317.95	745.23	115.88	429.09	884.85
$n=1$	Rigorous	185.47	519.34	946.38	266.07	642.72	1046.00
	Approximate	146.74	473.84	912.85	220.17	589.55	1018.41
	FEM	159.79	509.56	959.50	246.78	630.47	1092.79
$n=2$	Rigorous	301.71	737.16	1290.00	399.07	853.33	1466.70
	Approximate	224.63	570.58	1016.04	379.43	899.58	1346.93
	FEM	278.06	688.69	1494.05	379.62	945.56	1791.53

from the proposed theory demonstrate comparatively good agreement with results obtained from the solid element using the FEM.

7.9 Forced Vibration of Cellular Plates

We consider forced vibrations of the current circular cellular plate. The general solutions of Eqs. (7.4)–(7.6) are assumed to be the form, as similar to the preceding chapter,

$$w(r,\phi,t)=\sum_{m=1}\sum_{n=0} f_{mn}(r,\phi)\,w_{mn}(t) \tag{7.59}$$

$$\theta_r(r,\phi,t)=\sum_{m=1}\sum_{n=0} g_{rmn}(r,\phi)\,\theta_{rmn}(t) \tag{7.60}$$

$$\theta_\phi(r,\phi,t)=\sum_{m=1}\sum_{n=0} g_{\phi mn}(r,\phi)\,\theta_{\phi mn}(t) \tag{7.61}$$

in which w_{mn}, θ_{rmn}, and $\theta_{\phi mn}$ are unknown functions of time t.

Then, the Galerkin equations of Eqs. (7.4)–(7.6) may be rewritten as

$$\begin{aligned}\delta\, w_{mn}:\ &\sum_{\bar m=1}\sum_{\bar n=0}(\ddot w_{\bar m\bar n}F_{1mn\bar m\bar n}+2h_{\bar m\bar n}\omega_{\bar m\bar n}\dot w_{\bar m\bar n}F_{1mn\bar m\bar n}\\&-w_{\bar m\bar n}A_{1mn\bar m\bar n}-\theta_{r\bar m\bar n}A_{2mn\bar m\bar n}-\theta_{\phi\bar m\bar n}A_{3mn\bar m\bar n})-P_{mn}=0\end{aligned} \tag{7.62}$$

$$\delta\,\theta_{rmn}:\ \sum_{\bar m=1}\sum_{\bar n=0}(\ddot\theta_{r\bar m\bar n}F_{2mn\bar m\bar n}+w_{\bar m\bar n}B_{1mn\bar m\bar n}+\theta_{r\bar m\bar n}B_{2mn\bar m\bar n}+\theta_{\phi\bar m\bar n}B_{3mn\bar m\bar n})=0 \tag{7.63}$$

$$\delta\,\theta_{\phi mn}:\quad \sum_{\bar{m}=1}\sum_{\bar{n}=0}(\ddot{\theta}_{\phi\bar{m}\bar{n}}F_{3mn\bar{m}\bar{n}}+w_{\bar{m}\bar{n}}C_{1mn\bar{m}\bar{n}}+\theta_{r\bar{m}\bar{n}}C_{2mn\bar{m}\bar{n}}+\theta_{\phi\bar{m}\bar{n}}C_{3mn\bar{m}\bar{n}})=0 \tag{7.64}$$

in which h_{mn} is damping constant of the circular cellular plate, which depends on the mode indexes m and n. The dynamic response is obtained by solving the coupled Eqs. (7.62)–(7.64). The coefficients in these equations have nondiagonal terms $m \neq \bar{m}$ and/or $n \neq \bar{n}$ due to the effect of the voids. Therefore, the solution will be based on numerical computation, such as step-by-step integration methods.

For practical use, we consider an approximate solution in the closed form. Assuming that the behavior of a circular cellular plate is now dominated by only the diagonal terms in the coefficients, the coupled Eqs. (7.62)–(7.64) become of uncoupled with respect to m and $\bar{m}$ and n and $\bar{n}$, respectively. Furthermore, neglecting the rotatory inertia's terms in the uncoupled equations, we have the following differential equation in terms of only w_{mn}:

$$\ddot{w}_{mn}+2a_{mn}\dot{w}_{mn}+b_{mn}w_{mn}=Q_{mn}(t) \tag{7.65}$$

in which a_{mn}, b_{mn}, and $Q_{mn}(t)$ are defined as

$$a_{mn}\equiv h_{mn}\omega_{mn} \tag{7.66}$$

$$b_{mn}\equiv-\frac{A^*_{mn}}{F_{1mnmn}} \tag{7.67}$$

$$Q_{mn}(t)\equiv\frac{P_{mn}}{F_{1mnmn}} \tag{7.68}$$

where A^*_{mn} are defined as

$$\begin{aligned}A^*_{mn}=A_{1mnmn}-\frac{1}{\Delta_{mn}}[&A_{2mnmn}(C_{3mnmn}B_{1mnmn}-B_{3mnmn}C_{1mnmn})\\&+A_{3mnmn}(B_{2mnmn}C_{1mnmn}-B_{1mnmn}C_{2mnmn})]\end{aligned} \tag{7.69}$$

in which Δ_{mn} is given by

$$\Delta_{mn}=B_{2mnmn}C_{3mnmn}-B_{3mnmn}C_{2mnmn} \tag{7.70}$$

The general solution of Eq. (7.65) is

$$\begin{aligned}w_{mn}=&\,\mathrm{e}^{-a_{mn}t}(\bar{C}_1\sin\alpha_{0mn}t+\bar{C}_2\cos\alpha_{0mn}t)\\&+\frac{1}{\alpha_{0mn}}\int_0^t\mathrm{e}^{-a_{mn}(t-\tau)}\sin\alpha_{0mn}(t-\tau)Q_{mn}(\tau)\,\mathrm{d}\tau\end{aligned} \tag{7.71}$$

in which $\bar{C}_1$ and $\bar{C}_2$ are constants, and α_{0mn} are defined as

$$\alpha_{0mn} \equiv \sqrt{b_{mn} - a_{mn}^2} \tag{7.72}$$

Thus, the dynamic deflections of a circular cellular plate are determined by substituting Eq. (7.71) into Eq. (7.59).

7.10 Numerical Results for Dynamic Responses

In order to examine the closed-form approximate solution proposed here, numerical computations are carried out for the previous cellular plates, as shown in Fig. 7.5. The damping constants h_{mn} are 0.03 for all modes. The following two types of external lateral loads without the additional mass are assumed: firstly, a harmonic and uniformly distributed force $p = p^* \sin(\omega_p t)$, in which $p^* = 9.807 \times 10^6$ N/m^2 and $\omega_p = 62.83$ rad/s (10 cps), and secondly, test loads checking the disturbing effect of the vibration of floors caused by people walking and every other usage, as shown in Fig. 7.8. The test loads consist of four types of walking, tripping, bending and stretching, and heel, as already stated in the preceding chapters. They act as a concentrated load at the midspan. The current natural functions used are the same ones as used in the free vibrations.

Figure 7.9 represents the time history of the dynamic deflections at the midpoint of simply supported cellular plates of Type P1 subjected to the harmonic load. In this figure, the broken lines indicate values obtained from the approximate solution and the solid lines with circles indicate values obtained from FEM. Figure 7.10 shows the time history of the dynamic deflections at the midpoint of clamped cellular plates of P2, subjected to the test loads. The numerical results for all cases show that the approximate solution proposed here is applicable to the dynamic analyses of circular cellular plates in practical use.

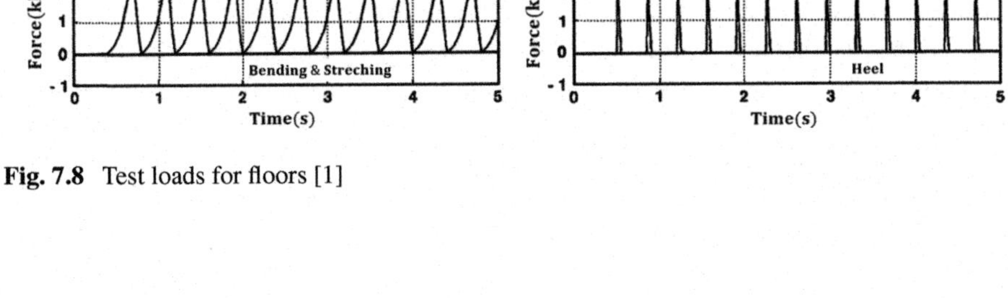

Fig. 7.8 Test loads for floors [1]

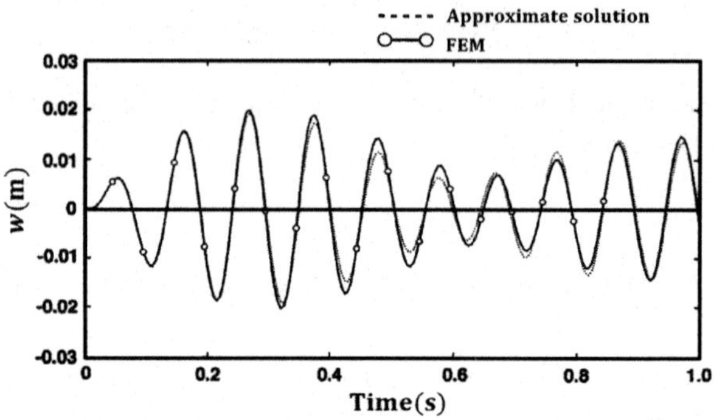

Fig. 7.9 Dynamic deflections of a simply supported plate P1 subjected to uniform load [1]

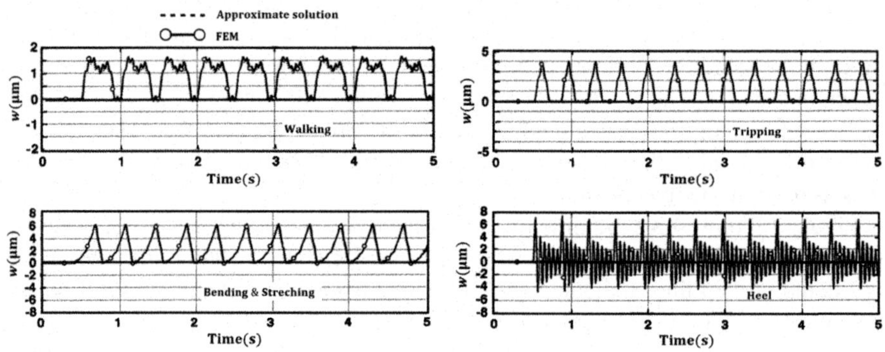

Fig. 7.10 Dynamic deflections of a clamped plate P4 subjected to test loads [1]

For these numerical models, the calculations using the present theory are remarkably fast as compared to those using FEM. Since the series in the theory proposed here converges very rapidly, the consideration of nine terms gives accuracy sufficient for all practical purposes.

7.11 Conclusions

The general and simplified analysis methods for an isotropic circular cellular plate with arbitrarily disposed voids have been proposed by considering the transverse shear deformation along with frame deformation. It is clarified from numerical computations that the simplified theory proposed here is usable in the preliminary and final stages of designs of such a cellular plate. The present theory is applicable to a circular cellular plate with arbitrarily shape voids by subdividing approximately the irregular void into many fan parallelepipeds.

Appendix 1: Translational Formulations

The coefficients and rotational angles are translated as follows:

$$\begin{aligned}
&w_{,x} = w_{,r}; \quad w_{,xx} = w_{,rr};\\
&w_{,y} = \frac{1}{r}w_{,\phi}; \quad w_{,yy} = \frac{1}{r}w_{,r} + \frac{1}{r^2}w_{,\phi\phi}; \quad w_{,xy} = \frac{1}{r}w_{,r\phi} - \frac{1}{r^2}w_{,\phi};\cdots
\end{aligned} \tag{7.73}$$

$$\begin{aligned}
&\theta_x = \theta_r; \quad \theta_y = \frac{1}{r}\theta_\phi; \quad \theta_{x,x} = \theta_{r,r}; \quad \theta_{x,y} = \frac{1}{r}\theta_{r,\phi} - \frac{1}{r^2}\theta_\phi; \quad \theta_{x,xx} = \theta_{r,rr};\\
&\theta_{x,xy} = \frac{1}{r}\theta_{r,r\phi} - \frac{1}{r}\left(\frac{1}{r}\theta_\phi\right)_{,r} - \frac{1}{r^2}\theta_{r,\phi} + \frac{1}{r^3}\theta_\phi; \quad \theta_{x,xxx} = \theta_{r,rrr};\\
&\theta_{y,x} = \left(\frac{1}{r}\theta_\phi\right); \quad \theta_{y,y} = \frac{1}{r}\theta_r + \frac{1}{r^2}\theta_{\phi,\phi};\\
&\theta_{y,yy} = \frac{1}{r}\left(\frac{1}{r}\theta_\phi\right)_{,r} + \frac{2}{r^2}\theta_{r,\phi} + \frac{1}{r^3}\theta_{\phi,\phi\phi} - \frac{1}{r^3}\theta_\phi;\\
&\theta_{y,xy} = \frac{1}{r}\left[\theta_{r,r} + \left(\frac{1}{r}\theta_\phi\right)_{,r\phi}\right] - \frac{1}{r^2}\left(\theta_r + \frac{1}{r}\theta_{\phi,\phi}\right);\\
&\theta_{x,yy} = \frac{1}{r}\theta_{r,r} + \frac{1}{r^2}\theta_{r,\phi\phi} - \frac{1}{r^2}\theta_r - \frac{2}{r^3}\theta_{\phi,\phi}; \quad \theta_{y,xx} = \left(\frac{1}{r}\theta_\phi\right)_{,rr};\cdots
\end{aligned} \tag{7.74}$$

Appendix 2: Coefficients and Load Terms

The coefficients in Eqs. (7.36)–(7.38) are defined as

$$\begin{aligned}
A_{1mn\bar{m}\bar{n}} = \int_0^{2\pi}\int_0^{r_0} \Big[&Gh_0\kappa_r\alpha_{Gr,r} f_{\bar{m}\bar{n},r} + Gh_0\kappa_r\alpha_{Gr} f_{\bar{m}\bar{n},rr}\\
&+ Gh_0\kappa_\phi\alpha_{G\phi,\phi}\frac{1}{r^2} f_{\bar{m}\bar{n},\phi} + Gh_0\kappa_\phi\alpha_{G\phi}\left(\frac{1}{r} f_{\bar{m}\bar{n},r} + \frac{1}{r^2} f_{\bar{m}\bar{n},\phi\phi}\right)\Big] f_{mn}\,\mathrm{d}r r\,\mathrm{d}\phi
\end{aligned} \tag{7.75}$$

$$\begin{aligned}
A_{2mn\bar{m}\bar{n}} = \int_0^{2\pi}\int_0^{r_0} \Big(&Gh_0\kappa_r\alpha_{Gr,r} g_{r\bar{m}\bar{n}} + Gh_0\kappa_r\alpha_{Gr} g_{r\bar{m}\bar{n},r}\\
&+ Gh_0\kappa_\phi\alpha_{G\phi}\frac{1}{r} g_{r\bar{m}\bar{n}}\Big) f_{mn}\,\mathrm{d}r r\,\mathrm{d}\phi
\end{aligned} \tag{7.76}$$

$$A_{3mn\bar{m}\bar{n}} = \int_0^{2\pi}\int_0^{r_0} \left(Gh_0\kappa_\phi\alpha_{G\phi,\phi}\frac{1}{r^2} g_{\phi\bar{m}\bar{n}} + Gh_0\kappa_\phi\alpha_{G\phi}\frac{1}{r^2} g_{\phi\bar{m}\bar{n},\phi}\right) f_{mn}\,\mathrm{d}r r\,\mathrm{d}\phi \tag{7.77}$$

$$B_{1mn\bar{m}\bar{n}} = -\int_0^{2\pi}\int_0^{r_0} Gh_0\kappa_r\alpha_{Gr} f_{\bar{m}\bar{n},r} g_{rmn}\, \mathrm{d}r r\, \mathrm{d}\phi, \tag{7.78}$$

$$\begin{aligned}
B_{2mn\bar{m}\bar{n}} = \int_0^{2\pi}\int_0^{r_0} &\left\{ D_0\mathrm{d}_{,r} g_{r\bar{m}\bar{n},r} + D_0\mathrm{d}_{,r}\nu\frac{1}{r} g_{r\bar{m}\bar{n}} \right.\\
&+ D_0\mathrm{d}\left[g_{r\bar{m}\bar{n},rr} + \nu\left(\frac{1}{r} g_{r\bar{m}\bar{n},r} - \frac{1}{r^2} g_{r\bar{m}\bar{n}}\right)\right]\\
&+ \frac{(1-\nu)}{2} D_0\mathrm{d}_{,\phi}\frac{1}{r^2} g_{r\bar{m}\bar{n},\phi} + \frac{(1-\nu)}{2} D_0\mathrm{d}\\
&\left.\times\left(\frac{2}{r} g_{r\bar{m}\bar{n},r} + \frac{1}{r^2} g_{r\bar{m}\bar{n},\phi\phi} - \frac{2}{r^2} g_{r\bar{m}\bar{n}}\right) - Gh_0\kappa_r\alpha_{Gr} g_{r\bar{m}\bar{n}} \right\} g_{rmn}\, \mathrm{d}r r\, \mathrm{d}\phi
\end{aligned} \tag{7.79}$$

$$\begin{aligned}
B_{3mn\bar{m}\bar{n}} = \int_0^{2\pi}\int_0^{r_0} &\left[D_0\mathrm{d}_{,r}\nu\frac{1}{r^2} g_{r\bar{m}\bar{n},\phi} + D_0\,\mathrm{d}\nu\left(-\frac{2}{r^3} g_{\phi\bar{m}\bar{n},\phi} + \frac{1}{r^2} g_{\phi\bar{m}\bar{n},r\phi}\right)\right.\\
&+ \frac{(1-\nu)}{2} D_0\mathrm{d}_{,\phi}\left(-\frac{2}{r^3} g_{\phi\bar{m}\bar{n}} + \frac{1}{r^2} g_{\phi\bar{m}\bar{n},r}\right)\\
&\left.+\frac{(1-\nu)}{2} D_0\mathrm{d}\left(\frac{1}{r^2} g_{\phi\bar{m}\bar{n},r\phi} - \frac{4}{r^3} g_{\phi\bar{m}\bar{n},\phi}\right)\right] g_{rmn}\, \mathrm{d}r r\, \mathrm{d}\phi
\end{aligned} \tag{7.80}$$

$$C_{1mn\bar{m}\bar{n}} = -\int_0^{2\pi}\int_0^{r_0} Gh_0\kappa_\phi\alpha_{G\phi}\frac{1}{r} f_{\bar{m}\bar{n},\phi} g_{\phi mn}\, \mathrm{d}r r\, \mathrm{d}\phi \tag{7.81}$$

$$\begin{aligned}
C_{2mn\bar{m}\bar{n}} = \int_0^{2\pi}\int_0^{r_0} &\left\{ D_0\mathrm{d}_{,\phi}\left(\frac{1}{r^2} g_{r\bar{m}\bar{n}} + \nu\frac{1}{r} g_{r\bar{m}\bar{n},r}\right)\right.\\
&+ D_0\mathrm{d}\left[\frac{2}{r^2} g_{r\bar{m}\bar{n},\phi} + \nu\left(\frac{1}{r} g_{r\bar{m}\bar{n},r\phi} - \frac{1}{r^2} g_{r\bar{m}\bar{n},\phi}\right)\right] + \frac{(1-\nu)}{2} D_0\mathrm{d}_{,r}\frac{1}{r} g_{r\bar{m}\bar{n},\phi}\\
&\left.+\frac{(1-\nu)}{2} D_0\mathrm{d}\times\left(\frac{1}{r} g_{r\bar{m}\bar{n},r\phi} - \frac{1}{r^2} g_{r\bar{m}\bar{n},\phi}\right)\right\} g_{\phi mn}\, \mathrm{d}r r\, \mathrm{d}\phi,
\end{aligned} \tag{7.82}$$

$$
\begin{aligned}
C_{3mn\bar{m}\bar{n}} = \int_0^{2\pi}\int_0^{r_0} \Bigg\{ & D_0 \mathrm{d}_{,\phi} \frac{1}{r^3} g_{\phi\,\bar{m}\bar{n},\phi} + D_0 \mathrm{d}\left[-\frac{2}{r^3} g_{\phi\,\bar{m}\bar{n}} + \frac{1}{r^2} g_{\phi\,\bar{m}\bar{n},r} \right. \\
& \left. + \frac{1}{r^3} g_{\phi\,\bar{m}\bar{n},\phi\phi} + \nu\left(\frac{2}{r^3} g_{\phi\,\bar{m}\bar{n}} - \frac{1}{r^2} g_{\phi\,\bar{m}\bar{n},r} \right) \right] \\
& + \frac{(1-\nu)}{2} D_0 \mathrm{d}_{,r}\left(-\frac{2}{r^2} g_{\phi\,\bar{m}\bar{n}} + \frac{1}{r} g_{\phi\,\bar{m}\bar{n},r} \right) \\
& + \frac{(1-\nu)}{2} D_0 \mathrm{d}\left(\frac{4}{r^3} g_{\phi\,\bar{m}\bar{n}} - \frac{3}{r^2} g_{\phi\,\bar{m}\bar{n},r} + \frac{1}{r} g_{\phi\,\bar{m}\bar{n},rr} \right) \\
& - G h_0 \kappa_\phi \alpha_{G\phi} \frac{1}{r} g_{\phi\,\bar{m}\bar{n}} \Bigg\} g_{\phi\,mn}\, \mathrm{d}r r\, \mathrm{d}\phi
\end{aligned} \tag{7.83}
$$

And the terms of external loads in Eq. (7.36) are

$$
P_{mn} = -\int_0^{2\pi}\int_0^{r_0} p\, f_{mn}\, \mathrm{d}r r\, \mathrm{d}\phi \tag{7.84}
$$

They may be calculated easily by the method shown in Appendix 3.

Appendix 3: Calculation Including the Extend Dirac Function $D(r - r_i)$

The integral calculation including the extended Dirac function $D(r - r_i)$ can be written as [4]

$$
\int_0^{r_0} D(r - r_i)\, f(r)\, \mathrm{d}r = \int_{r_i-(b_{ri,j}/2)}^{r_i+(b_{ri,j}/2)} f(\xi)\, \mathrm{d}\xi \tag{7.85}
$$

in which ξ is a supplementary variable of r. The n-th derivatives of the extended Dirac functions can therefore be expressed as

$$
\int_0^{r_0} D^{(n)}(r - r_i)\, f(r)\, \mathrm{d}r = \int_{r_i-(b_{ri,j}/2)}^{r_i+(b_{ri,j}/2)} (-1)^n f^{(n)}(\xi)\, \mathrm{d}\xi \tag{7.86}
$$

in which the superscripts enclosed within parentheses indicate the differential order. For calculations including the extended Dirac function $D(\phi - \phi_j)$, similar expressions may be obtained.

When the conditions $b_{ri,\,j} \ll r_0$ and $b_{\phi\,i,\,j} \ll 2\pi\, r$ are satisfied, the specific functions $D(r - r_i)$ and $D(\phi - \phi_j)$ are approximately related to the Dirac delta functions, $\delta(r - r_i)$ and $\delta(\phi - \phi_j)$, by

$$D(r - r_i) \approx b_{ri,j}\,\delta(r - r_i) \tag{7.87}$$

$$D(\phi - \phi_j) \approx b_{\phi i,j}\,\delta(\phi - \phi_j) \tag{7.88}$$

Appendix 4: The Governing Equations for Solid Plates Without Voids Based on Mindlin-Reissner Hypothesis

In Sect. 7.2, we had stated that the governing equations of circular cellular plates based on Mindlin-Reissner hypothesis may easily transform into rectangular solid (normal) plates without voids based on the same Mindlin-Reissner hypothesis by changing α_h and d to 1 and $\kappa_x G_x$ and $\kappa_y G_y$ to κG. We introduce the governing equations of circular solid plates based on Mindlin-Reissner hypothesis for practical use. From Eqs. (7.4) to (7.9), we have

$$\begin{aligned}\delta w:\quad & m_0\alpha_m\ddot{w} - [\kappa G h_0(w_{,r}+\theta_r)]_{,r} - (\kappa G h_0)_{,\phi}\left(\frac{1}{r^2}w_{,\phi} + \frac{1}{r^2}\theta_\phi\right)\\ & - \kappa G h_0\left(\frac{1}{r}w_{,r} + \frac{1}{r^2}w_{,\phi\phi} + \frac{1}{r}\theta_r + \frac{1}{r^2}\theta_{\phi,\phi}\right) - p + c\dot{w} = 0\end{aligned} \tag{7.89}$$

$$\begin{aligned}\delta\theta_r:\quad & I_p\ddot{\theta}_r + \kappa G h_0(w_{,r}+\theta_r) - D_0\left[\theta_{r,rr} + \nu\left(\frac{1}{r}\theta_{r,r} - \frac{2}{r^3}\theta_{\phi,\phi} + \frac{1}{r^2}\theta_{\phi,r\phi} - \frac{1}{r^2}\theta_r\right)\right]\\ & - \frac{(1-\nu)}{2}D_0\left(\frac{2}{r}\theta_{r,r} + \frac{1}{r^2}\theta_{r,\phi\phi} + \frac{1}{r^2}\theta_{\phi,r\phi} - \frac{1}{r^2}\theta_r - \frac{4}{r^3}\theta_{\phi,\phi}\right) = 0\end{aligned} \tag{7.90}$$

$$\begin{aligned}\delta\theta_\phi:\quad & \frac{1}{r}I_p\ddot{\theta}_\phi + \kappa G h_0\left(\frac{1}{r}w_{,\phi} + \frac{1}{r}\theta_\phi\right)\\ & - D_0\left[-\frac{2}{r^3}\theta_\phi + \frac{1}{r^2}\theta_{\phi,r} + \frac{2}{r^2}\theta_{r,\phi} + \frac{1}{r^3}\theta_{\phi,\phi\phi}\right.\\ & \left. + \nu\left(\frac{1}{r}\theta_{r,r\phi} + \frac{2}{r^3}\theta_\phi - \frac{1}{r^2}\theta_{\phi,r} - \frac{1}{r^2}\theta_{r,\phi}\right)\right]\\ & - \frac{(1-\nu)}{2}D_0\left(\frac{1}{r}\theta_{r,r\phi} + \frac{4}{r^3}\theta_\phi - \frac{3}{r^2}\theta_{\phi,r} - \frac{1}{r^2}\theta_{r,\phi} + \frac{1}{r}\theta_{\phi,rr}\right) = 0\end{aligned} \tag{7.91}$$

$$w = w^* \quad\text{or}\quad \kappa G h_0(w_{,r}+\theta_r) = v_r^* \tag{7.92}$$

$$\theta_r = \theta_r^* \quad\text{or}\quad D_0\left[\theta_{r,r} + \nu\left(\frac{1}{r}\theta_r + \frac{1}{r^2}\theta_{\phi,\phi}\right)\right] = m_r^* \tag{7.93}$$

$$\theta_\phi = \theta_\phi^* \quad\text{or}\quad \frac{1-\nu}{2}D_0\left(\frac{1}{r}\theta_{r,\phi} - \frac{2}{r^2}\theta_\phi + \frac{1}{r}\theta_{\phi,r}\right) = m_{r\phi}^* \tag{7.94}$$

at $r = r_0$, in which p are lateral loads; w^*, θ_r^*, and θ_ϕ^* are prescribed on the geometrical boundary conditions; and v_r^*, m_r^*, and $m_{r\phi}^*$ are prescribed on the mechanical boundary conditions at $r = r_0$.

References

1. Takabatake, H., Kajiwara, K., Takesako, R.: A simplified analysis of circular cellular plates. Comput. Struct. **61**(5), 789–804 (1996)
2. Takabatake, H., Morimoto, H., Fujiwara, T., Honma, T.: Simplified analysis of circular plates including voids. Comput. Struct. **58**(2), 263–275 (1996)
3. Szilard, R.: Theory and Analysis of Plates-Classical and Numerical Methods. Prentice-Hall, Englewood Cliffs, New Jersey (1974)
4. Takabatake, H.: Static analyses of elastic plates with voids. Int. J. Solids Struct. **28**(2), 179–196 (1991)

Chapter 8
Static and Dynamic Analyses of Rectangular Plates with Stepped Thickness

8.1 Introduction

Building slabs are usually supported by subdividing beams to thin the slab thickness. Recently, a building slab that is replaced the subdividing beams with stepped thickness is used frequently to decrease cost and simplify construction. In such a building slab with stepped thickness, the upper surface of slabs is, in general, flat in the whole surface, and the variation of the thickness is restricted on the lower surface. Therefore, analysis for building slabs with eccentrically stepped thickness is necessary but is lacking. A simplified solution is necessary to use in the preliminary and final stages of the design.

For such demand, this chapter proposes a simplified analytical method for static and dynamic problems of a rectangular plate with arbitrarily and eccentrically stepped thickness. The discontinuous variation of the bending stiffness and mass of such a plate due to arbitrarily-stepped thickness is expressed rationally as a continuous function by means of an extended Dirac function which is introduced in previous chapters. The use of extended Dirac function makes the formulation simple.

The general governing equations for a rectangular plate with stepped thickness are proposed by using Hamilton's principle. Simplified solutions for the static and dynamic problems are presented by the use of the Galerkin method. Finally, the exactness of the proposed solutions is established from numerical results by comparing the results obtained from the proposed theory with the results obtained from the FEM code NASTRAN. The methodology proposed in this chapter is mainly based on Ref. [1].

H. Takabatake, *Simplified Analytical Methods of Elastic Plates*,
https://doi.org/10.1007/978-981-13-0086-8_8

8.2 Governing Equations of Rectangular Plates with Stepped Thickness

We consider a rectangular plate with arbitrarily-stepped thickness, as shown in Fig. 8.1. The Cartesian coordinate system *x, y, z* is employed. Ridge lines of each stepped thickness are assumed to be parallel to the axis *x* or *y*. The midpoint, width, and varied thickness of the *i*-th stepped thickness being parallel to the axis *y* are noted by x_i, b_{xi}, and h_{xi}, respectively, in which h_{xi} is stepped thickness measured from the lower surface of the reference slab, as shown in Fig. 8.1. Similarly, for the *j*-th stepped thickness being parallel to the axis *x*, the midpoint, width, and varied thickness are represented by y_j, b_{yj}, and h_{yj}, respectively. The size and position of each stepped thickness are arbitrary except for the above-mentioned assumption.

This chapter considers the bending of isotropic rectangular plates in small deformations based on the validity of the Kirchhoff-Love hypothesis. The governing equation is presented by means of the following Hamilton's principle:

$$\delta I = \delta \int_{t_1}^{t_2} (T - U - V)\mathrm{d}t = 0 \tag{8.1}$$

in which *T, U,* and *V* are the kinetic energy, strain energy, and potential energy produced by external loads, respectively. δ indicates the variational operator taken during the indicated time interval from t_1 to t_2.

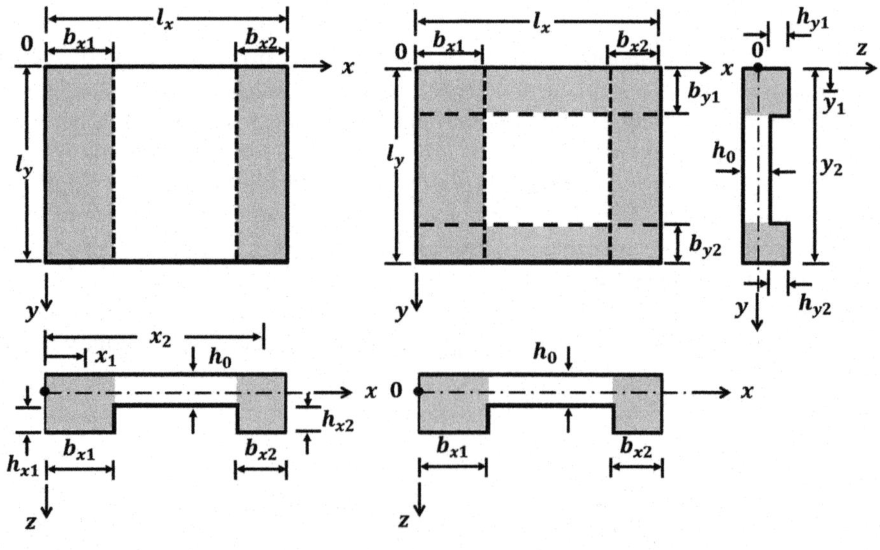

Fig. 8.1 Coordinates of rectangular plate with arbitrarily-stepped thickness

The location of the middle surface in the plate with stepped thickness varies due to the variation of thickness. The variation of the slab's thickness in usual building slabs is relatively small. For simplicity, we assume that the middle surface is located on a surface that bisects the thickness at each point. The location of the middle surface varies discontinuously at the boundary line between stepped thickness and reference thickness. This engineering assumption makes simple the formulation of current plate. The validity of this assumption will be examined from the numerical comparison with results obtained from the FEM. We also use the extended Dirac function to express rationally the discontinuous variation of the plate's stiffness due to stepped thickness.

Using the well-known linear strain–displacement relations and linear stress–displacement relations, the strain energy U is given by

$$U = \frac{1}{2}\int_0^{l_x}\int_0^{l_y}(M_x\kappa_x + M_y\kappa_y + 2M_{xy}\kappa_{xy})\mathrm{d}x\ \mathrm{d}y \tag{8.2}$$

in which κ_x, κ_y, and κ_{xy} indicate the curvatures and twist of the middle surface, respectively, and are defined as

$$\kappa_x = -w_{,xx};\quad \kappa_y = -w_{,yy};\quad \kappa_{xy} = -w_{,xy} \tag{8.3}$$

On the other hand, M_x, M_y, and M_{xy} are the bending and twisting moments per unit width, respectively; they are expressed as

$$M_x = -D_0 d(x, y)(w_{,xx} + \nu w_{,yy}) \tag{8.4}$$

$$M_y = -D_0 d(x, y)(w_{,yy} + \nu w_{,xx}) \tag{8.5}$$

$$M_{xy} = -(1-\nu)D_0 d(x, y)w_{,xy} \tag{8.6}$$

in which D_0 is the flexural rigidity of a reference plate neglecting stepped thickness, as given by $D_0 = Eh_0^3/[12(1-\nu^2)]$ (in which h_0 is the thickness of the reference normal plates, as given in Fig. 8.1). On the other hand, the stiffness coefficient $d\,(x, y)$ is defined as

$$d(x, y) = 1 + \alpha_{xi}D(x - x_i) + \alpha_{yj}D(y - y_j) + \alpha_{xyi,j}D(x - x_i)D(y - y_j) \tag{8.7}$$

in which the nondimensioned constants α_{xj}, α_{yj}, and $\alpha_{xyi,j}$ are given as

$$\alpha_{xi} = 3\frac{h_{xi}}{h_0} + 3\left(\frac{h_{xi}}{h_0}\right)^2 + \left(\frac{h_{xi}}{h_0}\right)^3 \tag{8.8}$$

$$\alpha_{yj} = 3\frac{h_{yj}}{h_0} + 3\left(\frac{h_{yj}}{h_0}\right)^2 + \left(\frac{h_{yj}}{h_0}\right)^3 \tag{8.9}$$

$$\alpha_{xyi,j} = 3\frac{h_{xyi,j}}{h_0} + 3\left(\frac{h_{xyi,j}}{h_0}\right)^2 + \left(\frac{h_{xyi,j}}{h_0}\right)^3 - \alpha_{xi} - \alpha_{yj} \tag{8.10}$$

Also, $D(x - x_i)$ and $D(y - y_j)$ are the extended Dirac functions. The extended Dirac function $D(x - x_i)$ is defined as a function where the Dirac delta function $\delta(x - \xi)$ exists continuously in the x-direction through the i-th stepped thickness: the region from $x_i - b_{xi}/2$ to $x_i + b_{xi}/2$, in which the supplemental variable ξ can take values continuously from $x_i - b_{xi}/2$ to $x_i + b_{xi}/2$. Similarly, the extended Dirac function $D(y - y_j)$ is defined as a function where the Dirac delta function $\delta(y - \eta)$ exists continuously in the y-direction through the j-th stepped thickness, in which the supplemental variable η can take values continuously from $y_j - b_{yj}/2$ to $y_j + b_{yj}/2$.

Hence, substituting Eqs. (8.3)–(8.6) into Eq. (8.2), the strain energy U can be obtained as

$$U = \frac{1}{2}\int_0^{l_x}\int_0^{l_y} D_0 d[(\kappa_x)^2 + (\kappa_y)^2 + 2\nu\kappa_x\kappa_y + 2(1-\nu)(\kappa_{xy})^2]\mathrm{d}x\,\mathrm{d}y \tag{8.11}$$

Next, the variation of the potential energy produced by external lateral loads, p, may be written as

$$\delta V = -\int_0^{l_x}\int_0^{l_y}(p - c\dot{w})\delta w\mathrm{d}x\mathrm{d}y + \int_0^{l_y}(m_x^*\delta w_{,x} - m_{xy}^*\delta w_{,y} + v_x^*\delta w)\mathrm{d}y\Big|_0^{l_x}$$
$$+ \int_0^{l_x}(m_y^*\delta w_{,y} - m_{yx}^*\delta w_{,x} + v_y^*\delta w)\mathrm{d}x\Big|_0^{l_y} \tag{8.12}$$

in which c is damping coefficient, and v_x^*, v_y^*, m_x^*, m_y^*, m_{xy}^*, and m_{yx}^* are external vertical forces, external moments, and external twist moments prescribed on the mechanical boundary edges, respectively.

The kinetic energy T may be given as

$$T = \frac{1}{2}\int_0^{l_x}\int_0^{l_y} m_0\alpha_h(\dot{w})^2\mathrm{d}x\,\mathrm{d}y \tag{8.13}$$

in which the notations α_h and m_0 are defined as

$$\alpha_h(x, y) = 1 + \left(\frac{h_{xi}}{h_0}\right)D(x - x_i) + \left(\frac{h_{yj}}{h_0}\right)D(y - y_j)$$
$$+ \left(\frac{h_{xyi,j} - h_{xi} - h_{yj}}{h_0}\right)D(x - x_i)D(y - y_j) \tag{8.14}$$

$$m_0 = \rho h_0 \tag{8.15}$$

m_0 is the mass per unit area of the reference plate and ρ is the mass density.

Substituting Eqs. (8.11)–(8.13) into Eq. (8.1), the differential equation of motion for a rectangular plate with arbitrarily-stepped thickness can be obtained as

$$\frac{m_0 \alpha_h \ddot{w}}{D_0} + \frac{c\dot{w}}{D_0} - \frac{p}{D_0} + (dw_{,xx})_{,xx} + (dw_{,yy})_{,yy}$$
$$+ \nu(dw_{,xx})_{,yy} + \nu(dw_{,yy})_{,xx} + 2(1-\nu)(dw_{,xy})_{,xy} = 0 \tag{8.16}$$

The associated boundary conditions are given as

$$w = 0 \quad \text{or} \quad (D_0 dw_{,xx})_{,x} + \nu(D_0 dw_{,yy})_{,y} + 2(1-\nu)(D_0 dw_{,xy})_{,y} - v_x^* = 0 \tag{8.17}$$

$$w_{,x} = 0 \quad \text{or} \quad D_0 d(w_{,xx} + \nu w_{,yy}) - m_x^* = 0 \tag{8.18}$$

at $x = 0$ and l_x; and

$$w = 0 \quad \text{or} \quad (1-\nu)D_0 dw_{,xy} = m_{xy}^* \tag{8.19}$$

at the corners. The similar boundary conditions corresponding to Eqs. (8.17)–(8.19) can be written as $y = 0$ and l_y.

The governing equations proposed here are general and can be applied to all rectangular plates in which the Kirchhoff-Love hypothesis is valid. As for uniform solid plates the coefficients α_h and d become all 1, the governing equations proposed here reduce to well-known general equations for a uniform plate, as given in Eq. (1.87).

8.3 Static Analysis

We present the static solutions by means of the Galerkin method. The static deflection w is also expressed by

$$w(x, y) = \sum_{m=1}\sum_{n=1} w_{mn} f_{xm}(x) f_{yn}(y) \tag{8.20}$$

in which w_{mn} are unknown displacement coefficients; and f_{xm} and f_{ym} are shape functions satisfying the specified boundary conditions.

Using the same procedure stated in the preceding chapters to the governing equations obtained here, the Galerkin equations for the static problem can be written as

$$\delta w_{\bar{m}\bar{n}}: \sum_{m=1}\sum_{n=1} w_{mn} A_{\bar{m}\bar{n}mn} = B_{\bar{m}\bar{n}} \tag{8.21}$$

in which the coefficients $A_{\bar{m}\bar{n}mn}$ and $B_{\bar{m}\bar{n}}$ are given as follows:

$$\begin{aligned}
A_{\bar{m}\bar{n}mn} = & F_{xm\bar{m}}(4,0;0)F_{yn\bar{n}}(0,0;0) + 2F_{xm\bar{m}}(2,0;0)F_{yn\bar{n}}(2,0;0) \\
& + F_{xm\bar{m}}(0,0;0)F_{yn\bar{n}}(4,0;0) \\
& + \sum_{i=1} \alpha_{xi}\big[F_{xm\bar{m}}(0,0;i)F_{yn\bar{n}}(4,0;0) + F_{xm\bar{m}}(2,2;i)F_{yn\bar{n}}(0,0;0) \\
& - 2(1-\nu)F_{xm\bar{m}}(1,1;i)F_{yn\bar{n}}(2,0;0) + \nu F_{xm\bar{m}}(2,0;i)F_{yn\bar{n}}(2,0;0) \\
& + \nu F_{xm\bar{m}}(0,2;i)F_{yn\bar{n}}(2,0;0) + \sum_{j=1} \alpha_{yj}\big[F_{xm\bar{m}}(4,0;0)F_{yn\bar{n}}(0,0;j) \\
& + F_{xm\bar{m}}(0,0;0)F_{yn\bar{n}}(2,2;j) - 2(1-\nu)F_{xm\bar{m}}(2,0;0)F_{yn\bar{n}}(1,1;j) \\
& + \nu F_{xm\bar{m}}(2,0;0)F_{yn\bar{n}}(2,0;j) + \nu F_{xm\bar{m}}(2,0;0)F_{yn\bar{n}}(0,2;j) \\
& + \sum_{i=1}\sum_{j=1} \alpha_{xyi,j}\big\{F_{xm\bar{m}}(2,2;i)F_{yn\bar{n}}(0,0;j) + F_{xm\bar{m}}(0,0;i)F_{yn\bar{n}}(2,2;j) \\
& + 2(1-\nu)F_{xm\bar{m}}(1,1;i)F_{yn\bar{n}}(1,1;j) \\
& + \nu F_{xm\bar{m}}(0,2;i)[F_{yn\bar{n}}(2,0;j) + F_{yn\bar{n}}(0,2;j)]\big\}
\end{aligned} \tag{8.22}$$

$$B_{\bar{m}\bar{n}} = \int_0^{l_x}\int_0^{l_y} \frac{P}{D_0} f_{x\bar{m}} f_{y\bar{n}} \mathrm{d}x\,\mathrm{d}y \tag{8.23}$$

in which the notations $F_{xm\bar{m}}(k_1,k_2;0)$, $F_{xm\bar{m}}(k_1,k_2;i)$, $F_{yn\bar{n}}(k_1,k_2;0)$, and $F_{yn\bar{n}}(k_1,k_2;j)$ are given in general form by

$$F_{xm\bar{m}}(k_1,k_2;0) = \int_0^{l_x} f_{xm}^{(k_1)} f_{x\bar{m}}^{(k_2)} \mathrm{d}x \tag{8.24}$$

$$F_{xm\bar{m}}(k_1,k_2;i) = \int_0^{l_x} f_{xm}^{(k_1)} f_{x\bar{m}}^{(k_2)} D(x - x_i)\mathrm{d}x \tag{8.25}$$

$$F_{yn\bar{n}}(k_1,k_2;0) = \int_0^{l_y} f_{yn}^{(k_1)} f_{y\bar{n}}^{(k_2)} \mathrm{d}y \tag{8.26}$$

$$F_{yn\bar{n}}(k_1,k_2;j) = \int_0^{l_y} f_{yn}^{(k_1)} f_{y\bar{n}}^{(k_2)} \tag{8.27}$$

The superscripts k_1 and k_2 enclosed within parentheses of the superscript indicate the differential order. The integral calculation including the extended Dirac function is explained in Appendix. Thus, the unknown displacement coefficients w_{mn} in the static problem are obtained by solving a set of algebraic linear equations given by Eq. (8.21).

8.4 Numerical Results for Static Solution

In the preceding section, the static analytical method for a rectangular plate with stepped thickness has been presented by means of the Galerkin method. The exactness of the proposed method is examined from numerical calculations for respective four types of simply supported and clamped plates with stepped thickness, as shown in Fig. 8.2. Data used in the numerical calculations are as follows: Young's modulus $E = 2.06 \times 10^{10}$ N/m^2; Poisson's ratio $\nu = 0.17$; mass density $\rho = 2400$ kg/m^3; and variation of the stepped thickness $h_i/h_0 = 0.5$ and $h_j/h_0 = 0.5$. For simplicity, the lateral load is assumed to be uniformly distributed load p_0.

The following shape functions f_{xm} are used

$$f_{xm} = \sin\frac{m\pi x}{l_x} \tag{8.28}$$

for current simply supported and

$$f_{xm} = \cosh\left(\frac{\lambda_m x}{l_x}\right) - \cos\left(\frac{\lambda_m x}{l_x}\right) - \frac{\cosh(\lambda_m) - \cos(\lambda_m)}{\sinh(\lambda_m) - \sin(\lambda_m)}\left[\sinh\left(\frac{\lambda_m x}{l_x}\right) - \sin\left(\frac{\lambda_m x}{l_x}\right)\right] \tag{8.29}$$

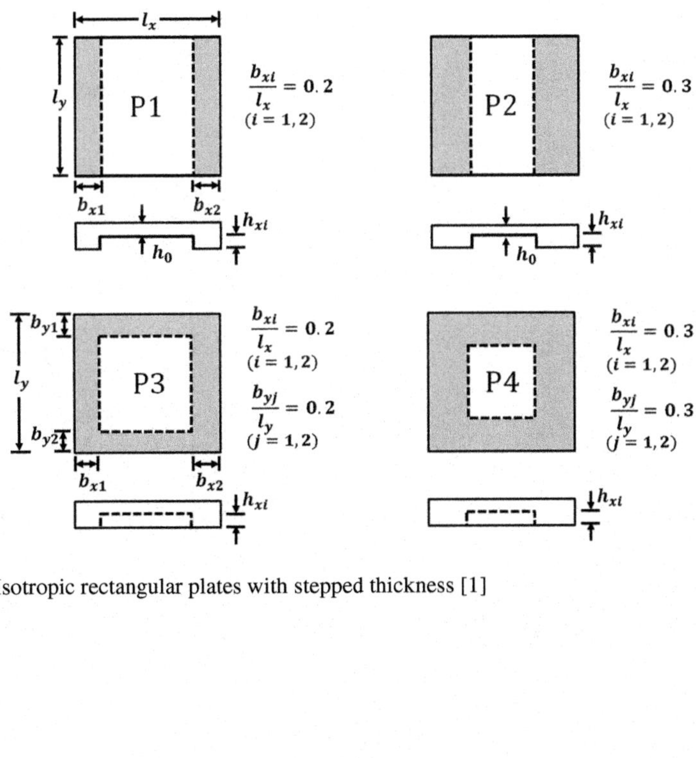

Fig. 8.2 Isotropic rectangular plates with stepped thickness [1]

for clamped plates. Here λ_m is well-known constant, as given after Eq. (1.86) in Sect. 1.8. The shape function f_{yn} in the y-direction can be expressed in the same form by replacing Eqs. (8.28) and (8.29) with the quantities in the y-direction.

Figures 8.3 and 8.4 show the distributions of the deflections and bending moments at $y = l_y/2$ for the clamped plates P1 and P3, respectively. In these figures, the solid lines represent the numerical results obtained from Eq. (8.21); the circles indicate results obtained by using the FEM code NASTRAN. The numerical results indicate that the results obtained from the present theory show good agreement with the results obtained from the finite element method. For current uniformly distributed loads, the consideration of five terms of m for f_{xm} gives sufficient accuracy.

Fig. 8.3 w and M_y for clamped plate P1 [1]

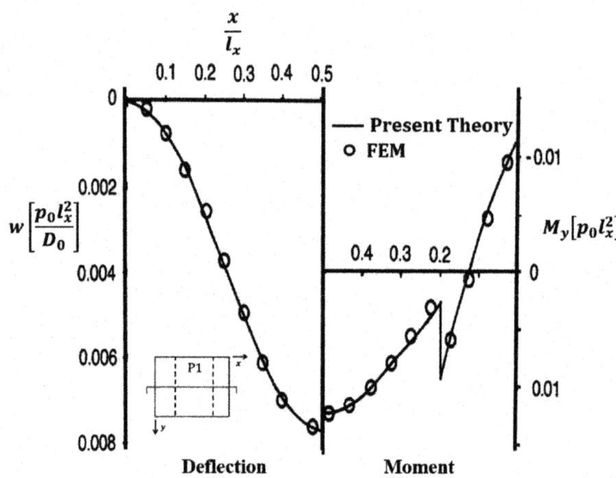

Fig. 8.4 w and M_y for clamped plate P3 [1]

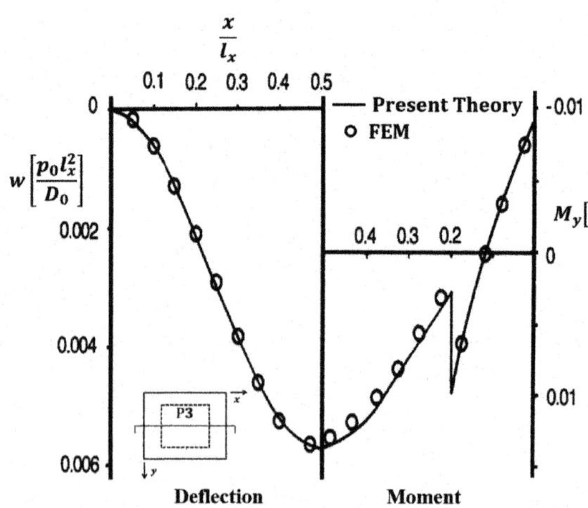

8.5 Free Transverse Vibrations of Plate with Stepped Thickness

We consider free transverse vibrations of rectangular plates with stepped thickness. Using the method of separation of variable, we assume that

$$w(x, y, t) = \Phi(t)W(x, y) \tag{8.30}$$

in which $W(x, y)$ is a function of x and y; and $\Phi(t)$ represents a function of time t. Applying Eq. (8.30) to the equation for free transverse vibrations obtained from Eq. (8.16), we have obtained the following equation for $W(x, y)$:

$$(dW_{,xx})_{,xx} + (dW_{,yy})_{,yy} + v(dW_{,xx})_{,yy} + v(dW_{,yy})_{,xx} + 2(1-v)(dW_{,xy})_{,xy} - \omega^2 \frac{m_0 \alpha_h}{D_0} W = 0 \tag{8.31}$$

in which ω is a constant.

The natural frequencies of a rectangular plate with stepped thickness are also presented by means of the Galerkin method. $W(x, y)$ is expressed as

$$W(x, y) = \sum_{m=1}\sum_{n=1} w_{mn} f_{xm}(x) f_{yn}(y) \tag{8.32}$$

where w_{mn} are the unknown coefficients; and $f_{xm}(x)$ and $f_{yn}(y)$ are natural functions satisfying the specified boundary conditions of the plates. The Galerkin equations of Eq. (8.31) can be expressed as

$$\delta w_{\bar{m}\bar{n}} : \sum_{m=1}\sum_{n=1} w_{mn}(A_{\bar{m}\bar{n}mn} - \lambda^2 B_{\bar{m}\bar{n}mn}) = 0 \tag{8.33}$$

in which the coefficient $A_{\bar{m}\bar{n}mn}$ is given by Eq. (8.22); and $B_{\bar{m}\bar{n}mn}$ and λ are defined as

$$B_{\bar{m}\bar{n}mn} = \int_0^{l_x}\int_0^{l_y} \alpha_h f_{xm} f_{yn} f_{x\bar{m}} f_{y\bar{n}} \mathrm{d}x\, \mathrm{d}y \tag{8.34}$$

$$\lambda^2 = \frac{\omega^2 m_0}{D_0} \tag{8.35}$$

Equation (8.33) presents a system of linear, homogeneous, simultaneous algebraic equations with respect to the unknown displacement coefficients w_{mn}. Solving Eq. (8.33) as eigenvalue problems for λ, the i-th natural frequency ω_i of a plate with stepped thickness, corresponding to the i-th value λ_i of λ, can be determined from Eq. (8.35) as

$$\omega_i = \lambda_i \sqrt{\frac{D_0}{m_0}} \tag{8.36}$$

Since the coefficient matrixes $A_{\bar{m}\bar{n}mn}$ and $B_{\bar{m}\bar{n}mn}$ have, in general, the values in the diagonal terms due to the existence of stepped thickness, Eq. (8.33) has coupled form. The natural frequencies of the plate with stepped thickness can be obtained from numerical computations.

Next, we consider the approximate expressions to obtain directly the natural frequencies. Although the natural frequencies are affected by the diagonal and nondiagonal terms in $A_{\bar{m}\bar{n}mn}$ and $B_{\bar{m}\bar{n}mn}$, the main behavior is considered to be dominated by only the diagonal terms. So, taking only the diagonal terms in $A_{\bar{m}\bar{n}mn}$ and $B_{\bar{m}\bar{n}mn}$ into consideration, Eq. (8.33) becomes an uncoupled form. Thus, the approximate value for the i-th natural frequency is obtained as

$$\omega_i \approx \sqrt{\frac{A_{mnmn}}{B_{mnmn}}} \sqrt{\frac{D_0}{m_0}} \tag{8.37}$$

8.6 Numerical Results for Natural Frequencies

The natural frequencies for a rectangular plate with stepped thickness have been presented by means of the Galerkin method. The method proposed here is examined from numerical calculations for the same models used at the static problem, as shown in Fig. 8.2.

Table 8.1 shows the numerical results for clamped plates P1–P4, in which the natural frequencies are represented in nondimensioned form by taking $(1/l_x^2)\sqrt{D_0/m_0}$ as a unit. The numerical results demonstrate that the differences between the rigorous results obtained from Eq. (8.33) and the approximate results obtained from Eq. (8.37) are negligible in practical use. And the results obtained from the proposed theory show excellent agreement when compared with results obtained from the FEM. Thus, the exactness of the approximate solution for the natural frequency is proven from the numerical calculations.

8.7 Forced Vibrations of Plate with Stepped Thickness

We consider forced vibrations of a plate with arbitrarily-stepped thickness. The general solution of Eq. (8.16) is assumed to be of the form

$$w(x, y, t) = \sum_{m=1}^{\infty} \sum_{n=1}^{\infty} f_{xm}(x) f_{yn}(y) \Phi_{mn}(t) \tag{8.38}$$

Table 8.1 Natural frequencies of clamped plates with stepped thickness [1]

Type	Analytical method	First (rad/s)	Second (rad/s)	Third (rad/s)	Fourth (rad/s)
P1	Rigorous method	46.98	84.47	95.10	135.19
	Approximate solution	46.96	84.33	95.29	135.07
	FEM	45.91	82.62	90.22	129.56
P2	Rigorous method	49.46	93.31	100.42	144.97
	Approximate solution	49.44	93.14	100.62	144.95
	FEM	49.31	91.66	91.91	136.94
P3	Rigorous method	54.32	100.49	100.49	148.92
	Approximate solution	54.29	100.58	100.58	148.81
	FEM	53.65	97.03	97.03	145.44
P4	Rigorous method	54.28	106.02	106.02	155.16
	Approximate solution	54.29	106.04	106.04	155.00
	FEM	54.78	101.34	101.34	151.72

in which $\Phi_{mn}(t)$ are unknown functions with respect to time t; and $f_{xm}(x)$ and $f_{yn}(y)$ are the known natural functions satisfying both Eq. (8.31) and the specified boundary conditions of current plate. Substituting Eq. (8.38) into Eq. (8.16) and using the relation of Eq. (8.31), it becomes

$$\sum_{m=1}\sum_{n=1} f_{xm} f_{yn} m_0 \alpha_h [\ddot{\Phi}_{mn}(t) + 2h_{mn}\omega_{mn}\dot{\Phi}_{mn}(t) + \omega_{mn}^2 \Phi_{mn}(t)] = p(x, y, t) \quad (8.39)$$

For more simplicity, we use that the relationships between the damping coefficient c_{mn} and damping constant h_{mn} of the current plate may be written as

$$\frac{c_{mn}}{m_0 \alpha_h} = 2h_{mn}\omega_{mn} \quad (8.40)$$

Since the mass coefficient α_h is a function of x and y, Eq. (8.39) cannot be transformed into an uncoupled form by means of the orthogonality relations for natural functions.

Multiplying both sides of Eq. (8.39) by $f_{x\bar{m}}$ and $f_{y\bar{n}}$ and integrating between 0–l_x and 0–l_y, Eq. (8.39) reduces to

$$\sum_{m=1}\sum_{n=1} K_{\bar{m}\bar{n}mn} [\ddot{\Phi}_{mn}(t) + 2h_{mn}\omega_{mn}\dot{\Phi}_{mn}(t) + \omega_{mn}^2 \Phi_{mn}(t)] = \frac{1}{m_0} Q_{\bar{m}\bar{n}}(t) \quad (8.41)$$

in which $K_{\bar{m}\bar{n}mn}$ and $Q_{\bar{m}\bar{n}}(t)$ are defined as

$$K_{\bar{m}\bar{n}mn} = \int_0^{l_x}\int_0^{l_y} \alpha_h(x, y) f_{xm} f_{y\bar{m}} f_{yn} f_{y\bar{n}} \mathrm{d}x\mathrm{d}y \tag{8.42}$$

$$Q_{\bar{m}\bar{n}}(t) = \int_0^{l_x}\int_0^{l_y} p(x, y, t) f_{x\bar{m}} f_{y\bar{n}} \mathrm{d}x\mathrm{d}y \tag{8.43}$$

However, Eq. (8.41) is a coupled form set of the equations of motion for each specified boundary conditions of the plates. $\Phi_{\bar{m}\bar{n}}$ $(\bar{m} = 1, 2, \ldots; \bar{n} = 1, 2, \ldots)$, and Eq. (8.41) can be solved by using step-by-step integration methods.

8.8 Approximate Solutions for Forced Vibrations

In the preceding section, the dynamic analysis for a damped plate with stepped thickness has been obtained by applying numerical computations, such as step-by-step integration methods to Eq. (8.41). For practical use, we present an approximation solution in uncoupled closed form.

The coefficients $K_{\bar{m}\bar{n}mn}$ in Eq. (8.41) have nondiagonal terms $m \neq \bar{m}$ and/or $n \neq \bar{n}$ due to the coefficient of the variation of thickness, $\alpha_h(x, y)$. Assuming that the behavior of a plate with stepped thickness is now dominated by only the diagonal terms in $K_{\bar{m}\bar{n}mn}$, Eq. (8.41) can be changed to an uncoupled form. The effectiveness of the assumption used here had been numerically demonstrated in the presentation of the approximate solutions for natural frequencies of the current plate. Then, Eq. (8.41) can be approximated as

$$K_{mnmn}[\ddot{\Phi}_{mn}(t) + 2h_{mn}\,\omega_{mn}\,\dot{\Phi}_{mn}(t) + \omega_{mn}^2\,\Phi_{mn}(t)] = \frac{1}{m_0} Q_{mn}(t) \tag{8.44}$$

The general solution of Eq. (8.44) can be expressed as

$$\Phi_{mn}(t) = \exp(-h_{mn}\omega_{mn}t)[C_1 \sin\omega_{Dmn}t + C_2 \cos\omega_{Dmn}t] + \frac{1}{K_{mnmn}m_0\omega_{Dmn}} \int_0^t \exp[-h_{mn}\omega_{mn}(t-\tau)] \sin\omega_{Dmn}(t-\tau) Q_{mn}(\tau)\mathrm{d}\tau \tag{8.45}$$

in which C_1 and C_2 are constants determined from the initial conditions; and ω_{mn} and ω_{Dmn} indicate the natural frequencies of the undamped and damped plates with stepped thickness, respectively. The well-known relationship between ω_{mn} and ω_{Dmn} is

$$\omega_{Dmn} = \omega_{mn}\sqrt{1 - h_{mn}^2} \tag{8.46}$$

Thus, the dynamic deflections of current plate are determined by substituting Eq. (8.45) into Eq. (8.38).

Next, we propose the dynamic solutions of the current plate, subjected to the following harmonic external load from Eq. (8.45):

$$p(x, y, t) = p^*(x, y)\sin\omega_p t \tag{8.47}$$

in which $p^*(x, y)$ is a function of x and y and is determined from the distribution of the external loads; and ω_p denotes the frequency of the external loads. When at the initial time $t = 0$, the following relations are effective:

$$w(r, \phi, 0) = 0, \quad \dot{w}(r, \phi, 0) = 0 \tag{8.48}$$

then $\Phi_{mn}(t)$ can be obtained as

$$\begin{aligned}\Phi_{mn} = {} & \frac{1}{K_{mnmn}m_0\omega_{Dmn}}\frac{Q_{mn}^*}{2}\left\{\frac{h_{mn}\omega_{mn}\cos\omega_p t + (\omega_p + \omega_{Dmn})\sin\omega_p t}{(h_{mn}\omega_{mn})^2 + (\omega_p + \omega_{Dmn})^2}\right. \\ & - \frac{h_{mn}\omega_{mn}\cos\omega_p t + (\omega_p - \omega_{Dmn})\sin\omega_p t}{(h_{mn}\omega_{mn})^2 + (\omega_p - \omega_{Dmn})^2} \\ & + \exp(-h_{mn}\omega_{mn}t)\left[\frac{(\omega_p + \omega_{Dmn})\sin\omega_{Dmn}t - h_{mn}\omega_{mn}\cos\omega_{Dmn}t}{(h_{mn}\omega_{mn})^2 + (\omega_p + \omega_{Dmn})^2}\right. \\ & \left.\left. + \frac{(\omega_p - \omega_{Dmn})\sin\omega_{Dmn}t + h_{mn}\omega_{mn}\cos\omega_{Dmn}t}{(h_{mn}\omega_{mn})^2 + (\omega_p - \omega_{Dmn})^2}\right]\right\}\end{aligned} \tag{8.49}$$

in which the notation Q_{mn}^* is defined as

$$Q_{mn}^* = \int_0^{l_x}\int_0^{l_y} p*(x, y) f_{xm} f_{yn} \mathrm{d}x\ \mathrm{d}y \tag{8.50}$$

Thus, the closed-form approximate solution for a damped rectangular plate with arbitrarily-stepped thickness subject to harmonic external loads has been presented in general form. Setting $h_{mn} \to 0$, $\omega_{Dmn} \to \omega_{mn}$ and $\exp(-h_{mn}\omega_{Dmn}t) \to 1$ in Eq. (8.49), the closed-form solution for an undamped plate with stepped thickness reduces to

$$\Phi_{mn} = \frac{Q_{mn}^*}{K_{mnmn}m_0\omega_{mn}}\frac{\omega_p\sin\omega_{mn}t - \omega_{mn}\sin\omega_p t}{\omega_p^2 - \omega_{mn}^2} \tag{8.51}$$

However, the solution for a uniform plate is easily obtained by replacing K_{mnmn} with 1.

8.9 Numerical Results for Dynamic Responses

Let us examine the exactness of the closed-form approximate solution proposed here for an isotropic rectangular plate with stepped thickness from numerical computations for four kinds of rectangular plates with stepped thickness, as shown in Fig. 8.2. The damping constants h_{mn} are assumed to be 0.03 for all modes. The external lateral load without the additional mass is assumed to be a harmonic and uniformly distributed force given in Eq. (8.47), in which $p^* = 9806\ \text{N/m}^2$ and $\omega_p = 62.83$ rad/s. The natural functions use Eqs. (8.28) and (8.29) for simply supported and clamped plates with stepped thickness, respectively.

Figure 8.5 demonstrates the time history of the dynamic deflections at the midpoint of clamped plates of P3. In this figure, the solid lines represent values obtained from Eq. (8.41), being the rigorous equation, in which the step-by-step integration methods use the Wilson-θ method; the broken lines indicate values obtained from the closed-form approximate solution Eq. (8.49); and the solid lines with circles indicate values obtained from the FEM. The difference between solid lines and broken lines is too small to plot. The few discrepancies between the results obtained from the proposed theories and from the FEM appear from 0 to 0.3 s. However, the numerical results show that the approximate solution proposed here is applicable to the dynamic analyses of plates with stepped thickness, in practical use.

Table 8.2 shows the maximum dynamic deflections at the midspan of the clamped plates with stepped thickness. The ratios in this table are defined as the value obtained from the present theory to the FEM.

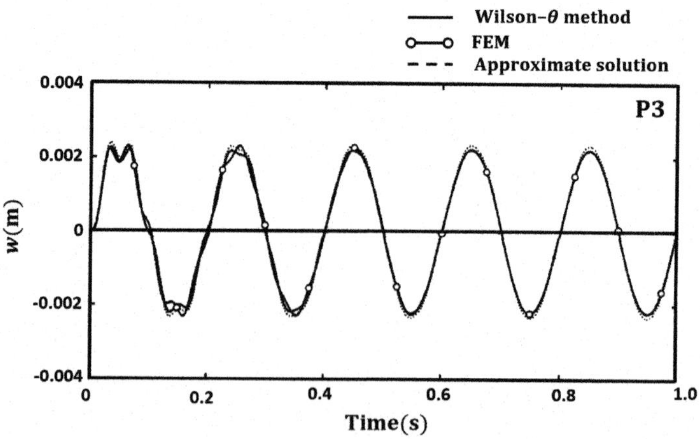

Fig. 8.5 Dynamic deflections for clamped plate P3 [1]

Table 8.2 Maximum dynamic deflections of clamped plates with stepped thickness [1]

Type	Analytical method	Maximum dynamic deflection (m)	Ratio
P1	Wilson-θ method	0.003516	0.981
	Approximate solution	0.003673	1.025
	FEM	0.003584	–
P2	Wilson-θ method	0.002881	1.106
	Approximate solution	0.003035	1.165
	FEM	0.002606	–
P3	Wilson-θ method	0.002287	0.992
	Approximate solution	0.002415	1.048
	FEM	0.002305	–
P4	Wilson-θ method	0.002022	1.034
	Approximate solution	0.002162	1.106
	FEM	0.001955	–

Maximum deflection ratio = present theory/FEM

For the sake of simplicity, we introduced in the formulation, as stated in Sect. 8.2, a simplified assumption that the middle surface of plate with stepped thickness is always located at the surface bisected the thickness at each point. The effectiveness of this assumption is examined from numerical calculation in which the stepped thickness ratios h_{xi}/h_0 and h_{yj}/h_0 in clamped plates P1 and P4, subjected to static uniformly distributed loads, are sequentially changed. Table 8.3 shows the ratios of maximum deflection obtained from the present theory to one obtained from the FEM for clamped plates P1 and P4, subjected to static uniformly distributed loads. It follows that the assumption proposed here is applicable to building slabs with thickness ratio h_{xi}/h_0 and/or h_{yj}/h_0 of about 0.5. However, when the size of the stepped thickness increases, the situation of the middle surface varies due to the local variation of stiffness. Thus, the deflection obtained from the theory proposed here is evaluated to be small when the step thickness ratio is 1.5 times or more.

8.10 Conclusions

The general analysis methods and closed-form approximate solutions for the static and dynamic problems of an isotropic rectangular plate with arbitrarily and eccentrically stepped thickness have been proposed by the use of the extended Dirac function. The exactness of the closed-form approximate solutions proposed here has been proven by comparing them with the numerical results obtained from the FEM.

Table 8.3 Relationships between stepped thickness ratios and maximum deflection ratios [1]

P1		P4	
h_{xi}/h_0	Maximum deflection ratio	$h_{xi}/h_0 = h_{yj}/h_0$	Maximum deflection ratio
0.0	0.999	0.0	0.999
0.1	0.999	0.1	0.999
0.2	1.001	0.2	1.012
0.3	1.008	0.3	1.030
0.4	1.014	0.4	1.049
0.5	**1.018**	**0.5**	**1.069**
0.6	1.019	0.6	1.087
0.7	1.016	0.7	1.102
0.8	1.009	0.8	1.113
0.9	1.000	0.9	1.093
1.0	0.987	1.0	1.121
1.2	0.954	1.2	1.111
1.4	0.912	1.4	1.085
1.5	0.889	1.5	1.068
2.0	0.765	2.0	0.945
2.5	0.638	2.5	0.942
3.0	0.522	3.0	0.640
3.5	0.423	3.5	0.532
4.0	0.341	4.0	0.430
4.5	0.275	4.5	0.349
5.0	0.223	5.0	0.284

Maximum deflection ratio = present theory/FEM

Appendix: Calculation Including Extended Dirac Function

The integral calculation including the extended Dirac function $D(x - x_i)$ can be written as

$$\int_0^{l_x} D(x - x_i)f(x)\mathrm{d}x = \int_{x_i-(b_{xi}/2)}^{x_i+(b_{xi}/2)} \left[\int_0^{l_x} \delta(x - \xi)f(x)\mathrm{d}x \right] \mathrm{d}\xi = \int_{x_i-(b_{xi}/2)}^{x_i+(b_{xi}/2)} f(\xi)\mathrm{d}\xi \quad (8.52)$$

in which ξ is a supplementary variable of x. The n-th derivatives of the extended Dirac functions can therefore be expressed as

$$\int_{0}^{l_x} D^{(n)}(x - x_i) f(x) \mathrm{d}x = \int_{x_{i-(b_{xi}/2)}}^{x_{i+(b_{xi}/2)}} (-1)^n f^{(n)}(\xi) \mathrm{d}\xi \tag{8.53}$$

in which the superscripts enclosed within parentheses indicate the differential order.

Reference

1. Takabatake, H., Imaizumi, T., Kunihiro, O.: Simplified analysis of rectangular plates with stepped thickness. J. Struct. Eng. ASCE **121**(1), 28–38 (1995)

Part III
Static and Dynamic Analysis of Special Plates

Chapter 9
Static and Dynamic Analyses of Rectangular Plates with Stepped Thickness Subjected to Moving Loads

9.1 Introduction

In the preceding Chap. 8, we have presented the static and dynamic analytical methodology for a rectangular plate with arbitrarily stepped thickness. A relatively thin stepped thickness in building slabs is used on to decrease cost and to simplify construction so that the problem of vibration impairment of the floor by a person becomes an important task in design. Since the load causing the floor vibration is the moving load, strictly speaking, it should be calculated by calculation of the moving load including the added mass due to the moving load. This handling is troublesome in practical design, so in the practical design, the analytical check to human response is judged from the response of slabs, subjected to a concentrated load of nonmoving foot impact at the midspan by means of FEM. The analytical method used FEM cannot solve the dynamic problem including the effect of additional mass due to moving loads.

Thus, a simplified analysis being practical and usable for building slabs, subjected to stationary and moving loads, is also demanded, so that the problems due to the vibration of building slabs may be removed in the preliminary stages and final stages of the design.

This chapter demonstrates a simplified analysis of an isotropic plate with relatively thin eccentrically stepped thickness, subjected to moving loads along with the effect of additional mass. First, the effect of additional mass due to moving loads is presented by modifying the governing equation proposed in Chap. 8. Second, the closed-form approximate solutions excluding the effect of additional mass due to moving loads are presented for rectangular plates with stepped thickness subjected to general moving loads. The accuracy of the proposed solutions is examined from numerical computation. The methodology proposed in this chapter is mainly based on Ref. [1].

H. Takabatake, *Simplified Analytical Methods of Elastic Plates*,
https://doi.org/10.1007/978-981-13-0086-8_9

9.2 Governing Equations of Plates with Stepped Thickness Including the Effect of Moving Loads with Additional Mass

We consider a rectangular plate with arbitrarily and eccentrically stepped thickness, as shown in Fig. 9.1. The Cartesian coordinate system *x, y, z* is also used. Ridgelines of each eccentrically stepped thickness are assumed to be parallel to the axis x or the axis y, as explained in the preceding chapter. The midpoint, width, and varied thickness of the i-th stepped thickness being parallel to the axis y are indicated by x_i, b_{xi}, and h_{xi}, respectively, in which h_{xi} is measured from the lower surface of the reference slab excluding stepped thickness. Similarly for the j-th stepped thickness being parallel to the axis x they are given by y_j, b_{yj}, and h_{yj}. In a part where the i-th and j-th stepped thickness being parallel to the x- and y-axes, respectively, cross each other the maximum value of their stepped thicknesses h_{xi} and h_{yj} is indicated with $h_{xyi,j}$. Namely, $h_{xyi,j} = \text{maximum}\,(h_{xi}, h_{yj})$.

The variation of slab's thickness in most building slabs is assumed to be relatively small. Therefore, for simplicity, we assume that the neutral surface of a plate is located in a surface that bisects the thickness at each point so that it varies discontinuously at the boundary line between eccentrically stepped thickness and reference thickness. It had been proven from discussion in Chap. 8 that this engineering assumption is practically effective within the range of $h_{xi}/h_0 = 1$ and $h_{yi}/h_0 = 1$, in which h_0 is the thickness of the reference slab.

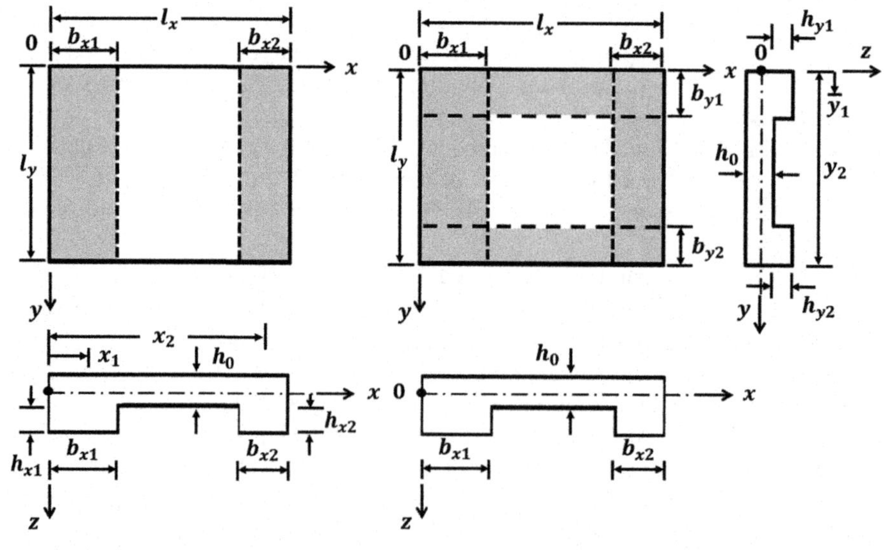

Fig. 9.1 Coordinates of a rectangular plate with arbitrarily stepped thickness [1]

We consider the bending problem of isotropic rectangular plates in small deformations based on the Kirchhoff-Love hypothesis. Using the above engineering assumption to the neutral surface, the flexural rigidity and mass of current plates with stepped thickness, which are function of x and y, are expressed by $D_0 d(x, y)$ and $m_0 \alpha_h(x, y)$, respectively, in which D_0 and m_0 are the flexural rigidity and mass of a reference plate neglecting stepped thickness, as given by $D_0 = Eh_0^3/[12(1 - \upsilon^2)]$ and $m_0 = \rho h_0$ which is mass per unit area, respectively. The coefficients of flexural rigidity and mass, $d(x, y)$ and $\alpha_h(x, y)$, respectively, are given as

$$d(x, y) = 1 + \alpha_{xi} D(x - x_i) + \alpha_{yj} D(y - y_j) + \alpha_{xyi,j} D(x - x_i) D(y - y_j) \quad (9.1)$$

$$\alpha_h(x, y) = 1 + \left(\frac{h_{xi}}{h_0}\right) D(x - x_i) + \left(\frac{h_{yj}}{h_0}\right) D(y - y_j) + \left(\frac{h_{xyi,j} - h_{xi} - h_{yj}}{h_0}\right) D(x - x_i) D(y - y_j) \quad (9.2)$$

in which constants α_{xi}, α_{yj}, and $\alpha_{xyi,j}$ are

$$\alpha_{xi} = 3\frac{h_{xi}}{h_0} + 3\left(\frac{h_{xi}}{h_0}\right)^2 + \left(\frac{h_{xi}}{h_0}\right)^3 \quad (9.3)$$

$$\alpha_{yj} = 3\frac{h_{yj}}{h_0} + 3\left(\frac{h_{yj}}{h_0}\right)^2 + \left(\frac{h_{yj}}{h_0}\right)^3 \quad (9.4)$$

$$\alpha_{xyi,j} = 3\frac{h_{xyi,j}}{h_0} + 3\left(\frac{h_{xyi,j}}{h_0}\right)^2 + \left(\frac{h_{xyi,j}}{h_0}\right)^3 - \alpha_{xi} - \alpha_{yj} \quad (9.5)$$

The extended Dirac functions $D(x - x_i)$ and $D(y - y_j)$ explained in Chap. 8 are significant only in prescribed regions. As stated before, the discontinuous variations of flexural rigidity and mass are expressed continuously by means of the extended Dirac function. Therefore, the current analysis is unnecessary to prepare an imaginary cutting at the discontinuities of rigidity and mass. Hence, although the assumption for the neutral surface is used on the discontinuities, the continuous condition of the deflection at the discontinuities is satisfied from natural functions used.

The governing equation of a rectangular plate with arbitrarily stepped thickness, including the effect of additional mass due to moving loads, can be given by modifying Eq. (8.16) as follows:

$$\frac{m_0 \alpha_h \ddot{w}}{D_0} + \frac{\dot{\tilde{m}} \dot{w}}{D_0} + \frac{\tilde{m} \ddot{w}}{D_0} + \frac{c\dot{w}}{D_0} - \frac{p}{D_0} + (dw_{,xx})_{,xx} + (dw_{,yy})_{,yy} + \nu(dw_{,xx})_{,yy} + \nu(dw_{,yy})_{,xx} + 2(1 - \nu)(dw_{,xy})_{,xy} = 0 \quad (9.6)$$

The associated boundary conditions

$$w = \bar{w} \quad \text{or} \quad (D_0 dw_{,xx})_{,x} + \nu(D_0 dw_{,yy})_{,x} + 2(1 - \nu)(D_0 dw_{,xy})_{,y} + v_x^* = 0 \quad (9.7)$$

$$w_{,x} = \bar{w}_{,x} \quad \text{or} \quad D_0 d\left(w_{,xx} + \nu w_{,yy}\right) + m_x^* = 0 \quad (9.8)$$

at $x = 0$ and l_x and

$$w = 0 \quad \text{or} \quad -(1-\nu)D_0 d w_{,xy} = m_{xy} \tag{9.9}$$

at the corners. Here, $\tilde{m}$ is additional mass due to moving loads and variable with respect to x, y, and t, in general. w is deflection on the middle surface of current plate. $\bar{w}$ and $\bar{w}_{,x}$ are displacement and rotation, respectively, prescribed from the geometrical boundary conditions. m_x^* and v_x^* are moment and vertical force, respectively, prescribed from the mechanical boundary conditions. The similar boundary conditions can be given at $y=0$ and l_y.

For uniform solid plates, the coefficients $d(x, y)$ and $\alpha_h(x, y)$ become all 1 and the governing equations are reduced to a general equation, including the effect of additional mass, for a uniform plate.

9.3 Forced Vibration of a Plate with Stepped Thickness

We assume that the general solution of Eq. (9.6) is the form

$$w(x, y, t) = \sum_{m=1}^{\infty} \sum_{n=1}^{\infty} f_{xm}(x) f_{yn}(y) \Phi_{mn}(t) \tag{9.10}$$

where $\Phi_{mn}(t)$ are unknown functions with respect to time t; and $f_{xm}(x)$ and $f_{yn}(y)$ are the natural functions satisfying the specified boundary conditions of current plate.

Employing the Galerkin method to Eq. (9.6), we have the following expression with respect to Φ_{mn}:

$$\begin{aligned}
&\sum_{m=1} \sum_{n=1} K_{\bar{m}\bar{n}mn} [\ddot{\Phi}_{mn}(t) + 2h_{mn}\omega_{mn}\dot{\Phi}(t) + \omega_{mn}^2 \Phi_{mn}(t)] \\
&\quad + \sum_{m=1} \sum_{n=1} [K_{\bar{m}\bar{n}m}^{(1)} \ddot{\Phi}_{mn}(t) + K_{\bar{m}\bar{n}m}^{(2)} \dot{\Phi}_{mn}(t)] \frac{1}{m_0} = \frac{1}{m_0} Q_{\bar{m}\bar{n}}(t)
\end{aligned} \tag{9.11}$$

where the notations $K_{\bar{m}\bar{n}mn}$, $K_{\bar{m}\bar{n}mn}^{(1)}$, $K_{\bar{m}\bar{n}mn}^{(2)}$, and $Q_{\bar{m}\bar{n}}(t)$ are defined as

$$K_{\bar{m}\bar{n}mn} = \int_0^{l_x} \int_0^{l_y} \alpha_h(x, y) f_{xm} f_{x\bar{m}} f_{yn} f_{y\bar{n}} \, dx \, dy \tag{9.12}$$

$$K_{\bar{m}\bar{n}mn}^{(1)}(t) = \int_0^{l_x} \int_0^{l_y} \bar{m}(x, y, t) f_{xm} f_{x\bar{m}} f_{yn} f_{y\bar{n}} \, dx \, dy \tag{9.13}$$

$$K^{(2)}_{\bar{m}\bar{n}mn}(t) = \int_0^{l_x}\int_0^{l_y} \dot{\bar{m}}(x, y, t) f_{xm} f_{x\bar{m}} f_{yn} f_{y\bar{n}}\, \mathrm{d}x\, \mathrm{d}y \tag{9.14}$$

$$Q_{\bar{m}\bar{n}}(t) = \int_0^{l_x}\int_0^{l_y} p(x, y, t) f_{x\bar{m}} f_{y\bar{n}}\, \mathrm{d}x\, \mathrm{d}y \tag{9.15}$$

and h_{mn} are damping constants of the current plate. The coefficients $K^{(1)}_{\bar{m}\bar{n}mn}$ and $K^{(2)}_{\bar{m}\bar{n}mn}$ are variable with respect to time. Therefore, Eq. (9.11) can be solved by step-by-step integration methods.

The explicit expression of Eq. (9.12) is written as

$$\begin{aligned} K_{\bar{m}\bar{n}mn} = {} & \int_0^{l_x}\int_0^{l_y} f_{xm} f_{x\bar{m}} f_{yn} f_{y\bar{n}} \mathrm{d}x\, \mathrm{d}y + \sum_{i=1} \frac{h_{xi}}{h_0} \int_0^{l_x}\int_0^{l_y} D(x - x_i) f_{xm} f_{x\bar{m}} f_{yn} f_{y\bar{n}} \mathrm{d}x\, \mathrm{d}y \\ & + \sum_{j=1} \frac{h_{yj}}{h_0} \int_0^{l_x}\int_0^{l_y} D(y - y_j) f_{xm} f_{x\bar{m}} f_{yn} f_{y\bar{n}}\, \mathrm{d}x\, \mathrm{d}y \\ & + \sum_{i=1}\sum_{j=1} \frac{h_{xyi,j} - h_{xi} - h_{yj}}{h_0} \int_0^{l_x}\int_0^{l_y} D(x - x_i) D(y - y_j) f_{xm} f_{x\bar{m}} f_{yn} f_{y\bar{n}}\, \mathrm{d}x\, \mathrm{d}y \end{aligned} \tag{9.16}$$

The use of the extended Dirac function is powerful in the integral calculation, as stated in the preceding chapters. Namely, for a function $f(x)$, the integral calculation including the extended Dirac function $D(x - x_i)$ can be given by

$$\int_0^{l_x} D(x - x_i) f(x) \mathrm{d}x = \int_{x_i-(b_{xi}/2)}^{x_i+(b_{xi}/2)} \left[\int_0^{l_x} \delta(x - \xi) f(x) \mathrm{d}x\right] \mathrm{d}\xi = \int_{x_i-(b_{xi}/2)}^{x_i+(b_{xi}/2)} f(\xi) \mathrm{d}\xi \tag{9.17}$$

in which $\delta(x - \xi)$ is the Dirac delta function and ξ is a supplementary variable of x. The n-th derivatives of the characteristic functions can be obtained by the following expression:

$$\int_0^{l_x} D^{(n)}(x - x_i) f(x) \mathrm{d}x = \int_{x_i-(b_{xi}/2)}^{x_i+(b_{xi}/2)} (-1)^n f^{(n)}(\xi) \mathrm{d}\xi \tag{9.18}$$

where the superscripts enclosed within parentheses indicate the differential order. The reduced integral is easily calculated by means of Chebyshev's formula.

9.4 Approximate Solution Excluding the Effect of Additional Mass Due to Moving Loads

Equation (9.11) is coupled differential equation with the variable coefficients because of the effect of additional mass due to moving loads. The dynamic analysis including the effect of additional mass due to moving loads must solve Eq. (9.11) in coupled form by the step-by-step integration method, in which the location and mass of moving mass move according to time history of moving loads. The computation is very complicated, in general.

Then, in order to present a closed-form approximate solution as for practical use, we introduce the following assumptions: (1) The effect of additional mass due to moving loads is negligible and (2) only the diagonal terms in $K_{\bar{m}\bar{n}mn}$ are considered. Thus, Eq. (9.11) is approximated in the following uncoupled equation:

$$\delta\Phi_{mn}:\quad K_{mnmn}[\ddot{\Phi}_{mn}(t)+2h_{mn}\omega_{mn}\dot{\Phi}_{mn}(t)+\omega_{mn}^2\Phi_{mn}(t)]=\frac{1}{m_0}Q_{mn}(t) \tag{9.19}$$

Equation (9.19) agrees with Eq. (8.39) in Chap. 8. The general solution of Eq. (9.19) is obtained as

$$\Phi_{mn}(t)=\exp(-h_{mn}\omega_{mn}t)[C_1\sin\omega_{Dmn}t+C_2\cos\omega_{Dmn}t] + \frac{1}{K_{mnmn}m_0\omega_{Dmn}}\int_0^t \exp[-h_{mn}\omega_{mn}(t-\tau)]\sin\omega_{Dmn}(t-\tau)Q_{mn}(\tau)d\tau \tag{9.20}$$

where the constants C_1 and C_2 are determined from the initial conditions; ω_{mn} is the natural frequency of current plate. The well-known following relation is also effective: $\omega_{Dmn}=\omega_{mn}\sqrt{1-h_{mn}^2}$.

The Duhamel integral in Eq. (9.20) can be calculated approximately as follows:

$$\Phi_{mn}(t)=A_{mn}(t)\sin\omega_{Dmn}t-B_{mn}(t)\cos\omega_{Dmn}t \tag{9.21}$$

in which $A_{mn}(t)$ and $B_{mn}(t)$ are given by

$$A_{mn}(t)=\frac{1}{K_{mnmn}m_0\omega_{Dmn}}\int_0^t \exp[-h_{mn}\omega_{mn}(t-\tau)]\cos\omega_{Dmn}\tau\, Q_{mn}(\tau)\mathrm{d}\tau \tag{9.22}$$

$$B_{mn}(t)=\frac{1}{K_{mnmn}m_0\omega_{Dmn}}\int_0^t \exp[-h_{mn}\omega_{mn}(t-\tau)]\sin\omega_{Dmn}\tau\, Q_{mn}(\tau)\mathrm{d}\tau \tag{9.23}$$

$A_{mn}(t)$ can be expressed approximately with the following incremental form

$$A_{mn}(t) \approx A_{mn}(t-\Delta t)\exp(-h_{mn}\omega_{mn}\Delta t) + \frac{1}{K_{mnmn}m_0\omega_{Dmn}}\int_{t-\Delta t}^{t} \exp[-h_{mn}\omega_{mn}(t-\tau)]\cos\omega_{Dmn}\tau\, Q_{mn}(\tau)\,d\tau \tag{9.24}$$

Here Δt is incremental time. The first term on the right side represents the value of A_{mn} at time $t - \Delta t$ and Similarly,

$$B_{mn}(t) \approx B_{mn}(t-\Delta t)\exp(-h_{mn}\omega_{mn}\Delta t) + \frac{1}{K_{mnmn}m_0\omega_{Dmn}}\int_{t-\Delta t}^{t} \exp[-h_{mn}\omega_{mn}(t-\tau)]\sin\omega_{Dmn}\tau\, Q_{mn}(\tau)d\tau \tag{9.25}$$

We consider a concentrated moving load $p(x, y, t)$ advancing along from the left support to the right one at $y=\eta$ with constant velocity v_0, given by

$$p(x, y, t) = \delta(x - v_0 t)\delta(y - \eta)\bar{p}(t) \tag{9.26}$$

in which $\bar{p}(t)$ is the amplitude of the moving load, depending on time. The substitution of Eq. (9.26) into Eq. (9.15) becomes

$$Q_{\bar{m}\bar{n}}(t) = f_{\bar{n}}(y - \eta) f_{\bar{m}}(v_0 t)\bar{p}(t) \tag{9.27}$$

9.5 Numerical Results

The exactness of the solution given in Eq. (9.20) for an isotropic rectangular plate with eccentrically stepped thickness is examined from numerical computations for four kinds of simply supported and clamped rectangular plates with eccentrically stepped thickness, as shown in Fig. 9.2. The data used in the numerical calculation are as follows: span length $l_x = l_y = 6$m; the slab's thickness $h_0 = 0.12$ m; Young's modulus $E=2.06 \times 10^{10}$ N/m^2; Poisson's ratio $\nu=0.17$; and the mass density $\rho=2400$ kg/m^3; the variation of the stepped thickness $h_i/h_0 = 0.5$ and $h_j/h_0 = 0.5$. The damping constants h_{mn} are 0.03 for all modes.

For the external lateral loads, we use test loads checking the disturbing effect of the vibration of floors caused by people walking and other everyday usage, as shown in Fig. 9.3. The test loads consist of four loading types: walking, tripping, bending and stretching, and heel. These loading types are created for a weight of 65 kg. The velocity for moving loads is $v_0 = 1$ m/s except for $v_0 = 2$ m/s for tripping. Kushida [3] discussed the relationships among these criteria and guidelines for evaluation of habitability to building vibrations by using the nonmoving test loads.

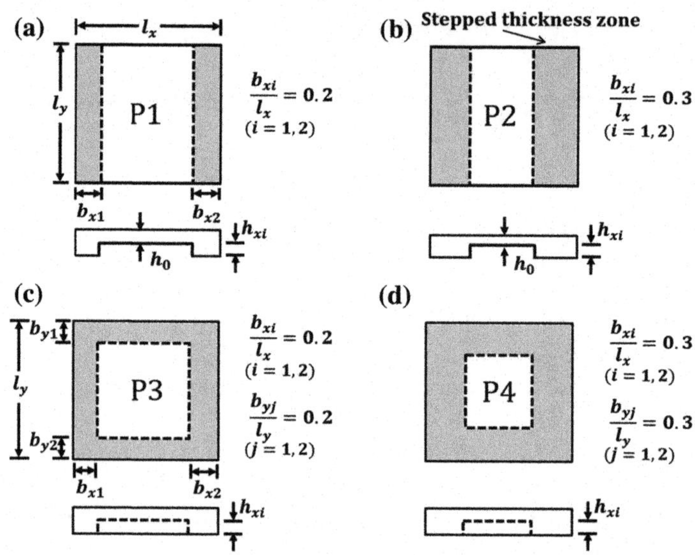

Fig. 9.2 Isotropic rectangular plates with stepped thickness **a** P1, **b** P2, **c** P3, and **d** P4 [1]

The current natural functions f_{xm} take the following well-known natural functions for the simply supported beam and clamped beam, respectively,

$$f_{xm} = \sin \frac{m\pi x}{l_x} \tag{9.28}$$

$$f_{xm} = \cosh\left(\frac{\lambda_m x}{l_x}\right) - \cos\left(\frac{\lambda_m x}{l_x}\right) - \frac{\cosh(\lambda_m) - \cos(\lambda_m)}{\sinh(\lambda_m) - \sin(\lambda_m)}\left[\sinh\left(\frac{\lambda_m x}{l_x}\right) - \sin\left(\frac{\lambda_m x}{l_x}\right)\right] \tag{9.29}$$

in which λ_m represents well-known constants as given $\lambda_1 = 4.7300$; $\lambda_2 = 7.8532$; $\lambda_3 = 10.9956$; $\lambda_4 = 14.1372$; $\lambda_m \approx (2m + 1)\pi/2$ for m large. The natural functions f_{yn} are given as well.

Firstly, we present numerical results for nonmoving test load. Figure 9.4 shows the time history of the dynamic deflections at the midspan of the clamped plates of type P1, subjected to the nonmoving test loads at the midspan. In this figure, the solid lines represent the numerical result obtained from Eq. (9.11) by means of the Wilson-θ method; the broken lines denote values obtained from the approximate solution Eq. (9.20); and the solid lines with circles indicate values obtained from the FEM code NASTRAN. The finite element used is an isotropic rectangular plate element with 20 × 20 divisions for the whole plate. The difference between solid lines (rigorous theory) and broken lines (approximate theory) is too small to plot. Table 9.1 shows the maximum dynamic deflections and the maximum deflection ratios obtained from the proposed method compared to values obtained from FEM. The numerical results demonstrate that the results obtained from the proposed method are in relatively good agreement with the results obtained from FEM.

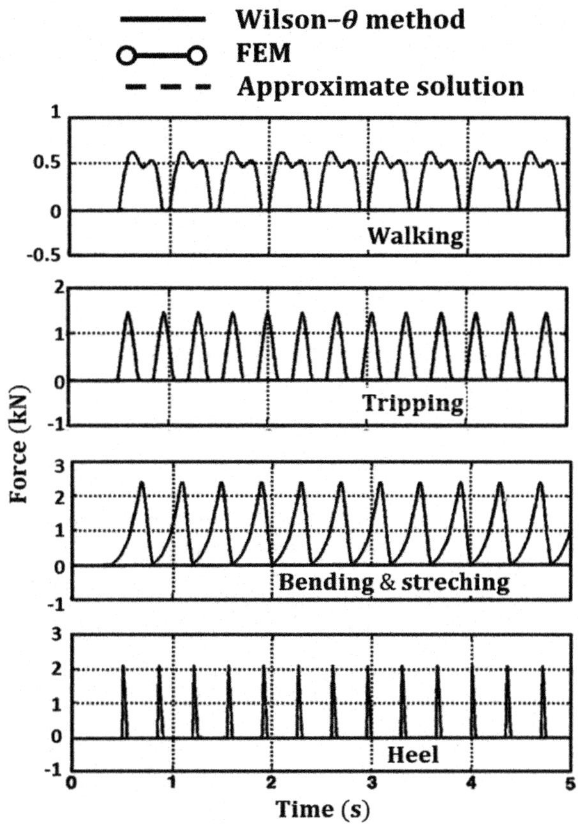

Fig. 9.3 Test loads for building slab [1]

Secondly, we present numerical results for moving test load. The moving load is obtained by considering the effect of the movement on the four nonmoving test loads given in Fig. 9.3. Figures 9.5 and 9.6 represent the effect of moving additional mass due to moving loads on the dynamic deflections at the midspan of type P1, subjected to various moving loads of type of walking and heel, respectively. In these figures, the axis of abscissa indicates the variation of weight of the additional mass. These results are obtained from solving Eq. (9.11) by means of the Wilson-θ method. The moving loads move along $\eta = l_y/2$ in Eq. (9.26) and $v_0 = 1$ m/s except for $v_0 = 2$ m/s for tripping. It follows from these figures that the effect of additional mass due to moving loads increases significantly as the additional mass due to moving loads becomes heavier. The effect of additional mass due to moving loads must be considered for plates subjected to heavyweight moving loads, such as motorcar and airplane. However, the effect is negligible on the vibration of building floors caused by people walking and other everyday usage.

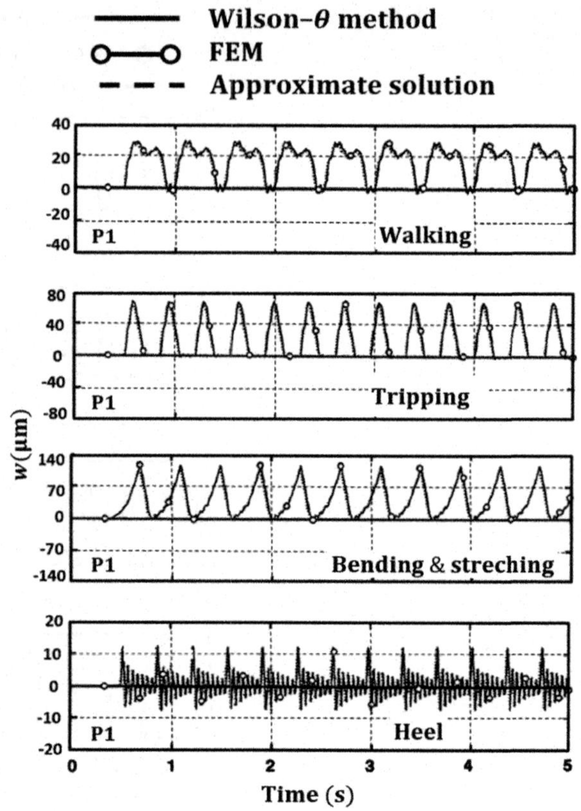

Fig. 9.4 Dynamic deflections for a clamped plate P1 subjected to an nonmoving load [1]

Table 9.1 Maximum dynamic deflections of clamped plates with stepped thickness subjected to an nonmoving load [1]

Type	Load	Wilson-θ method (μm)		Approximate (μm)		FEM (μm)
P1	Walking	28.99	(1.02)	27.39	(0.97)	28.51
	Tripping	66.71	(1.04)	62.97	(0.99)	64.14
	Bending and stretching	112.00	(0.97)	105.70	(0.91)	116.10
	Heel	125.00	(1.07)	116.40	(1.02)	119.40
P4	Walking	20.11	(1.09)	16.94	(0.92)	18.57
	Tripping	46.24	(0.98)	38.96	(0.83)	47.47
	Bending and stretching	75.49	(1.09)	63.45	(0.91)	69.69
	Heel	84.69	(0.99)	73.33	(0.86)	85.58

(Maximum deflection ratio) = present theory/FEM

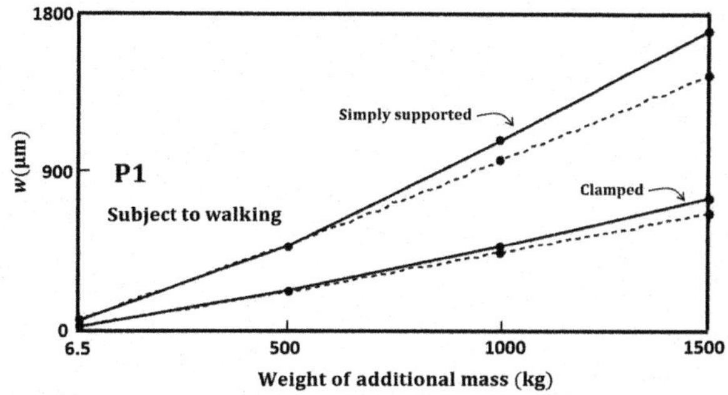

Fig. 9.5 The effect of additional mass due to moving loads for plate P1 subjected to walking: straight line includes the effect of additional mass; dash line excludes the effect of additional mass [1]

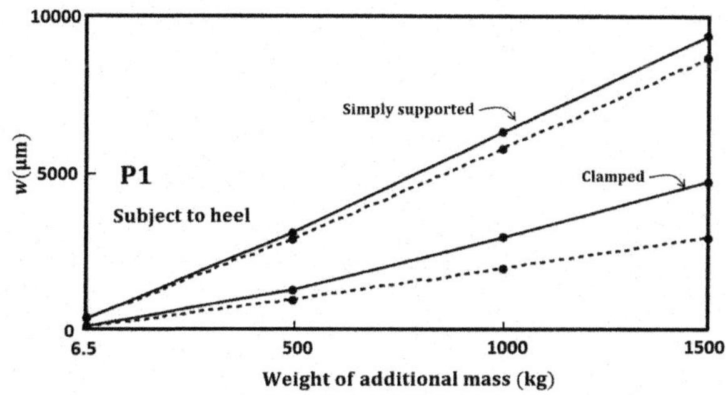

Fig. 9.6 The effect of additional mass due to moving loads for plate P1 subjected to heel: straight line includes the effect of additional mass; dash line excludes the effect of additional mass [1]

Thirdly, in order to present practical use to building slabs, the effect of additional mass due to moving loads is neglected in the following numerical computations. Figures 9.7 and 9.8 represent the time histories of the dynamic deflections, excluding the effect of moving additional mass, at the midspan of the clamped plates of types P1 and P4, subjected to moving test loads, in which $\eta = l_y/2$ in Eq. (9.26) and $v_0 = 1$ m/s except for $v_0 = 2$ m/s of tripping. Table 9.2 shows the maximum dynamic deflections and the maximum deflection ratios, in which the maximum deflection obtained is divided by the maximum deflection obtained from FEM The numerical results demonstrate that the analytical method proposed here is also applicable to the dynamic analyses of current plates.

Lastly, for these numerical models the series in the theory proposed here converges very rapidly. The consideration of each of the nine terms for m and n gives sufficient accuracy for all practical purposes.

9.6 Conclusions

The general analysis methods for an isotropic rectangular plate with arbitrarily and eccentrically stepped thickness, subjected to moving loads along with the effect of additional mass, have been presented by extending the theory for nonmoving external load proposed in Chap. 8 to one for moving loads. The effect of additional mass

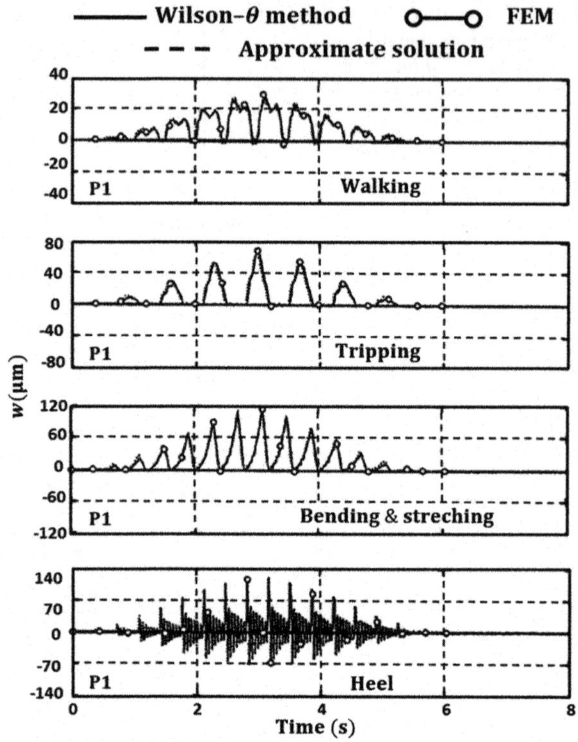

Fig. 9.7 Dynamic deflections for a clamped plate P1 subjected to a moving load [1]

due to moving loads becomes significant when the amplitude of additional mass is heavyweight. However, the effect has been clarified to be negligible on the vibration of floors caused by people walking and other every usage because the additional mass is relatively small compared with the weigh of the plate.

Also, the approximate solutions proposed here, excluding the effect of additional mass due to moving loads, have been demonstrated numerically to be usable in the preliminary stage. Therefore, by comparing the dynamic response obtained from the proposed theory with criteria and assessment for human response, such as the sensitive curve of Meister [2], International Standard ISO 6897, and other standards in each country, trouble in building slabs is avoidable in the preliminary stage of the design.

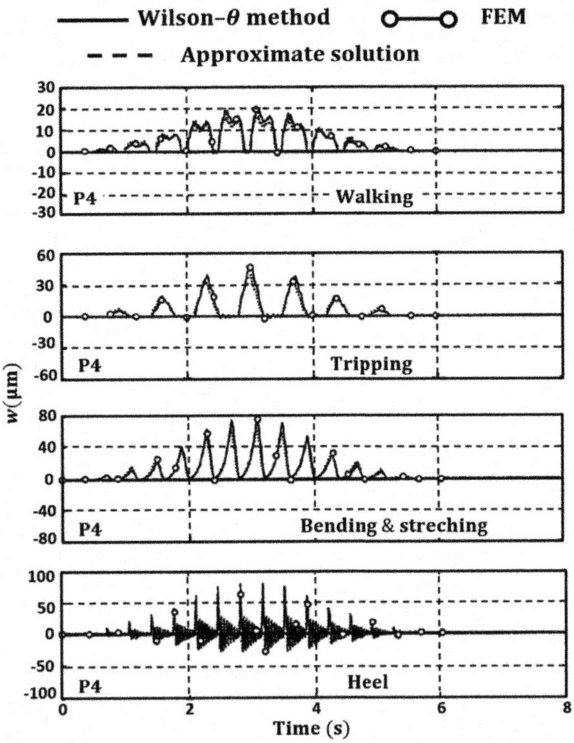

Fig. 9.8 Dynamic deflections for a clamped plate P4 subjected to a moving load [1]

Table 9.2 Maximum dynamic deflections of clamped plates with stepped thickness subjected to a moving load [1]

Type	Load	Wilson-θ method (μm)		Approximate (μm)		FEM (μm)
P1	Walking	28.73	(0.98)	27.19	(0.93)	29.49
	Tripping	65.05	(0.99)	62.60	(0.95)	65.88
	Bending and Stretching	111.20	(0.98)	105.10	(0.92)	114.20
	Heel	121.10	(1.02)	116.40	(0.98)	119.40
P4	Walking	20.06	(1.00)	16.88	(0.84)	20.09
	Tripping	46.82	(1.00)	38.58	(0.82)	47.04
	Bending and Stretching	74.67	(0.99)	62.94	(0.84)	75.66
	Heel	79.27	(1.04)	68.19	(0.90)	76.23

(Maximum deflection ratio) = present theory/FEM

References

1. Takabatake, H.: Dynamic analysis of rectangular plates with stepped thickness subjected to moving loads including additional mass. J. Sound Vib. **213**(5), 829–842 (1998)
2. Meister, F.J.: Die empfindlichkeit des menschen gegen erschutterungen. Forschung auf dem Gebiete des Ingenieurwesens **6**(2), 116–120 (1935)
3. Kushida, H.: A study of design method in consideration of vibration characteristics and optimum environment. J. Archit. Planning Environ. **404** (1989) (in Japanese)

Chapter 10
Static and Dynamic Analyses of Rectangular Floating Plates Subjected to Moving Loads

10.1 Introduction

Very large floating structures, such as a floating international airport, are built in the sea near the megacity for the convenience of transportation. Those structures consisted of a floating plate, a mooring structure, an upper structure for operations, and accommodations for access. The behavior of floating plates subjected to external loads is different from that for normal structures built on land. The vertical loads acting on the floating structure are supported by its footing. The horizontal loads produced by wind, tides, and waves are supported by the mooring structures. Since the safety of such a mega floating structure depends on the structure form against the action of wind loads, live loads (including moving loads, and dead loads), wave action, etc., it is necessary to sufficiently study rational shape determination at the preliminary design stage. A literature survey of research on the hydroelastic analysis of pontoon-type very large floating structures is presented by Watanabe et al. [1].

The structural form of very large floating structures is changed many times at the preliminary design stage so that a simple analytical method for the mega structures is demanded from the structural engineers.

Pontoon-type very large floating structures are built up by connecting many separate structural parts. Therefore, the structural behavior is affected by the stiffness of both the structural parts and the connections. Furthermore, since the very large floating structures are constructed in relatively shallow coastal regions, the vertical supporting force varies due to changes in the tide under the floating plate.

Generally, the analytical method of such very large floating plates is treated as thin plates on an elastic foundation (Winkler type). In the previous works proposed by many researchers, a simplified closed-form solution under both small deformation and the Kirchhoff-Love hypothesis or Mindlin-Reissner hypothesis is presented. However, the general description for a rectangular plate including local variations in structural stiffness due to factors such as voids, stiffened beams, and notches is lacking.

H. Takabatake, *Simplified Analytical Methods of Elastic Plates*,
https://doi.org/10.1007/978-981-13-0086-8_10

Meanwhile, the buoyancy forces on the floating plates are often modeled by an elastic foundation. The interaction between the plate and the foundation can be modeled in a number of ways that are taken from various engineering problems. The Winkler model, which regards the foundation as a series of separate uncoupled spring, so that it is the simplest model for the elastic foundation. The interactions between the separate springs in the Winker model are taken into account by two-parameter elastic models. Two-parameter elastic models have two independent elastic constants and have been approached along two distinct lines. The first approach is to improve the discontinuous behavior of Winkler model by providing mechanical interaction between the individual spring elements, as proposed by Filonenko-Borodich [2, 3], Hetényi [4], Pasternak [5], and Kerr [6]. On the other hand, the second approach is based on elastic continuum model, as proposed by Reissner [7] and Vlasov and Leontiev [8]. Selvadural [9] published about elastic analysis of soil-foundation interaction.

For many very large floating structures, such as a floating international airport, the effect of moving masses must be considered. The dynamic behavior of structures under the influence of moving loads has attracted the attention of many researchers over the last few decades. Fryba [10] provided a complete formulation for the moving mass on a Kirchhoff-Love plate. Rofooei and Nikkhoo [11] presented the dynamic response of a thin rectangular plate under a moving mass.

Moreover, since the floating plates are always subjected to the influence of waves, the hydroelastic behavior of floating plates and the scattering of the surface wave are important subjects. In the case of an airport, the platform has tight restrictions on the maximum permissible vibration amplitudes and loads from the viewpoint of navigational instruments mounted on a floating runaway. Therefore, the investigation of vibration reduction for these floating platform systems has been presented by many researchers. Zhao et al. [12] investigated the water-induced response and the vibration reduction of an elastic floating plate in water waves. Sturova [13] discussed the effect of periodic surface pressure on a rectangular elastic plate floating on shallow water. Sturova [14] presented the unsteady behavior of a thin elastic Euler beam floating on shallow water. Tkacheva [15] clarified the diffraction of surface waves at an oblique incidence by a floating elastic plate. Sturova [16] proposed a method of solving the linear time-varying problem of a floating elastic beam plate on shallow water. For very large floating structures constructed in relatively shallow coastal areas, the dynamic lateral deflection is considered to be influenced by the variation of floating stiffness due to the variation of the tide.

This chapter is to present a general and simple but exact analytical method that can be used at the preliminary design stage to determine the optimal structural form of a huge floating structure, such as a floating international airport. The closed-form approximate solution for the dynamic problem is proposed under both the small deformation and the Kirchhoff-Love hypothesis. The methodology proposed in this chapter is mainly based on Ref. [17], proposed by the author.

10.2 Governing Equations of a Rectangular Plate on an Elastic Foundation

We consider a rectangular very large floating elastic plate with arbitrarily located variations in stiffness and mass placed on an elastic foundation (variable spring stiffness), as shown in Fig. 10.1. In the analysis, we use the following assumptions:

(1) The floating structure is regarded as the rectangular thin plate with arbitrarily variable structural stiffness and variable mass on the elastic foundation of the Winkler model with variable spring stiffness. The elastic foundation with variable spring stiffness is a Winkler-type model representing the variation of buoyancy forces produced by the change in tide in shallow coastal seas.
(2) Although the influence of water wave action on the floating platform is important in the structural design of very large floating structures, the action of water waves is dealt with an external load, and the effect of the interaction between the structure and wave action is neglected.
(3) The deformation behavior of floating structures is the small deformation and the Kirchhoff-Love hypothesis can be applied.
(4) Employing Cartesian coordinate system *x, y, z*, the neutral plane is located on the middle surface of the floating plate.
(5) The local variation in structural stiffness and mass is generated by the existence of stiffened beams, voids, and notches, as shown in Fig. 10.2. For simplicity, the variation of such structural stiffness and mass is assumed to occur due to the uneven thickness of the plates. Also, the unevenness is assumed to be symmetric with respect to the middle plane of the plate and is parallel to the *x*- and/or *y*-directions of the rectangular floating plate.

Thus, expressing the upper and lower heights measured from the middle surface for each variation in stiffness with h^*_{1x}, h^*_{2x}, h^*_{1xy}, h^*_{2xy}, h^*_{1y}, and h^*_{2y}, respectively, the following expressions for plates with stiffened beams can be obtained as

$$h^*_{1x} = \frac{h_0}{2} + \frac{h_{xi}}{2} D(x - x_i) \quad \text{for the upper height} \tag{10.1}$$

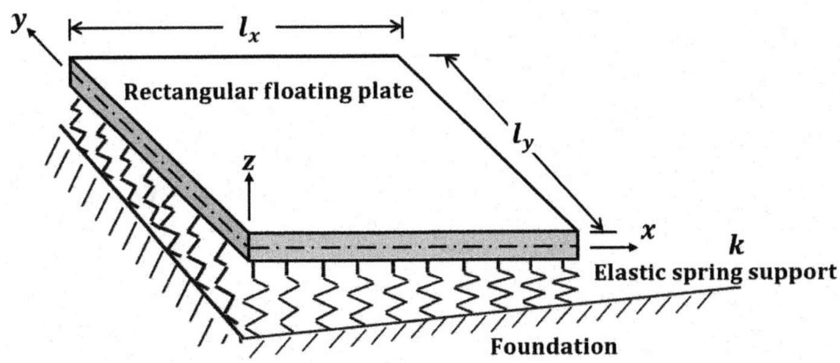

Fig. 10.1 Schematic model [17]

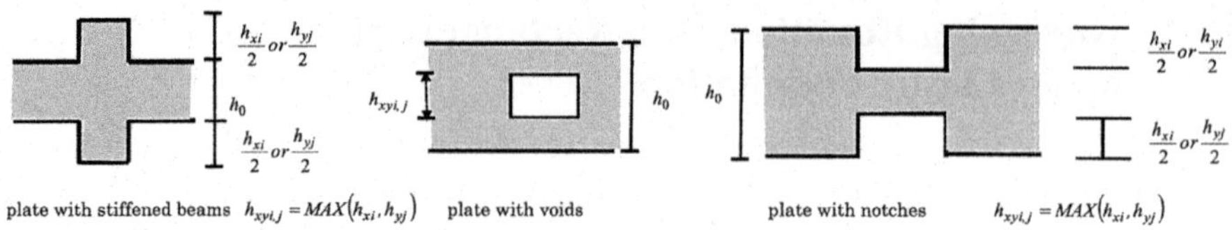

Fig. 10.2 Various floating plates [17]

$$h_{2x}^{*} = -\frac{h_0}{2} - \frac{h_{xi}}{2} D(x - x_i) \quad \text{for the lower height} \tag{10.2}$$

for the i-th variation of stiffness depending on only the x-direction, and

$$h_{1xy}^{*} = \frac{h_0}{2} + \frac{h_{xyi,j}}{2} D(x - x_i) D(y - y_j) \quad \text{for the upper height} \tag{10.3}$$

$$h_{2xy}^{*} = -\frac{h_0}{2} - \frac{h_{xyi,j}}{2} D(x - x_i) D(y - y_j) \quad \text{for the lower height} \tag{10.4}$$

for the i-, j-th variation of stiffness depending on both the x- and y-directions. The j-th variation of the stiffness depending on only the y-direction can be similarly expressed as

$$h_{1y}^{*} = \frac{h_0}{2} + \frac{h_{yj}}{2} D(y - y_j) \tag{10.5}$$

$$h_{2y}^{*} = -\frac{h_0}{2} - \frac{h_{yj}}{2} D(y - y_j) \tag{10.6}$$

In the above equations, h_{xi}, h_{yj}, and $h_{xyi,j}$ represent the varied height of the plate at the i-th, j-th, and i-, j-th stiffness which apart from the reference thickness h_0. Also, $D(x - x_i)$ and $D(y - y_j)$ are the extended Dirac functions, already mentioned in the previous chapters. The i-th extended Dirac function $D(x - x_i)$, in which x takes values from $x_i - b_{xi}/2$ to $x_i + b_{xi}/2$, acts on a function $f(x)$ as follows:

$$\int_0^{l_x} D(x - x_i) f(x) \mathrm{d}x = \int_{x_i - (b_{xi}/2)}^{x_i + (b_{xi}/2)} \left[\int_0^{l_x} \delta(x - \xi) f(x) \mathrm{d}x \right] \mathrm{d}\xi = \int_{x_i - (b_{xi}/2)}^{x_i + (b_{xi}/2)} f(\xi) \mathrm{d}\xi \tag{10.7}$$

where $\delta(x - \xi)$ is the Dirac delta function and ξ is the variable of integration. The action of the n-th derivative of the extended Dirac function is also calculated by

$$\int_0^{l_x} D(x-x_i)^{(n)} f(x)\mathrm{d}x = \int_{x_i-(b_{xi}/2)}^{x_i+(b_{xi}/2)} (-1)f^{(n)}(\xi)\mathrm{d}\xi$$
$$= \int_0^{l_x} D(x-x_i)(-1)^n f^{(n)}(x)\mathrm{d}x \tag{10.8}$$

If the length b_{xi} of the i-th stiffness in the x-direction is sufficiently smaller than the span l_x, the following simplification is effective:

$$\int_0^{l_x} D(x-x_i)f(x)\mathrm{d}x = b_{xi}f(x_i) \tag{10.9}$$

The bending and twisting moments acting on the plate element are defined as

$$M_x = \int_{h_{2y}^*}^{h_{1y}^*} \sigma_x z\,\mathrm{d}z; \quad M_y = \int_{h_{2x}^*}^{h_{1x}^*} \sigma_y z\,\mathrm{d}z \tag{10.10}$$

$$M_{xy} = \int_{h_{2y}^*}^{h_{1y}^*} \tau_{xy} z\,\mathrm{d}z; \quad M_{yx} = \int_{h_{2x}^*}^{h_{1x}^*} \tau_{yx} z\,\mathrm{d}z \tag{10.11}$$

Under the Kirchhoff-Love hypothesis, the relationships between stress and displacement for the isotropic plates are given by

$$\sigma_x = -\frac{Ez}{1-\nu^2}(w_{,xx} + \nu\, w_{,yy}) \tag{10.12}$$
$$\sigma_y = -\frac{Ez}{1-\nu^2}(w_{,yy} + \nu\, w_{,xx}) \tag{10.13}$$
$$\tau_{xy} = \tau_{yx} = -2Gzw_{,xy} \tag{10.14}$$

in which $w(x, y, t)$ is the transverse displacement of the middle surface. Substituting Eqs. (10.12)–(10.14) into Eqs. (10.10) and (10.11), we have

$$M_x = -D_0 d(x, y)(w_{,xx} + \nu\, w_{,yy}) \tag{10.15}$$
$$M_y = -D_0 d(x, y)(w_{,yy} + \nu\, w_{,xx}) \tag{10.16}$$
$$M_{yx} = M_{xy} = -(1-\nu)D_0 d(x, y)w_{,xy} \tag{10.17}$$

where $D_0 = Eh_0^3/[12(1-\nu^2)]$. D_0 is the flexural stiffness of the uniform (reference) plate.

On the other hand, the i-, j-th stiffness coefficient $d(x, y)$ is defined as

$$d(x, y) = 1 + \alpha_{xi} D(x - x_i) + \alpha_{yj} D(y - y_j) + \alpha^*_{xyi,j} D(x - x_i) D(y - y_j) \quad (10.18)$$

in which the stiffness constants α_{xi}, α_{yj}, and $\alpha^*_{xyi,j}$ are given for each variation of the stiffness, as shown in Fig. 10.2, as follows:

$$\alpha_{xi} = 3\left(\frac{h_{xi}}{h_0}\right) + \left(\frac{h_{xi}}{h_0}\right)^2 + \left(\frac{h_{xi}}{h_0}\right)^3 \quad (10.19)$$

$$\alpha_{yj} = 3\left(\frac{h_{yj}}{h_0}\right) + \left(\frac{h_{yj}}{h_0}\right)^2 + \left(\frac{h_{yj}}{h_0}\right)^3 \quad (10.20)$$

$$\alpha^*_{xyi,j} = 3\left(\frac{h_{xyi,j}}{h_0}\right) + 3\left(\frac{h_{xyi,j}}{h_0}\right)^2 + \left(\frac{h_{xyi,j}}{h_0}\right)^3 - \alpha_{xi} - \alpha_{yj} \quad (10.21)$$

for plates with the *i*-, *j*-th stiffened beam, and

$$\alpha_{xi} = 0; \quad \alpha_{yj} = 0; \quad \alpha^*_{xyi,j} = -\left(\frac{h_{xyi,j}}{h_0}\right)^3 \quad (10.22)$$

for plates with the *i*-, *j*-th void, and

$$\alpha_{xi} = -3\left(\frac{h_{xi}}{h_0}\right) + \left(\frac{h_{xi}}{h_0}\right)^2 - \left(\frac{h_{xi}}{h_0}\right)^3 \quad (10.23)$$

$$\alpha_{yj} = -3\left(\frac{h_{yj}}{h_0}\right) + \left(\frac{h_{yj}}{h_0}\right)^2 - \left(\frac{h_{yj}}{h_0}\right)^3 \quad (10.24)$$

$$\alpha^*_{xyi,j} = -3\left(\frac{h_{xyi,j}}{h_0}\right) + 3\left(\frac{h_{xyi,j}}{h_0}\right)^2 - \left(\frac{h_{xyi,j}}{h_0}\right)^3 - \alpha_{xi} - \alpha_{yj} \quad (10.25)$$

for plates with the *i*-, *j*-th notch.

The distribution of mass in the floating plate is influenced by the variation of structural stiffness, and it can be expressed in general form by the use of the extended Dirac function as follows:

$$m(x, y) = m_0 \alpha_h(x, y) \quad (10.26)$$

in which m_0 is the constant mass of a uniform plate per unit area, which is given by $m_0 = \rho h_0$. Also, in order to represent the variation of mass the mass coefficient $\alpha_h(x, y)$ is introduced as

$$\alpha_h = 1 + \alpha_{hxi} D(x - x_i) + \alpha_{hyj} D(y - y_j) + \alpha^*_{hxyi,j} D(x - x_i) D(y - y_j) \quad (10.27)$$

The mass coefficients $\alpha_h(x, y)$ corresponding to the three types of stiffness can be given by

$$\alpha_{hxi} = \frac{h_{xi}}{h_0}; \quad \alpha_{hyj} = \frac{h_{yj}}{h_0}; \quad \alpha^*_{hxyi,j} = \frac{h^*_{xyi,j}}{h_0} \tag{10.28}$$

$$h^*_{xyi,j} = h_{xyi,j} - h_{xi} - h_{yj} \tag{10.29}$$

for the *i*-, *j*-th part of the stiffened beams,

$$\alpha_{hxi} = 0; \quad \alpha_{hyj} = 0; \quad \alpha^*_{hxyi,j} = -\frac{h_{xyi,j}}{h_0} \tag{10.30}$$

for the *i*-, *j*-th part of the voids, and

$$\alpha_{hxi} = -\frac{h_{xi}}{h_0}; \quad \alpha_{hyj} = -\frac{h_{yj}}{h_0}; \quad \alpha^*_{hxyi,j} = -\frac{h^*_{xyi,j}}{h_0} \tag{10.31}$$

$$h^*_{xyi,j} = h_{xyi,j} - h_{xi} - h_{yj} \tag{10.32}$$

for the *i*-, *j*-th part of the notches.

On the other hand, the moving mass $\tilde{m}(x, y, t)$ of the moving load produced by a jet airplane on the current floating plate is variable with respect to both the location x and y and the time t. Hence, the moving mass is represented in a general expression as

$$\tilde{m}(x, y, t) = \tilde{m}_0(t)\,\delta(x - x_0(t))\,\delta(y - y_0(t)) \tag{10.33}$$

where $\tilde{m}_0$ is the magnitude of the moving load and depends on the time t. $x_0(t)$ and $y_0(t)$ indicate the coordinate values of the moving mass at time t.

The kinetic energy T, including the effect of the moving mass $\tilde{m}(x, y, t)$, may be written as

$$T = \frac{1}{2}\iint m_0\alpha_h(\dot{w})^2 \mathrm{d}x\mathrm{d}y + \frac{1}{2}\iint \tilde{m}(x, y, t)(\dot{w})^2 \mathrm{d}x\mathrm{d}y \tag{10.34}$$

The differential equations of motion for the floating plates, including the effect of moving loads can be obtained from Hamilton's principle as

$$\begin{aligned} &m_0\alpha_h\ddot{w} + \dot{\tilde{m}}\dot{w} + \tilde{m}\ddot{w} + c\dot{w} + (D_0 d\, w_{,xx})_{,xx} + (D_0 d\, w_{,yy})_{,yy} + \nu(D_0 d\, w_{,xx})_{,yy} \\ &+ \nu(D_0 d\, w_{,yy})_{,xx} + 2(1-\nu)(D_0 d\, w_{,xy})_{,xy} - p + k_c w = 0 \end{aligned} \tag{10.35}$$

in which k_c is the modulus of the subgrade reaction corresponding to a Winkler-type foundation and it is determined from the reaction of the seabed. The interaction between the structure and the water is an important item on very large floating plates. The additional water action produced by the interaction can be regarded as an additional external load. If the additional external load is previously given by applying another suitable method to a part of the structure, the effect of the interaction is reflected on external load. So, this chapter neglects the interaction between the structure and the water.

This chapter considers the dynamic problem of very large floating structures which are constructed in the coastal water. Even if the structure rises to the water surface, the behavior is considered to be local and has not a great influence. Since the structure is assumed to be always buried in the water, we do not consider tensionless elastic foundations. The moving load is explicitly given by $\tilde{m}(x, y, t)g$, in which g is the acceleration due to gravity.

The associated boundary conditions are given as

$$w = w^* \quad \text{or} \quad (D_0 d\, w_{,xx})_{,x} + \nu(D_0 d\, w_{,yy})_{,x} + 2(1-\nu)(D_0 d\, w_{,xy})_{,y} + v_x^* - k_c w = 0 \tag{10.36}$$

$$w_{,x} = w_{,x}^* \quad \text{or} \quad D_0 d(w_{,xx} + \nu\, w_{,yy}) + m_x^* = 0 \tag{10.37}$$

at $x = 0$ and l_x;

$$w = w^* \quad \text{or} \quad (D_0 d\, w_{,yy})_{,y} + \nu(D_0 d\, w_{,xx})_{,y} + 2(1-\nu)(D_0 d\, w_{,xy})_{,x} + v_y^* - k_c w = 0 \tag{10.38}$$

$$w_{,y} = w_y^* \quad \text{or} \quad D_0 d(w_{,yy} + \nu w_{,xx}) + m_y^* = 0 \tag{10.39}$$

at $y = 0$ and l_y; and

$$w = w^* \quad \text{or} \quad -2(1-\nu) D_0 d\, w_{,xy} = 2m_{xy}^* \tag{10.40}$$

at the corners. In the above expressions, w^*, w_x^*, and w_y^* represent the displacement components prescribed on the geometrical boundary condition. v_x^*, v_y^*, m_x^*, m_y^*, and m_{xy}^* are the vertical edge forces, the end moments and the end torsional moment, respectively, per unit length prescribed on the mechanical boundary conditions.

10.3 Free Transverse Vibrations

Assuming that the ground modulus k_c on the Winkler-type model varies linearly with respect to the x- and y-directions, respectively, as shown in Fig. 10.3, we have

$$k_c(x, y) = k_0(1 + \eta_x\, x)(1 + \eta_y\, y) \tag{10.41}$$

where k_0 is a constant corresponding to a uniform ground modulus. η_x and η_y are the gradients of the ground modulus in the x- and y-directions, respectively, produced by the variation of the floating force. For a uniform floating force, the gradients η_x and η_y take zero.

Employing the method of separation of variables for the dynamic deflection $w(x, y, t)$, we have

$$w(x, y, t) = W(x, y)e^{i\omega t} \tag{10.42}$$

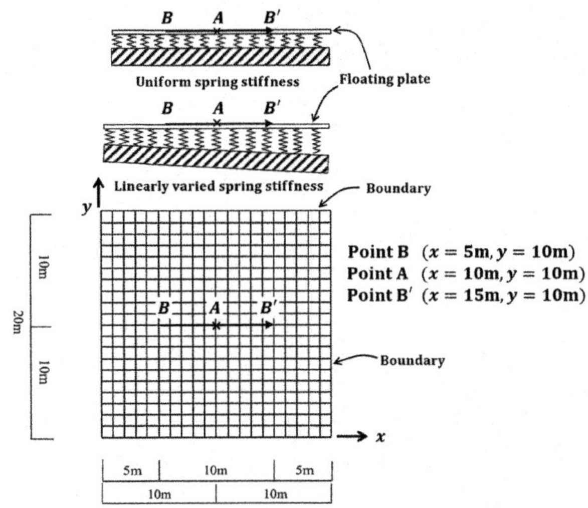

Fig. 10.3 Floating plate and plate elements used in NASTRAN [17]

where W is a function of x and y. Hence, the free transverse vibration can be obtained from Eq. (10.35) as

$$(D_0 d\, W_{,xx})_{,xx} + (D_0 d\, W_{,yy})_{,yy} + \nu(D_0 d\, W_{,xx})_{,yy} + \nu(D_0 d\, W_{,yy})_{,xx} + 2(1-\nu)(D_0 d\, W_{,xy})_{,xy} + k_c W - \omega^2 m_0 \alpha_h W(x, y) = 0 \tag{10.43}$$

where ω is a constant.

The natural frequencies are presented by means of the Galerkin method, where the function W is expressed with an unknown displacement coefficient w_{mn} and the shape function f_{mn} satisfying the specified boundary conditions as

$$W(x, y) = \sum_{m=1} \sum_{n=1} w_{mn}\, f_{mn}(x, y) \tag{10.44}$$

Hence, the Galerkin equation of Eq. (10.43) becomes

$$\delta\, w_{\bar{m}\bar{n}} : \sum_{m=1} \sum_{n=1} w_{mn}[A_{\bar{m}\bar{n}mn} - \lambda^2 B_{\bar{m}\bar{n}mn}] = 0 \tag{10.45}$$

in which $A_{\bar{m}\bar{n}mn}$, $B_{\bar{m}\bar{n}mn}$, and λ^2 are defined as

$$A_{\bar{m}\bar{n}mn} = D_0 \int_0^{l_x} \int_0^{l_y} \big[(d\, f_{mn,xx})_{,xx} + (d\, f_{mn,yy})_{,yy} + \nu (d\, f_{mn,xx})_{,yy} + \nu (d\, f_{mn,yy})_{,xx}$$

$$+ 2(1 - \upsilon)(d f_{mn,xy})_{,xy} \big] f_{\bar{m}\bar{n}} \mathrm{d}x\mathrm{d}y + k_0 \int_0^{l_x} \int_0^{l_y} (f_{mn} + \eta_x x\, f_{mn} + \eta_y y\, f_{mn}$$

$$+ \eta_x \eta_y xy\, f_{mn}) f_{\bar{m}\bar{n}} \mathrm{d}x\mathrm{d}y \tag{10.46}$$

$$B_{\bar{m}\bar{n}mn} = \int_0^{l_x} \int_0^{l_y} \alpha_{\mathrm{h}}\, f_{mn}\, f_{\bar{m}\bar{n}} \mathrm{d}x\, \mathrm{d}y \tag{10.47}$$

$$\lambda^2 = \omega^2 m_0 \tag{10.48}$$

Since Eq. (10.45) has a coupled form with respect to m and n, the natural frequencies ω must be obtained by solving Eq. (10.45) as a generalized eigenvalue problem.

Considering only the diagonal terms of the coefficient matrices $A_{\bar{m}\bar{n}mn}$ and $B_{\bar{m}\bar{n}mn}$ in Eq. (10.45), we can given the simplified solution in uncoupled form for the m-, n-th natural frequency ω_{mn} as

$$\omega_{mn} = \frac{1}{\sqrt{m_0}} \sqrt{\frac{A_{mnmn}}{B_{mnmn}}} \tag{10.49}$$

The shape function f_{mn} is given in the Appendix.

10.4 Forced Transverse Vibrations

We consider transverse forced vibrations of the current rectangular floating plate. We assume that the general solutions of Eq. (10.35) can be expressed in the following form

$$w\,(x,\, y,\, t) = \sum_{m=1}^{\infty} \sum_{n=1}^{\infty} f_{mn}\,(x,\, y) \Phi_{mn}(\mathrm{t}) \tag{10.50}$$

in which Φ_{mn} is an unknown function of time t.

Then, using the above equation and noting that the functions $f_{mn}(x, y)$ satisfy both the equation of free vibration and the specified boundary condition, Eq. (10.35) can be obtained as

$$\sum_{m=1} \sum_{n=1} f_{mn} \big[m_0 \alpha_{\mathrm{h}} \ddot{\Phi}_{mn}(t) + c_{mn} \dot{\Phi}_{mn}(t) + \omega_{mn}^2 m_0 \alpha_h \Phi_{mn}(t) \big] = p(x,\, y,\, t) \tag{10.51}$$

where c_{mn} is the m-, n-th value of the damping coefficient. The m-, n-th damping coefficient c_{mn} is related to the m-, n-th damping constant h^*_{mn} as

$$\frac{c_{mn}}{m_0 \alpha_h} = 2h^*_{mn}\omega_{mn} \tag{10.52}$$

The damping constant h^*_{mn} depends on the coordinate axes x and y due to the variation of the structural mass distribution. Using the above equation, Eq. (10.51) can be rewritten as

$$\sum_{m=1}\sum_{n=1} f_{mn} m_0 \alpha_h [\ddot{\Phi}_{mn}(t) + 2h^*_{mn}\omega_{mn}\dot{\Phi}_{mn}(t) + \omega^2_{mn}\Phi_{mn}(t)] = p(x, y, t) \tag{10.53}$$

Since the above equation includes variables t and x and y, we translate it to the equation including variable only t in order to solve. So, multiplying $f_{\bar{m}\bar{n}}$ by the above equation and integrating the result from 0 to l_x, l_y, respectively, Eq. (10.53) can be translated as

$$\sum_{m=1}\sum_{n=1} \{K_{\bar{m}\bar{n}mn}[\ddot{\Phi}_{mn}(t) + 2h_{mn}\omega_{mn}\dot{\Phi}_{mn}(t) + \omega^2_{mn}\Phi_{mn}(t)]\} = \frac{1}{m_0} Q_{\bar{m}\bar{n}}(t) \tag{10.54}$$

in which $K_{\bar{m}\bar{n}mn}$ and $Q_{\bar{m}\bar{n}}$ are defined as

$$K_{\bar{m}\bar{n}mn} = \int_0^{l_x}\int_0^{l_y} \alpha_h\, f_{\bar{m}\bar{n}} f_{mn} \mathrm{d}x\, \mathrm{d}y \tag{10.55}$$

$$Q_{\bar{m}\bar{n}}(t) = \int_0^{l_x}\int_0^{l_y} p\,(x, y, t) f_{\bar{m}\bar{n}}\, \mathrm{d}x\, \mathrm{d}y \tag{10.56}$$

In the process conducted in Eq. (10.54), we use the assumption that the damping constant h^*_{mn} is the damping constant h_{mn}, which indicates the whole damping constant for the current floating plate by the process of integration of the whole structure. This approximation is effective because it is used in integral calculation considering the entire structure. Since Eq. (10.54) takes a coupled form with respect to $\bar{m}$ and $\bar{n}$, it must be solved by numerical computation such as in the step-by-step integration methods.

The external term $Q_{\bar{m}\bar{n}}(t)$ can be calculated if the moving load is given. As a concrete calculation example, we consider a concentrated traveling load $p(x, y, t)$ advancing along the x direction with a constant velocity v_0 at a constant distance η in the y-direction, as given by

$$p(x, y, t) = \delta(x - v_0 t)\delta(y - \eta)\bar{p}(t) \tag{10.57}$$

in which $\bar{p}(t)$ depends on time. In this case, $Q_{\bar{m}\bar{n}}$ is given by

$$Q_{\bar{m}\bar{n}(t)} = f_{\bar{m}}(v_0 t) f_{\bar{n}}(\eta) \bar{p}(t) \tag{10.58}$$

10.5 Approximate Solutions for Forced Transverse Vibration

Since Eq. (10.54) takes a coupled form, the calculation is complicated. Therefore, for practical use, we consider an approximate solution in the closed form for forced transverse vibration. Assuming that the behavior of the current floating plate is now dominated by only the diagonal terms of the coefficients, Eq. (10.54) becomes the following uncoupled differential equation in terms of only Φ_{mn}:

$$\delta\,\Phi_{mn} : K_{mnmn}\left[\ddot{\Phi}_{mn} + 2h_{mn}\,\omega_{mn}\,\dot{\Phi}_{mn} + \omega_{mn}^2\,\Phi_{mn}\right] = \frac{Q_{mn}}{m_0} \tag{10.59}$$

The general solution of the above equation can be written as

$$\begin{aligned}\Phi_{mn}(t) &= \exp(-h_{mn}\,\omega_{mn}\,t)[C_1 \sin\omega_{Dmn}t + C_2\cos\omega_{Dmn}t] \\ &+ \frac{1}{\omega_{Dmn}}\int_0^t \exp[-h_{mn}\omega_{mn}(t-\tau)]\sin\omega_{Dmn}(t-\tau)\frac{Q_{mn}(\tau)}{m_0 K_{mnmn}}\mathrm{d}\tau\end{aligned} \tag{10.60}$$

in which $\omega_{Dmn} = \omega_{mn}\sqrt{1-h_{mn}^2}$.

Equation (10.60) can be expressed in an explicit form when moving load is defined. So, we consider the special case where the external load $p(x, y)$ is the following harmonic unmoving load with a constant frequency ω_p:

$$p(x, y, t) = p^*(x, y)\sin\omega_p t \tag{10.61}$$

in which $p^*(x, y)$ indicates the magnitude of the external lateral load, depending on x and y. The closed-form solution for $\Phi_{mn}(t)$ can be given from Eq. (10.60) as

$$\begin{aligned}\Phi_{mn} = \frac{1}{K_{mnmn}m_0\omega_{Dmn}}\frac{Q_{mn}^*}{2}&\left\{\frac{h_{mn}\omega_{mn}\cos\omega_p t + (\omega_p+\omega_{Dmn})\sin\omega_p t}{(h_{mn}\omega_{mn})^2 + (\omega_p+\omega_{Dmn})^2}\right. \\ &- \frac{h_{mn}\omega_{mn}\cos\omega_p t + (\omega_p-\omega_{Dmn})\sin\omega_p t}{(h_{mn}\omega_{mn})^2 + (\omega_p-\omega_{Dmn})^2} \\ &+ \exp(-h_{mn}\omega_{mn}t)\left[\frac{(\omega_p+\omega_{Dmn})\sin\omega_{Dmn}t - h_{mn}\omega_{mn}\cos\omega_{Dmn}t}{(h_{mn}\omega_{mn})^2 + (\omega_p+\omega_{Dmn})^2}\right. \\ &\left.\left.+\frac{(\omega_p-\omega_{Dmn})\sin\omega_{Dmn}t + h_{mn}\omega_{mn}\cos\omega_{Dmn}t}{(h_{mn}\omega_{mn})^2 + (\omega_p-\omega_{Dmn})^2}\right]\right\}\end{aligned} \tag{10.62}$$

in which Q^*_{mn} is defined as

$$Q^*_{\bar{m}\bar{n}} \equiv \int_0^{l_x} \int_0^{l_y} p^*(x, y) f_{\bar{m}\bar{n}} \, \mathrm{d}x \, \mathrm{d}y \tag{10.63}$$

Thus, the closed-form approximate solution has been presented for damped floating plates with arbitrarily positioned variable stiffness and mass. For undamped floating plates, the following conversions $h_{mn} \to 0$, we have $\omega_{Dmn} \to \omega_{mn}$ and $\exp(-h_{mn}\omega_{Dmn}t) \to 1$ in Eq. (10.62), the closed-form solution for the undamped floating plates can be translated to more simple form

$$\Phi_{mn} = \frac{Q^*_{mn}}{K_{mnmn} m_0 \omega_{mn}} \frac{(\omega_p \sin \omega_{mn} t - \omega_{mn} \sin \omega_p t)}{\omega_p^2 - \omega_{mn}^2} \tag{10.64}$$

10.6 Numerical Results

The dynamic analytical method for a general rectangular floating plate with a local variation in both the structural stiffness and the spring stiffness in addition to the variation in mass, including the mass of a moving load has been proposed. We use FEM code NASTRAN to evaluate the numerical results obtained using the theory proposed here. Conventional FEM code cannot directly calculate the response due to moving mass. Therefore, we use a special method where the time-history data of the moving load is previously given to all of the nodal points at which the moving load passes through because all external forces in FEM code NASTRAN must be given as known values at the corresponding nodal points. This method is very troublesome and requires very long computation times.

Numerical calculation is carried out for two examples of floating plates. The first example is for a square plate of side 20 m. The second example is for a very large floating plate of length 5000 m and width 1000 m which is typical of the size of an international airport. Both the rectangular floating plates are considered to be three kind of structural characteristics: a uniform floating plate, a floating plate with voids, and a floating plate with stiffened beams. The boundary conditions are assumed to be a simply supported plate. Since the floating plates with notches may be treated in the same way as the case with voids, the demonstration for the notches is neglected in this section.

Example 1: Numerical Results for 20 m x 20 m Floating Plate

We assume the floating structure and moving load in Example 1 as follows: The moving load moves from $x = 5$ m to $x = 15$ m along the midline at $y = 10$ m. The magnitude of the moving load in the downward direction is

$$p = [-10(x - 5) + 100] \times 9.8067 \text{ (kN)} \tag{10.65}$$

The floating plate is composed of steel. The original floating plate is converted into an equivalent uniform stiffness with an equivalent thickness of 0.2 m. The modulus of the floating reaction is 9.8067 kN/m^2 for the uniform floating support and $(1 + 0.0098x) \times 9.8067$ kN/m^2 for the linearly varying floating support.

Figure 10.3 shows the plane of the floating plate, the wake of the moving load, and the supporting condition for floating. In this figure, the vertical and horizontal meshes marked in the floating plate display the element division used in NASTRAN. Figure 10.4 represents the location of voids in the floating plate. The horizontal projection of the voids is rectangular at 2 m × 2 m, and the height of the voids is 0.05 m or 0.1 m. Figure 10.5 shows the floating plate with stiffened beams. The total height of the stiffened beam is either 0.25 m or 0.3 m, including the 0.2 m thickness of the reference plate.

Firstly, in order to examine easily the exactness of the proposed analytical method, we consider the static numerical results obtained from the unmoving and static load

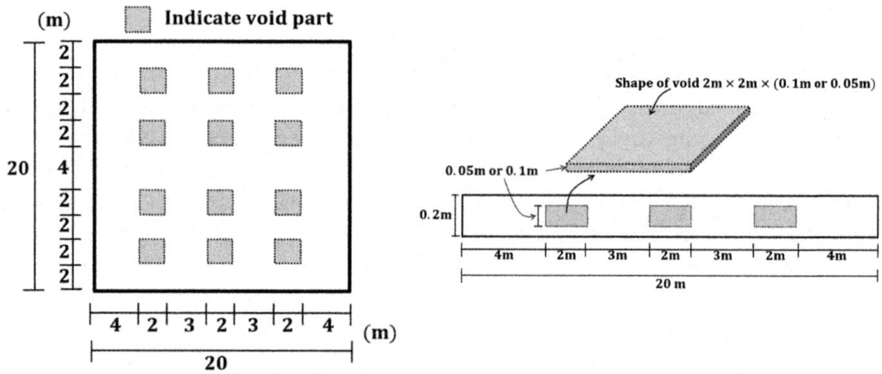

(a) 2D (b) Section

Fig. 10.4 A floating plate with voids

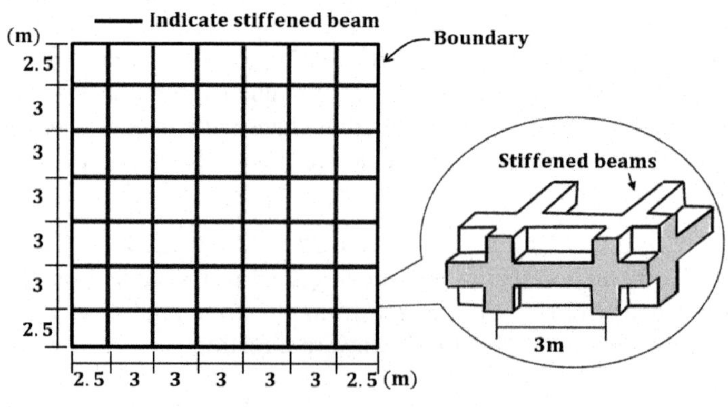

Fig. 10.5 Floating plate with stiffened beams [17]

located at the starting point B of $x = 5$ m and $y = 10$ m. Table 10.1 gives the results for deflections obtained using the proposed theory and NASTRAN at two positions B and A ($x = 5$ m and 10 m at $y = 10$ m, respectively) for the uniform floating plate subject to the concentrated load at the point B. Tables 10.2 and 10.3 represents the numerical results of deflections for the floating plate with voids and stiffened beams, respectively. These Tables 10.1, 10.2, and 10.3 illustrate the difference of static deflections for static loads on floating plates with both variable structural stiffness and variable spring stiffness. It demonstrates from the numerical results that the theory proposed here has good agreement with the results obtained from NASTRAN for all cases.

Secondly, we examine the exactness of the present theory for free transverse vibration. Tables 10.4, 10.5, and 10.6 demonstrate the excellent relationships between the present theory and the NASTRAN results for the natural frequencies at relatively low modes. Thus, the exactness of the present theory for the free transverse vibrations of floating plates is also examined from these tables.

Thirdly, we consider the dynamic problem for these floating plates. Figures 10.6, 10.7 and 10.8 compare the results obtained from the present theory and NASTRAN for the time history of dynamic lateral displacements at $x = 5$ m and $y = 10$ m. For other locations, there is also good agreement between the results obtained using

Table 10.1 Relationship between the present theory and NASTRAN results for static deflections of normal floating plate [17]

Location	Uniform spring stiffness			Variable spring stiffness		
	Theory (m)	NASTRAN (m)	Theory/ NASTRAN	Theory (m)	NASTRAN (m)	Theory/ NASTRAN
Point B	0.0318	0.0331	0.96	0.0311	0.0324	0.96
Point A	0.0336	0.0358	0.94	0.0329	0.0349	0.94

Table 10.2 Relationship between the present theory and NASTRAN results for static deflections of floating plate with voids [17]

Location	Uniform spring stiffness			Variable spring stiffness		
	Theory (m)	NASTRAN (m)	Theory/ NASTRAN	Theory (m)	NASTRAN (m)	Theory/ NASTRAN
Point B	0.0350	0.0336	1.04	0.0342	0.0328	1.04
Point A	0.0370	0.0363	1.02	0.0354	0.0361	0.98

Table 10.3 Relationship between the present theory and NASTRAN results for static deflections of floating plate with stiffened beams [17]

Location	Uniform spring stiffness			Variable spring stiffness		
	Theory (m)	NASTRAN (m)	Theory/ NASTRAN	Theory (m)	NASTRAN (m)	Theory/ NASTRAN
Point B	0.0246	0.0232	1.06	0.0242	0.0228	1.06
Point A	0.0166	0.0196	0.85	0.0257	0.0247	1.04

Table 10.4 Relationship between the present proposed theory and NASTRAN results for natural frequencies (rad/s) of uniform floating plate [17]

Mode		Uniform spring stiffness			Variable spring stiffness		
m	*n*	Theory (rad/s)	NASTRAN (rad/s)	Theory/ NASTRAN	Theory (rad/s)	NASTRAN (rad/s)	Theory/ NASTRAN
1	1	4.947	4.933	1.00	5.010	4.996	1.00
1	2	12.23	12.17	1.00	12.26	12.20	1.00
1	3	24.43	24.31	1.00	24.44	24.32	1.00
2	2	19.55	19.31	1.01	19.56	19.33	1.01
2	3	31.75	31.25	1.02	31.76	31.26	1.02
3	3	43.95	42.89	1.02	43.96	42.89	1.02

Table 10.5 Relationship between the present theory and NASTRAN results for natural frequencies (rad/s) of floating plate with voids [17]

Mode		Uniform spring stiffness			Variable spring stiffness		
m	*n*	Theory (rad/s)	NASTRAN (rad/s)	Theory/ NASTRAN	Theory (rad/s)	NASTRAN (rad/s)	Theory/ NASTRAN
1	1	4.845	4.930	0.98	4.907	4.993	0.98
1	2	11.91	12.16	0.98	11.94	12.19	0.98
1	3	23.92	24.29	0.98	23.93	24.30	0.98
2	2	19.04	19.30	0.99	19.05	19.32	0.99
2	3	31.10	31.23	1.00	13.11	31.24	1.00
3	3	43.07	42.86	1.00	43.08	42.87	1.00

Table 10.6 Relationship between the present theory and NASTRAN results for natural frequencies (rad/s) of floating plate with stiffened beams [17]

Mode		Uniform spring stiffness			Variable spring stiffness		
m	*n*	Theory (rad/s)	NASTRAN (rad/s)	Theory/ NASTRAN	Theory (rad/s)	NASTRAN (rad/s)	Theory/ NASTRAN
1	1	5.600	5.519	1.01	5.649	5.568	1.01
1	2	13.99	14.63	0.96	14.01	13.72	1.02
1	3	28.12	29.58	0.95	28.13	27.41	1.03
2	2	22.21	22.94	0.97	22.22	21.66	1.03
2	3	36.19	37.48	0.97	36.19	35.01	1.03
3	3	49.93	51.40	0.97	49.94	47.83	1.04

the present theory and those obtained using NASTRAN. It is also clarified that the exactness of the present theory is proven from dynamic problems with all floating plate considered here.

Example 2: Numerical Results for 5000-m-Long and 1000-m-Wide Very Large Floating Plate

We examine the exactness of the present theory for a very large floating plate representing an international airport subjected to moving loads. The floating plate has a long side of 5000 m and a short side of 1000 m, as shown in Fig. 10.9. The floating plate is assumed to be one continuous plate made up of steel material, the equivalent thickness of the floating plate is 2.289 m. The moving load represents a Boeing 747-200B jet airplane. The moving load p is assumed to be

$$p = \left[-\frac{377.8}{3300}(x - 900) + 377.8 \right] \times 9.8067 \text{ (kN)} \tag{10.66}$$

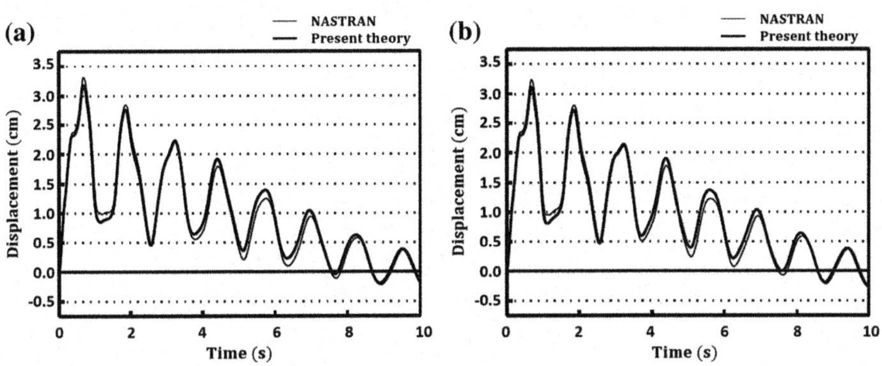

Fig. 10.6 Time history of dynamic lateral displacement at the point B of normal floating plate for **a** uniform spring stiffness and **b** variable spring stiffness [17]

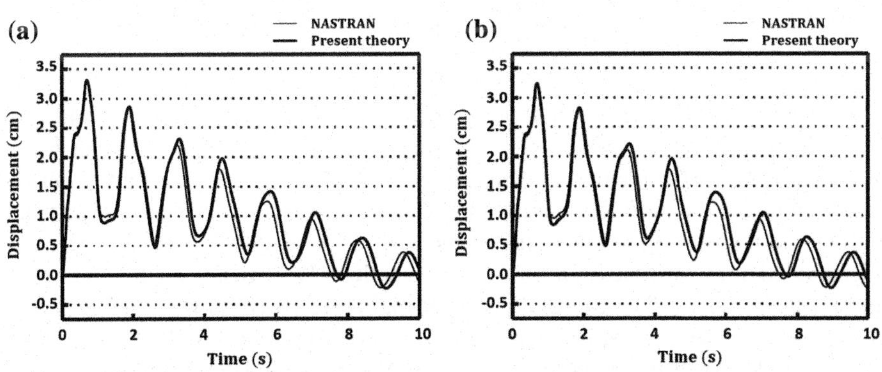

Fig. 10.7 Time history of dynamic lateral displacement at the point B of floating plate with voids for **a** uniform spring stiffness and **b** variable spring stiffness [17]

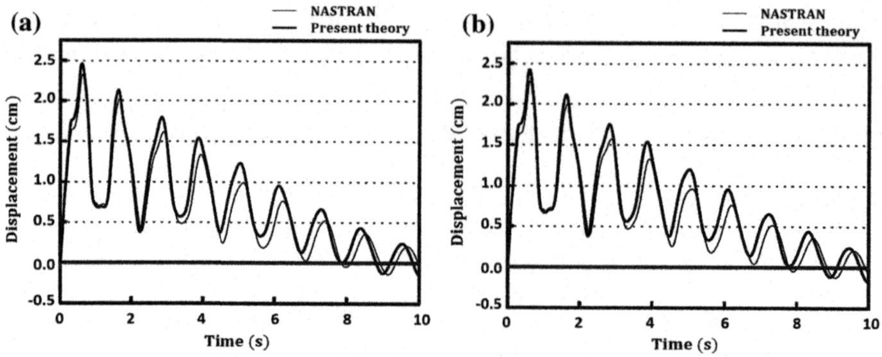

Fig. 10.8 Time history of dynamic lateral displacement at the point B of floating plate with stiffened beams for **a** uniform spring stiffness and **b** variable spring stiffness [17]

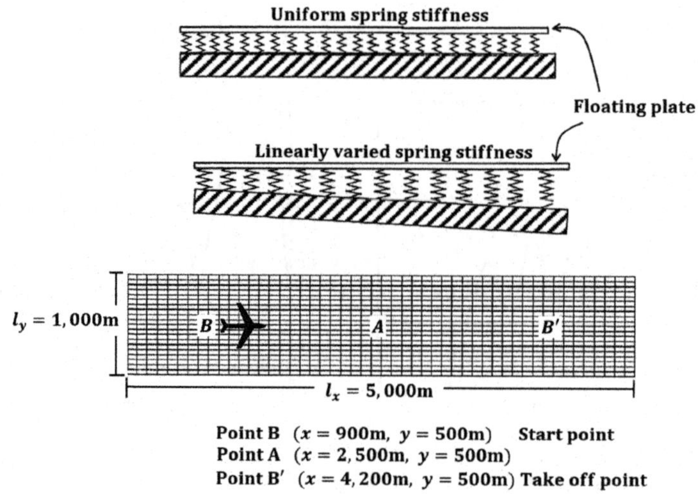

Fig. 10.9 Very large floating plate and plate elements used in NASTRAN [17]

and moves from the starting point at $x = 900$ m to the takeoff point at $x = 4200$ m along the midline at $y = 500$ m. These points are marked in Fig. 10.9 with B and B', respectively. The magnitude of the moving load decreases gradually from the starting point to the takeoff point due to the running lift of the airplane. The spring stiffness corresponding to the spring of the elastic foundation for the floating force is examined for two cases: uniform and variable. The variable coefficient varies linearly from 1 at $x = 0$ m. to 1.5 at $x = 5000$ m. Figure 10.10 shows the time history of the dynamic lateral displacements at the point B ($x = 900$ m and $y = 500$ m). Figure 10.11 demonstrates the time history of the dynamic lateral displacements at the point A ($x = 2500$ m and $y = 500$ m). Table 10.7 compares the effect of between the uniform and variable spring stiffnesses on the dynamic lateral displacement.

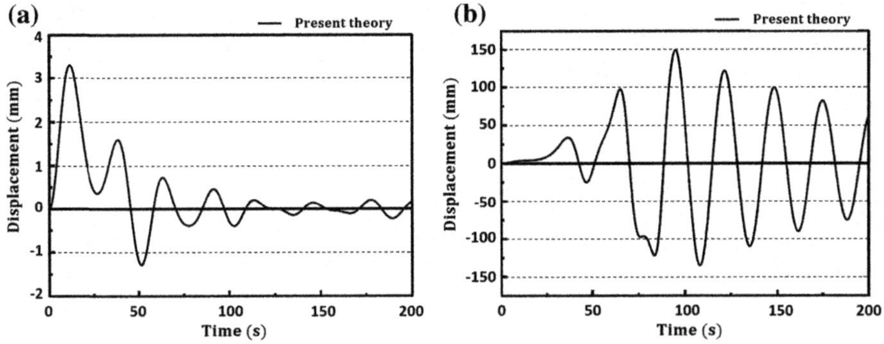

Fig. 10.10 Dynamic lateral displacement at the point B for **a** constant spring stiffness and **b** variable spring stiffness [17]

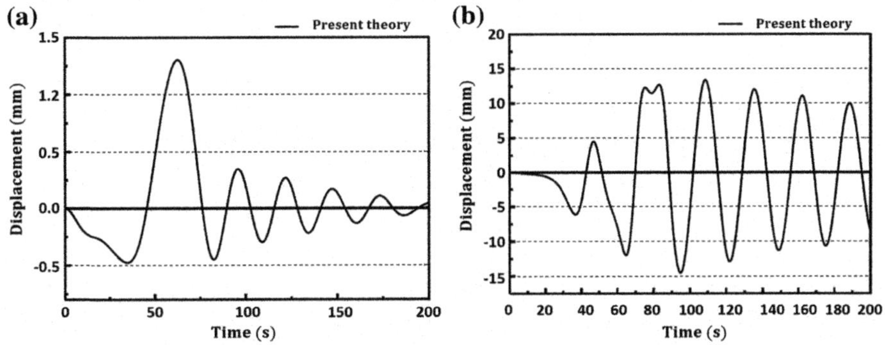

Fig. 10.11 Dynamic lateral displacement at the point A for **a** constant spring and **b** variable spring stiffness [17]

It follows from these numerical results that the dynamic response of the floating structure is remarkably influenced by the spring stiffness of the foundation. For that purpose, sufficient examination is required at the preliminary design stage using a simple but exact analytical method proposed here.

10.7 Conclusions

The design for very large floating plates, as for an international airport, is necessary to easily understand the structural behavior in the preliminary design stage. A general floating structure has variable structural stiffness and variable spring stiffness of foundation in addition of the effect of moving loads. This chapter has proposed a simple but exact analytical method for the dynamic problem of a general floating

Table 10.7 Relationship between maximum dynamic displacements for uniform and variable spring stiffness [17]

Location	Uniform spring stiffness		Variable spring stiffness	
	Time (s)	Displacement (m)	Time (s)	Displacement (m)
Point B ($x =$ 900 m)	11.6	3.3×10^{-3}	11.6	3.3×10^{-3}
Point A ($x =$ 2500 m)	62.2	1.3×10^{-3}	62.2	1.3×10^{-3}
Point B' ($x =$ 4200 m)	88.4	3.8×10^{-4}	88.4	3.8×10^{-4}

structure. The exactness of the proposed theory is proven from numerical computations.

Appendix: Shape Function of Beam on Elastic Foundation, Supported with Translational and Rotational Stiffnesses

In the theory the shape function of the current rectangular floating plates is substituted with one of a uniform beam on the Winkler-type elastic foundation. So, we present the dynamic problem of uniform beams on a uniform elastic foundation supported with elastic translational stiffness and rotational stiffness at the both ends, as shown in Fig. 10.12.

Where k_c is the translational stiffness of the elastic foundation, and k_a, k_b, R_a, and R_b are the translational stiffnesses and rotational stiffnesses of the supports at both ends, respectively. $\hat{m}_0$ is the mass per unit length of the uniform beam.

The equation of motion for the uniform beam may be given by

$$\hat{m}_0 \ddot{w} + [EI\, w'']'' = p(x, y) - k_c w \tag{10.67}$$

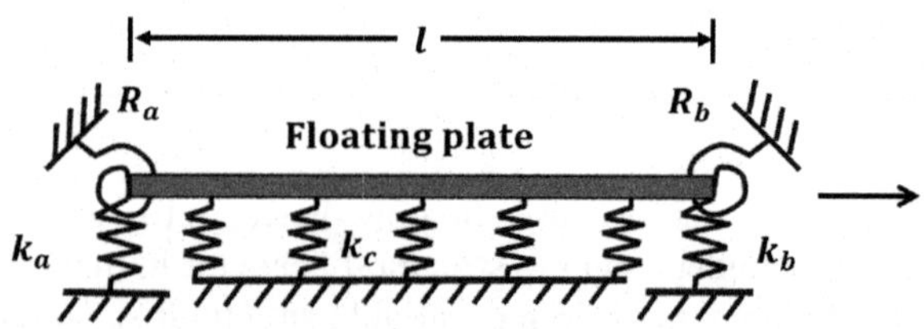

Fig. 10.12 Beam supported with translational stiffness and rotational stiffness on elastic foundations [17]

where *EI* is the bending stiffness of the uniform beam.

The equation for the transverse free vibration becomes from Eq. (10.67) as

$$\hat{m}_0 \ddot{w} + [EI\, w'']'' + k_c w = 0 \tag{10.68}$$

The boundary conditions are also obtained as

Elastic support

$$k_a w_a = -EI\, w''' \quad \text{at}\, x = 0 \tag{10.69}$$

$$k_b w_b = EI\, w''' \quad \text{at}\, x = l \tag{10.70}$$

Elastic restraint

$$R_a w_a' = EI\, w'' \quad \text{at}\, x = 0 \tag{10.71}$$

$$R_b w_b' = -EI\, w'' \quad \text{at}\ x = l \tag{10.72}$$

The various boundary conditions can be made up easily by changing the coefficients of the translational stiffness and rotational stiffness. For example,

When $k_a \to \infty$, the unmoving support condition is $w_a = 0$.
When $k_a \to 0$, the free support condition is $EI\, w''' = 0$.
When $R_a \to \infty$, the constrained condition for rotation is $w_a' = 0$.
When $R_a \to 0$, the free condition for rotation is $EI\ w'' = 0$.

Expressing the deflection $w(x, t)$ in the form

$$w(x, t) = \bar{w}(\mathrm{x})\, \mathrm{e}^{i\omega t} \tag{10.73}$$

the frequency equation for the current beam may be expressed as

$$\bar{w}'''' - k^4 \bar{w} = 0 \tag{10.74}$$

in which k^2 is defined as

$$k^2 = \sqrt{\frac{\omega^2 \hat{m}_0 - \bar{k}_c}{EI}} \tag{10.75}$$

The solution of Eq. (10.74) is given as

$$\begin{aligned}\bar{w}(x) &= C_1[\cos(kx) + \cosh(kx)] + C_2[\cos(kx) - \cosh(kx)] \\ &+ C_3[\sin(kx) + \sinh(kx)] + C_4[\sin(kx) - \sinh(kx)]\end{aligned} \tag{10.76}$$

in which $C_1 - C_4$ are arbitrary constants.

The boundary conditions (10.69)–(10.72) can be rewritten in the form

$$\alpha_a \bar{k}_{a1} \bar{w}_{(0)} + \bar{k}_{a2} \bar{w}'''_{(0)} = 0; \quad \alpha_b \bar{k}_{b1} \bar{w}_{(l)} - \bar{k}_{b2} \bar{w}'''_{(l)} = 0 \tag{10.77}$$

$$\beta_a \bar{R}_{a1} \bar{w}'_{(0)} - \bar{R}_{a2} \bar{w}''_{(0)} = 0; \quad \beta_b \bar{R}_{b1} \bar{w}'_{(l)} + \bar{R}_{b2} \bar{w}''_{(l)} = 0 \tag{10.78}$$

where α_a, α_b, β_a, and β_b are defined to be

$$\alpha_a = \frac{k_a}{EI}; \quad \alpha_b = \frac{k_b}{EI} \tag{10.79}$$

$$\beta_a = \frac{R_a}{EI}; \quad \beta_b = \frac{R_b}{EI} \tag{10.80}$$

The constants $\bar{k}_{a1}, \bar{k}_{a2}, \ldots, \bar{R}_{b1}$, and $\bar{R}_{b2}$ are constants and take the value 1 for the general boundary condition including the translational stiffness and rotational stiffness. When the translation or rotation degree is constrained, the constants $\bar{k}_{a1}$, $\bar{k}_{b1}$, $\bar{R}_{a1}$, and $\bar{R}_{b1}$ takes 1, and $\bar{k}_{a2}$, $\bar{k}_{b2}$, $\bar{R}_{a2}$, and $\bar{R}_{b2}$ are 0.

Conversely, when the translational or rotational is free, the constants $\bar{k}_{a1} \ldots$ and $\bar{R}_{b1}$ take 0, and $\bar{k}_{a2}, \ldots$ and $\bar{R}_{b2}$ are 1. These constants are the general boundary conditions and include extreme (classic) boundary conditions such as perfectly constrained or free for translation and rotation. Substituting Eq. (10.76) into Eqs. (10.77) and (10.78) and using the condition that the determinant $|A_{ij}|$ of the reduced equations is zero, the following algebraic equation for kl is obtained

$$\begin{bmatrix} A_{11} & A_{12} & A_{13} & A_{14} \\ A_{21} & A_{22} & A_{23} & A_{24} \\ A_{31} & A_{32} & A_{33} & A_{34} \\ A_{41} & A_{42} & A_{43} & A_{44} \end{bmatrix} \begin{bmatrix} C_1 \\ C_2 \\ C_3 \\ C_4 \end{bmatrix} = \begin{bmatrix} 0 \\ 0 \\ 0 \\ 0 \end{bmatrix} \tag{10.81}$$

where $A_{11} \sim A_{44}$ are defined as

$$\begin{cases} A_{11} = 2\alpha_a \bar{k}_{a1} \\ A_{12} = 0 \\ A_{13} = 0 \\ A_{14} = -2\bar{k}_{a2}\,(kl)^3 \end{cases}$$

$$\begin{cases} A_{21} = \alpha_b \bar{k}_{b1}\,[\cos kl + \cosh kl] - \bar{k}_{b2}\,(kl)^3[\sin kl + \sinh kl] \\ A_{22} = \alpha_b \bar{k}_{b1}\,[\cos kl - \cosh kl] - \bar{k}_{b2}\,(kl)^3[\sin kl - \sinh kl] \\ A_{23} = \alpha_b \bar{k}_{b1}\,[\sin kl + \sinh kl] - \bar{k}_{b2}\,(kl)^3[-\cos kl + \cosh kl] \\ A_{24} = \alpha_b \bar{k}_{b1}\,[\sin kl - \sinh kl] + \bar{k}_{b2}\,(kl)^3[\cos kl + \cosh kl] \end{cases}$$

$$\begin{cases} A_{31} = 0 \\ A_{32} = 2\bar{R}_{a2}\,(kl)^2 \\ A_{33} = 2\beta_a \bar{R}_{a1}(kl) \\ A_{34} = 0 \end{cases}$$

$$\begin{cases} A_{41} = \beta_B \bar{R}_{b1}(kl)\,[-\sin(kl) + \sinh(kl)] + \bar{R}_{b2}\,(kl)^2[-\cos(kl) + \cosh(kl)] \\ A_{42} = -\beta_B \bar{R}_{b1}(kl)\,[\sin(kl) + \sinh(kl)] - \bar{R}_{b2}\,(kl)^2[\cos(kl) + \cosh(kl)] \\ A_{43} = \beta_B \bar{R}_{b1}(kl)\,[\cos(kl) + \cosh(kl)] + \bar{R}_{b2}\,(kl)^2[-\sin(kl) + \sinh(kl)] \\ A_{44} = \beta_B \bar{R}_{b1}(kl)\,[\cos(kl) - \cosh(kl)] - \bar{R}_{b2}\,(kl)^2[\sin(kl) + \sinh(kl)] \end{cases} \tag{10.82}$$

The constants $k_1 \sim k_4$ take the following values for the four cases:
for $A_{11} \neq 0$ and $A_{32} \neq 0$

$$k_1 = -\frac{A_{14}}{A_{11}}; \quad k_2 = -\frac{A_{33}}{A_{32}}k_3; \quad k_3 = \frac{A_{32}}{A_{11}}\left(\frac{A_{14}A_{21} - A_{24}A_{11}}{-A_{33}A_{22} + A_{23}A_{32}}\right); \quad k_4 = 1 \tag{10.83}$$

for $A_{11} \neq 0$, $A_{32} = 0$, $A_{33} \neq 0$

$$k_1 = -\frac{A_{14}}{A_{11}}; \quad k_2 = \frac{A_{33}}{A_{11}}\left(\frac{A_{14}A_{21} - A_{24}A_{11}}{A_{33}A_{22} - A_{23}A_{32}}\right); \quad k_3 = -\frac{A_{32}}{A_{33}}k_2; \quad k_4 = 1 \tag{10.84}$$

for $A_{11} = 0$, $A_{14} \neq 0$, $A_{32} \neq 0$

$$k_1 = \frac{A_{14}}{A_{32}}\left(\frac{A_{33}A_{22} - A_{23}A_{32}}{A_{21}A_{14} - A_{11}A_{24}}\right); \quad k_2 = -\frac{A_{33}}{A_{32}}; \quad k_3 = 1; \quad k_4 = -\frac{A_{11}}{A_{14}}k_1 \tag{10.85}$$

for $A_{11} = 0$, $A_{14} \neq 0$, $A_{32} = 0$, $A_{33} \neq 0$

$$k_1 = A_{14}\left(\frac{A_{22} + k_3 A_{32}}{A_{21}A_{14} - A_{11}A_{24}}\right); \quad k_2 = 1; \quad k_3 = -\frac{A_{32}}{A_{33}}; \quad k_4 = -\frac{A_{11}}{A_{14}}k_1 \tag{10.86}$$

References

1. Watanabe, E., Utsunomiya, T., Wang, C.M.: Hydroelastic analysis of pontoon-type VLFS: a literature survey. Eng. Struct. **26**(2), 245–256 (2004)
2. Filonenko-Borodich, M.M.: Some approximate theories of the elastic foundation. Uchenyie Zapiski Moskovskogo Gosudarstvennogo Universiteta **46**, 3–18 (1940)
3. Filonenko-Borodich, M.M.: A very simple model of an elastic foundation capable of spreading the load. Sbornik Moskovkovo Elektro Instituta Transzheldorizdat **53** (1945)
4. Hetényi, M.: Beams on Elastic Foundation: Theory with Applications in the Fields of Civil and Mechanical Engineering, University of Michigan Press, Ann Arbor (1946)
5. Pasternak, P.L.: On a New Method of Analysis of an Elastic Foundation by Means of Two Foundation Constants. Gosudarstvennoe Izdatelstvo Literaturi po Stroitelstvu i Arkhitekture, Moscow (1954)
6. Kerr, A.D.: Elastic and viscoelastic foundation models. J. Appl. Mech. **31**(3), 491–498 (1964)
7. Reissner, E.: Deflection of plates on viscoelastic foundation. J. Appl. Mech. **80**, 144–145 (1958)
8. Vlasov, V. Z., Leont'ev, N. N.: Beams, Plates and Shells on Elastic Foundations. Israel Program for Scientific Translations, Jerusalem (1966)
9. Selvadurai, A.P.S.: Elastic analysis of soil-foundation interaction. Geotech. Eng. **17** (1979)
10. Frýba, L.: Vibration of Solids and Structures Under Moving Loads. Thomas Telford, London (1999)
11. Rofooei, F.R., Nikkhoo, A.: Application of active piezoelectric patches in controlling the dynamic response of a thin rectangular plate under a moving mass. Int. J. Solids Struct. **46**(11–12), 2429–2443 (2009)
12. Zhao, C., Zhang, J., Huang, W.: Vibration reduction of floating elastic plates in water waves. Mar. Struct. **20**(1–2), 71–99 (2007)
13. Sturova, I.V.: The effect of periodic surface pressure on a rectangular elastic plate floating on shallow water. J. Appl. Math. Mech. **70**(3), 378–386 (2006)
14. Sturova, I.V.: Effect of bottom topography on the unsteady behaviour of an elastic plate floating on shallow water. J Appl. Math. Mech. **72**(4), 417–426 (2008)
15. Tkacheva, L.A.: The diffraction of surface waves by a floating elastic plate at oblique incidence. J. Appl. Math. Mech. **68**(3), 425–436 (2004)
16. Sturova, I.V.: Unsteady behaviour of a heterogeneous elastic beam floating on shallow water. J. App. Math. Mech. **72**(6), 704–714 (2008)
17. Takabatake, H.: A simplified analysis of rectangular floating plates subjected to moving loads. Ocean Eng. **97**, 37–47 (2015)

Part IV
Effects of Dead Loads on Elastic Plates

Chapter 11
Effects of Dead Loads on Static and Dynamic Analyses of Rectangular Plates

11.1 Introduction

Plates in buildings are main structural members subject to large vertical loads. The theory of thin plates has progressed from the classical theories based on the Kirchhoff-Love and Mindlin-Reissner hypotheses, as shown in Timoshenko [1] and Volterra and Zachmanoglou [2], to more recent advanced theories.

The interesting subjects of recent advanced theories are classified as follows:

(1) To propose generalized analytical method of plates with various boundary conditions: Shimpi et al. [3]
(2) To propose higher-order shear deformation under the restrictions on limited variable: Wu et al. [4], Boscolo and Banerjee [5]
(3) Analytical method of plates with arbitrarily cutouts: Fang et al. [6], Lee and Chen [7]
(4) To propose generalized analytical method of various laminate composite plates: Tanveer and Singh [8]
(5) To propose simplified analytical method closed-form solution of various plates: Rao and Saheb [9], Takabatake [10–17].

However, the above-mentioned plate theories do not reflect an important characteristic of plates frequently used in building structures such as roof slabs. Such plates are subjected to only dead loads in the initial state. External loads, such as live loads, always act on the deformed state caused by the invariant dead loads. If the dead load is extremely heavy, then the initial stress is very large in the initial dead-load-only state. The difference between a heavy concrete slab and a steel lightweight slab is obvious in building structures. Thus, the author proposes that the behavior of plates used in building structures should be different depending on the invariant dead load. Although theoretically the reinforced concrete structures and steel structures should have the same degree of safety, the collapse due to snow loads on roofs is for less in reinforced concrete structures than in lightweight steel structures. The dead

H. Takabatake, *Simplified Analytical Methods of Elastic Plates*,
https://doi.org/10.1007/978-981-13-0086-8_11

loads of structures are considered to play an important part in this phenomenon that lightweight steel structures collapse more frequently.

Structures are always subject to dead loads, and initial stresses produced by these dead loads are conservative. When structures are subjected to live loads, the strain energy is the sum of the strain energy produced by only live loads and the additional strain energy produced by the conservative initial stresses. This additional strain energy has the effect of decreasing external disturbances such as displacements and stress couples produced by live loads. However, due to the unknown nature of this phenomenon, current trends in structural design do not consider the effect of dead loads. If the effect of dead loads is clarified, it will be possible to estimate the effective value of live loads which reflects the existence of dead loads. As a result, it will be possible to have the same safety factor for both heavy structures and lightweight structures and to design truly safe structures. Previous studies for linear and nonlinear plate problems do not take into consideration the effect of dead loads, which is an essential problem for plates used in structures.

The author [18–20] demonstrated the effect of dead loads on the static and dynamic problems of elastic beams and proposed the effect from a closed-form approximate solution. This new attention presented by the author became an important jumping-off point for extensions of elementary beam theory. It was also extended to the finite element method using a beam element that takes the effect of dead loads into account [21]. The existence of an initial bend in a beam due to a dead load has been suggested to increase the natural frequencies of lateral vibrations [22]. The author [23, 24] reported the effect of dead loads on the static and dynamic responses of a uniform elastic rectangular plate and clarified the physical factors governing this effect. Mostaghel and Yu [25] revealed that, based on the large deflection theory for thin plates and the principle of conservation of energy, preforming a thin plate into any shape has the effect of increasing its natural frequencies. Yu et al. [26] reported a phenomenon, whereby the natural frequencies increase when the plate is preformed into a shape with a specific mode of vibration. Zhou [27] proposed a finite element formulation for plates that takes into consideration the stiffening effect of dead loads. Recently, Durmaz and Daloğlu [28] created a map to obtain the precise variation of ground snow loads in the Eastern Black Sea Region, where dead loads were recognized to have a great influence on the structural behavior produced by live loads, such as snow loads.

This chapter clarifies the effect of dead loads in static and dynamic elastic plates. First, the advanced governing equation of plates, which takes into account the effect of dead loads, is presented from Hamilton's principle. The effect of dead loads is based on strain energy resulting from conservative initial stresses produced by dead loads. This strain energy supports part of the potential energy of the live loads. If the dead loads are large, then the initial stresses are large, and the energy supported by the initial stresses is large. Consequently, the displacements and stress couples produced by live loads are smaller than for those excluding the effect of dead loads. Second, from the results of numerical calculations using the Galerkin method, the effect of dead loads is shown for the case of simply supported and clamped rectangular plates subjected to static live loads. Third, in order to apply the effect of dead loads to design, the physical parameters which express this effect are explicitly shown by a closed-form solution for simply supported plates in the static problem. Fourth, a method that reflects the effect of dead loads on live loads is explained through an example. Fifth, the effect of dead load on the free transverse vibration of thin elastic plates is clarified by the use of the Galerkin method. Sixth, a closed-form approximate solution for dynamic deflections of plates is proposed from the governing equation including the effect of dead loads. It is shown that the numerical results obtained from the proposed solution agree with the numerical results obtained from the Galerkin method. Finally, a method that reflects the effect of dead loads to dynamic live loads is presented by an example. The methodology proposed in this chapter is mainly based on references [23, 24] presented by the author.

11.2 Governing Equations Including the Effect of Dead Loads for Plates

We consider a rectangular normal plate with Cartesian coordinate system, as shown in Fig. 11.1. It is assumed that both the external forces are transverse loads only and the axial forces are neglected. In the initial state, the plate causes deflections $\tilde{w}$ due to dead loads $\tilde{p}$ of the plates. This deformed state is defined as the reference state. When dynamic or static live loads $\bar{p}$ per unit area act on this reference state, the deflections $\bar{w}$ due to the live load $\bar{p}$ occur from this reference state. The deflections $\bar{w}$ are measured from the reference state. Deflections $\tilde{w}$ and $\bar{w}$ and transverse loads $\tilde{p}$ and $\bar{p}$ are defined as positive when they indicate positive in the coordinate axis z.

We assume that the Kirchhoff-Love hypothesis is valid and the stretching strains in the middle surface are negligible. The nonlinear strain–displacement relations of plates can be given from Washizu [29] as follows:

$$\left.\begin{aligned} \bar{\varepsilon}_x &= -z\bar{w}_{,xx} + \underline{\tfrac{1}{2}(\bar{w}_{,x})^2} \\ \bar{\varepsilon}_y &= -z\bar{w}_{,yy} + \underline{\tfrac{1}{2}(\bar{w}_{,y})^2} \\ \bar{\gamma}_{xy} &= -2z\bar{w}_{,xy} + \underline{\bar{w}_{,x}\bar{w}_{,y}} \end{aligned}\right\} \tag{11.1}$$

$$\left.\begin{aligned} \tilde{\varepsilon}_x &= -z\tilde{w}_{,xx} + \underline{\tfrac{1}{2}(\tilde{w}_{,x})^2} \\ \tilde{\varepsilon}_y &= -z\tilde{w}_{,yy} + \underline{\tfrac{1}{2}(\tilde{w}_{,y})^2} \\ \tilde{\gamma}_{xy} &= -2z\tilde{w}_{,xy} + \underline{\tilde{w}_{,x}\tilde{w}_{,y}} \end{aligned}\right\} \tag{11.2}$$

where $\bar{\varepsilon}_x$, $\bar{\varepsilon}_y$, $\bar{\gamma}_{xy}$ are strains produced by live loads $\bar{p}$, and $\tilde{\varepsilon}_x$, $\tilde{\varepsilon}_y$, $\tilde{\gamma}_{xy}$ are strains produced by dead loads $\tilde{p}$. z is element thickness with respect to reference. The underlined terms in Eqs. (11.1) and (11.2) represent nonlinear terms.

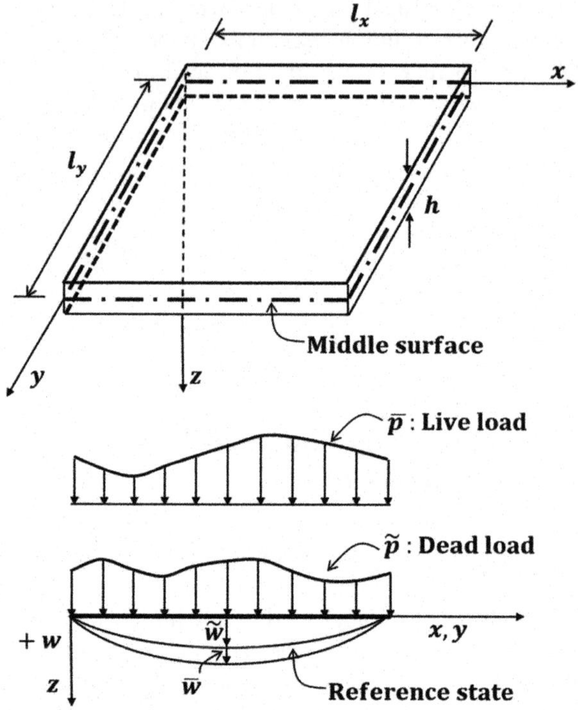

Fig. 11.1 Coordinates and load distributions of a plate [23, 24]

The equation of motion of plates including the effect of dead loads is formulated from the following Hamilton's principle:

$$\delta I = \delta \int_{t_1}^{t_2} (T - U - V)\mathrm{d}t = 0 \tag{11.3}$$

where T indicates kinetic energy, U is strain energy, and V is potential energy produced by external loads. δ denotes variation operator taken during the indicated time interval from t_1 to t_2.

The strain energy U can be given by

$$U = \bar{U} + \tilde{U} \tag{11.4}$$

in which $\bar{U}$ is the strain energy produced by only live loads and $\tilde{U}$ is the additional strain energy produced by the conservative initial bending stresses caused by only dead loads. These energies can be expressed by

$$\bar{U} = \frac{1}{2} \iiint (\bar{\sigma}_x \bar{\varepsilon}_x + \bar{\sigma}_y \bar{\varepsilon}_y + \bar{\tau}_{xy} \bar{\gamma}_{xy}) \mathrm{d}x\, \mathrm{d}y\, \mathrm{d}z \tag{11.5}$$

$$\tilde{U} = \iiint (\tilde{\sigma}_x \bar{\varepsilon}_x + \tilde{\sigma}_y \bar{\varepsilon}_y + \tilde{\tau}_{xy} \bar{\gamma}_{xy}) \mathrm{d}x\,\mathrm{d}y\,\mathrm{d}z \tag{11.6}$$

where the stresses $\bar{\sigma}_x, \bar{\sigma}_y, \bar{\tau}_{xy}$, and $\tilde{\sigma}_x, \tilde{\sigma}_y, \tilde{\tau}_{xy}$ are stresses produced by live loads and dead loads, respectively. Now, using linear stress–strain relations, we have

$$\left.\begin{aligned} \bar{\sigma}_x &= \tfrac{E}{1-\nu^2}(\bar{\varepsilon}_x + \nu\bar{\varepsilon}_y); \quad \bar{\sigma}_y = \tfrac{E}{1-\nu^2}(\bar{\varepsilon}_y + \nu\bar{\varepsilon}_x); \quad \bar{\tau} = G\bar{\gamma}_{xy} \\ \tilde{\sigma}_x &= \tfrac{E}{1-\nu^2}(\tilde{\varepsilon}_x + \nu\tilde{\varepsilon}_y); \quad \tilde{\sigma}_y = \tfrac{E}{1-\nu^2}(\tilde{\varepsilon}_y + \nu\tilde{\varepsilon}_x); \quad \tilde{\tau} = G\tilde{\gamma}_{xy} \end{aligned}\right\} \tag{11.7}$$

in which E is Young's modulus of elasticity, G is the modulus of rigidity, and ν is Poisson's ratio.

Since the deflections $\tilde{w}$ due to dead loads $\tilde{p}$ are previously known, we consider the linear governing equation including the effect of dead loads with respect to unknown displacements $\bar{w}$. Hence, the strain energies $\bar{U}$ and $\tilde{U}$ given in Eqs. (11.5) and (11.6) take into account up to the second order with respect to $\bar{w}$. Hence, the strain energy $\bar{U}$ is calculated using the linear strain–displacement relations, while the strain energy $\tilde{U}$ is calculated using the nonlinear strain–displacement relations. It should be noted that the effect of dead loads considered here is not the well-known stiffening effect of midplane stretching in plates. Because the current plate is considered only in the bending state, the midplane stretching can be neglected in the strain–displacement relations. The effect of dead loads in the plate is considered by using these different types of strain–displacement relation for $\bar{U}$ and $\tilde{U}$. The author [18] used similar treatments to present the effect of dead loads in static beams, too. Haroun [30] used similar treatments to present the effect of initial hoop stress due to hydrostatic pressures on liquid storage cylindrical tanks. Thus, the strain energy $\bar{U}$ can be obtained as

$$\bar{U} = \frac{D_0}{2} \iint \left\{(\bar{w}_{,xx} + \bar{w}_{,yy})^2 + 2(1-\nu)\left[(\bar{w}_{,xy})^2 - \bar{w}_{,xx}\bar{w}_{,yy}\right]\right\} \mathrm{d}x\,\mathrm{d}y \tag{11.8}$$

where D_0 is the flexural stiffness of the plates $\left(D_0 = E\,h_0^3/\left[12(1-\nu^2)\right]\right)$, in which h_0 is the thickness of the plate.

On the other hand, the strain energy $\tilde{U}$ is given as

$$\begin{aligned} \tilde{U} = D_0 \iint & \left[\tilde{w}_{,xx}\bar{w}_{,xx} + \tilde{w}_{,yy}\bar{w}_{,yy} + 2\tilde{w}_{,xy}\bar{w}_{,xy} \right. \\ & \left. +\nu(\tilde{w}_{,yy}\bar{w}_{,xx} + \tilde{w}_{,xx}\bar{w}_{,yy} - 2\tilde{w}_{,xy}\bar{w}_{,xy})\right]\mathrm{d}x\mathrm{d}y \\ & + \frac{Eh_0}{4(1-\nu^2)} \iint \left\{(\tilde{w}_{,x})^2(\bar{w}_{,x})^2 + (\tilde{w}_{,y})^2(\bar{w}_{,y})^2 + 2\tilde{w}_{,x}\tilde{w}_{,y}\bar{w}_{,x}\bar{w}_{,y} \right. \\ & \left. + \nu\left[(\tilde{w}_{,y})^2(\bar{w}_{,x})^2 + (\tilde{w}_{,x})^2(\bar{w}_{,y})^2 - 2\tilde{w}_{,x}\tilde{w}_{,y}\bar{w}_{,x}\bar{w}_{,y}\right]\right\}\mathrm{d}x\mathrm{d}y \end{aligned} \tag{11.9}$$

The strain energy $\tilde{U}$ is a function of both the unknown displacements $\bar{w}$ due to live loads and the known displacements $\tilde{w}$ due to dead loads.

Next, for simplicity, we assume that transverse loads act on the middle surface of plates and that the end moments and end forces at the boundaries are absent. The potential energy produced by the external loads can be given by

$$\bar{V} = -\iint (\bar{p} + \tilde{p})\, \bar{w}\, \mathrm{d}x\, \mathrm{d}y \tag{11.10}$$

Also, neglecting the effect of rotatory inertia, the kinetic energy T can be expressed as

$$T = \iint \frac{m_0}{2} (\dot{\bar{w}})^2\, \mathrm{d}x\, \mathrm{d}y \tag{11.11}$$

where m_0 is the mass per unit length ($m_0 = \rho h_0$) of the plate.

For the reference state where only the dead loads $\tilde{p}$ act, the following familiar governing equation and boundary conditions can be given

$$\nabla^4 \tilde{w} - \frac{\tilde{p}}{D_0} = 0 \tag{11.12}$$

and

$$\left.\begin{aligned}
&\left.\begin{aligned}
&\tilde{w} = 0 \;\text{ or }\; \tilde{w}_{,xxx} + (2-\nu)\tilde{w}_{,xyy} = 0\\
&\tilde{w}_{,x} = 0 \;\text{ or }\; \tilde{w}_{,xx} + \nu \tilde{w}_{,yy} = 0
\end{aligned}\right) \text{ at } x = 0 \text{ and } x = l_x;\\
&\left.\begin{aligned}
&\tilde{w} = 0 \;\text{ or }\; \tilde{w}_{,yyy} + (2-\nu)\tilde{w}_{,xxy} = 0\\
&\tilde{w}_{,y} = 0 \;\text{ or }\; \tilde{w}_{,yy} + \nu \tilde{w}_{,xx} = 0
\end{aligned}\right) \text{ at } y = 0 \text{ and } y = l_y;\\
&\tilde{w} = 0 \;\text{ or }\; \tilde{w}_{,xy} = 0 \qquad \text{at corners;}
\end{aligned}\right\} \tag{11.13}$$

where $\nabla^2 = \partial^2/\partial x^2 + \partial^2/\partial y^2$.

Substituting Eqs. (11.8)–(11.11) into Eq. (11.3) and considering Eqs. (11.12) and (11.13), the advanced governing equations of the plates subjected to live loads $\bar{p}$, which include the effect of dead loads, can be obtained as follows:

$$\frac{m_0 \ddot{\bar{w}}}{D_0} + \nabla^4 \bar{w} - \frac{\bar{p}}{D_0} + R(\bar{w}, \tilde{w}) = 0 \tag{11.14}$$

and

$$
\left.\begin{array}{ll}
\left.\begin{array}{l}
\bar{w}=0 \quad \text{or}\ \bar{w}_{,xxx}+(2-\nu)\bar{w}_{,xyy} \\
\qquad \underline{-\frac{6}{h_0^2}\bar{w}_{,x}[(\tilde{w}_{,x})^2+\nu(\tilde{w}_{,y})^2]} \\
\qquad \underline{-\frac{6(1-\nu)}{h_0^2}\tilde{w}_{,x}\tilde{w}_{,y}\bar{w}_{,y}}=0 \\
\bar{w}_{,x}=0 \ \text{or}\ \bar{w}_{,xx}+\nu\bar{w}_{,yy}=0
\end{array}\right) & \text{at} \quad x=0 \text{ and } x=l_x; \\
\left.\begin{array}{l}
\bar{w}=0 \quad \text{or}\ \bar{w}_{,yyy}+(2-\nu)\bar{w}_{,xxy} \\
\qquad \underline{-\frac{6}{h_0^2}\bar{w}_{,x}[(\tilde{w}_{,x})^2+\nu(\tilde{w}_{,y})^2]} \\
\qquad \underline{-\frac{6(1-\nu)}{h_0^2}\tilde{w}_{,x}\tilde{w}_{,y}\bar{w}_{,y}}=0 \\
\bar{w}_{,y}=0 \ \text{or}\ \bar{w}_{,yy}+\nu\bar{w}_{,xx}=0
\end{array}\right) & \text{at} \quad y=0 \text{ and } y=l_y; \\
\bar{w}=0 \ \text{or}\ \bar{w}_{,xy}=0 & \text{at corners,}
\end{array}\right\} \tag{11.15}
$$

where the term $R(\bar{w}, \tilde{w})$ is defined as

$$
\begin{aligned}
R(\bar{w}, \tilde{w}) = &-\frac{6}{h_0^2}[2\tilde{w}_{,x}\tilde{w}_{,xx}\bar{w}_{,x}+(\tilde{w}_{,x})^2\bar{w}_{,xx}+2\tilde{w}_{,y}\tilde{w}_{,yy}\bar{w}_{,y}+(\tilde{w}_{,y})^2\bar{w}_{,yy}] \\
&-\frac{6\nu}{h_0^2}[2\tilde{w}_{,y}\tilde{w}_{,xy}\bar{w}_{,x}+(\tilde{w}_{,y})^2\bar{w}_{,xx}+2\tilde{w}_{,x}\tilde{w}_{,xy}\bar{w}_{,y}+(\tilde{w}_{,x})^2\bar{w}_{,yy}] \\
&-\frac{6(1-\nu)}{h_0^2}[(\tilde{w}_{,x}\tilde{w}_{,y}\bar{w}_{,y})_{,x}+(\tilde{w}_{,x}\tilde{w}_{,y}\bar{w}_{,x})_{,y}]
\end{aligned} \tag{11.16}
$$

Since the static deflections $\tilde{w}$ produced by the dead loads are previously known, Eqs. (11.14) and (11.15) are linear with respect to unknown deflections $\bar{w}$. The underlined terms in Eq. (11.15) and $R(\bar{w}, \tilde{w})$ in Eq. (11.14) relate to the effect of dead loads. The expressions neglecting them reduce to the well-known equations for plates subjected to only live loads $\bar{p}$.

11.3 Formulation of Static Problem Including the Effect of Dead Loads

In the preceding section, we have obtained the advanced governing equations including the effect of dead loads for plates. This section considers the effect of dead loads on the static problem of plates. For the static problem, neglecting dynamic terms of Eq. (11.14), the equilibrium equation including the effect of dead loads for plates is given as

$$
\nabla^4\bar{w}-\frac{\bar{p}}{D_0}+R(\bar{w}, \tilde{w})=0 \tag{11.17}
$$

We examine the effect of dead loads on a uniform rectangular plate subject to static live loads by means of the Galerkin method. The Galerkin equation for the current plates may be obtained as follows:

$$\int_0^{l_x}\int_0^{l_y} Q \cdot \delta\bar{w}\, dxdy = 0 \tag{11.18}$$

where the notation Q indicates the equilibrium equation given in Eq. (11.17), and l_x and l_y are the dimensions of the plate in the x- and y-directions, respectively. The deflection $\bar{w}$ due to live loads can be given by power series expansions as follows:

$$\bar{w}\,(x, y) = \sum_{m=1}\sum_{n=1} \bar{w}_{mn}\, f_{mn} \begin{pmatrix} m=1,2,3,\ldots,\infty \\ n=1,2,3,\ldots,\infty \end{pmatrix} \tag{11.19}$$

where $\bar{w}_{mn}$ are unknown displacement coefficients and f_{mn} are the shape functions satisfying the prescribed boundary conditions of the plates. The following functions f_{mn} for simply supported and clamped rectangular plates can be represented as:

$$f_{mn} = \sin\frac{m\pi x}{l_x}\sin\frac{n\pi\, y}{l_y} \quad \text{(for a simply supported plate)} \tag{11.20}$$

$$f_{mn} = \sin\frac{\pi\, x}{l_x}\sin\frac{m\pi\, x}{l_x}\sin\frac{\pi\, y}{l_y}\sin\frac{n\pi\, y}{l_y} \quad \text{(for a clamped plate)} \tag{11.21}$$

Substituting Eq. (11.19) into Eq. (11.18), the Galerkin equation can be rewritten as

$$\delta\bar{w}_{mn} \quad : \int_0^{l_x}\int_0^{l_y} Q\, f_{mn} dxdy = 0 \begin{pmatrix} m=1,2,3,\ldots,\infty \\ n=1,2,3,\ldots,\infty \end{pmatrix} \tag{11.22}$$

Since the governing equation Q contains deflections $\tilde{w}$ due to dead loads $\tilde{p}$, it is necessary to determine beforehand deflections $\tilde{w}$ from Eqs. (11.12) and (11.13). The deflections $\tilde{w}$ can be expressed generally as follows:

$$\tilde{w}(x, y) = \sum_{m=1}^{\infty}\sum_{n=1}^{\infty} \tilde{w}_{mn}\, f_{mn}(x, y) \tag{11.23}$$

where $\tilde{w}_{mn}$ are known displacement coefficients when dead loads $\tilde{p}$ are prescribed. For simply supported plates, $\tilde{w}_{mn}$ and f_{mn} are easily determined from Navier's double Fourier series. For simplicity, the shape functions f_{mn} for clamped plates use approximately Eq. (11.21). Substituting Eqs. (11.19) and (11.23) into Eq. (11.22), the Galerkin equation reduces to the following linear, nonhomogeneous, algebraic equations with respect to unknown displacement coefficients $\bar{w}_{mn}$.

$$
\begin{aligned}
\delta\bar{w}_{mn}:\ & \sum_{\bar{m}=1}\sum_{\bar{n}=1}\bar{w}_{\bar{m}\bar{n}}\int_0^{l_x}\int_0^{l_y}\left[f_{\bar{m}\bar{n},xxxx}+2f_{\bar{m}\bar{n},xxyy}+f_{\bar{m}\bar{n},yyyy}\right.\\
& -\sum_{\tilde{m}=1}\sum_{\tilde{n}=1}\sum_{\hat{m}=1}\sum_{\hat{n}=1}\tilde{w}_{\tilde{m}\tilde{n}}\tilde{w}_{\hat{m}\hat{n}}\frac{6}{h_0^2}\{2f_{\tilde{m}\tilde{n},x}f_{\hat{m}\hat{n},xx}f_{\bar{m}\bar{n},x}\\
& +f_{\tilde{m}\tilde{n},x}f_{\hat{m}\hat{n},x}f_{\bar{m}\bar{n},xx}+2f_{\tilde{m}\tilde{n},y}f_{\hat{m}\hat{n},yy}f_{\bar{m}\bar{n},y}+f_{\tilde{m}\tilde{n},y}f_{\hat{m}\hat{n},y}f_{\bar{m}\bar{n},yy}\\
& +\nu[2f_{\tilde{m}\tilde{n},y}f_{\hat{m}\hat{n},xy}f_{\bar{m}\bar{n},x}+f_{\tilde{m}\tilde{n},y}f_{\hat{m}\hat{n},y}f_{\bar{m}\bar{n},xx}\\
& +2f_{\tilde{m}\tilde{n},x}f_{\hat{m}\hat{n},xy}f_{\bar{m}\bar{n},y}+f_{\tilde{m}\tilde{n},x}f_{\hat{m}\hat{n},x}f_{\bar{m}\bar{n},yy}]\\
& \left.+(1-\nu)[(f_{\tilde{m}\tilde{n},x}f_{\hat{m}\hat{n},y}f_{\bar{m}\bar{n},y})_{,x}+(f_{\tilde{m}\tilde{n},x}f_{\hat{m}\hat{n},y}f_{\bar{m}\bar{n},x})_{,y}]\}\right]f_{mn}\mathrm{d}x\mathrm{d}y\\
& =\int_0^{l_x}\int_0^{l_y}\frac{\bar{p}}{D_0}f_{mn}\mathrm{d}x\mathrm{d}y
\end{aligned}
\tag{11.24}
$$

Solving Eq. (11.24) for $\bar{w}_{\bar{m}\bar{n}}$ and applying the result to Eq. (11.19), the displacements $\bar{w}$ due to live loads are obtained. The stress couples $\bar{M}_x$, $\bar{M}_y$, and $\bar{M}_{xy}$ produced by live loads are obtained from the well-known relations $\bar{M}_x=-D_0(\bar{w}_{,xx}+\nu\bar{w}_{,yy})$, $\bar{M}_y=-D_0(\bar{w}_{,yy}+\nu\bar{w}_{,xx})$, and $\bar{M}_{xy}=-(1-\nu)D_0\bar{w}_{,xy}$.

11.4 Numerical Results to Static Live Loads

We examine numerically the effect of dead loads on simply supported and clamped plates. In the numerical computations, we use the following data. Two types of rectangular plates are considered: a reinforced concrete plate and a steel plate. The steel plate is considered an equivalent plate obtained from a steel structure. The dead loads $\tilde{p}$ and live loads $\bar{p}$ are uniformly distributed loads, in which $\bar{p}$ =5.88 kN/m^2 being a snow depth of 2 m with a snow density of 2.94 kN/m^3. Young's modulus E and Poisson's ratio ν are E=2.1 × 10^{11} N/m^2 and ν=0.3 for the steel plate and E=2.1 × 10^{10} N/m^2 and ν=0.17 for the reinforced concrete plate. The standard thicknesses h_0^* of the plates are 0.07 and 0.14 m for simply supported steel and reinforced concrete plates, respectively, and 0.05 and 0.1 m for the clamped steel and reinforced concrete plates, respectively. In addition, the standard length of the plate, l_x, in the x-direction is 5 m. The length of the plate, l_y, in the y-direction and the thickness h_0 are given by

$$
\left.\begin{aligned}
l_y&=\alpha_l l_x\ \ (\alpha_l=1\text{ and }2)\\
h_0&=\alpha_h h_0^*\ (\alpha_h=1\text{ and }2)
\end{aligned}\right\}
\tag{11.25}
$$

where the span ratio α_l and the thickness ratio α_h are parameters having values of 1 and 2. Numerical calculations are obtained by varying the dead load $\tilde{p}$ and two parameters α_l and α_h under a constant live load $\bar{p}$.

Figures 11.2 and 11.3 show the numerical results for rectangular plates with the variation of the thickness h_0 and dead loads $\tilde{p}$ under a constant live load $\bar{p}$. Figures 11.4 and 11.5 represent the numerical results for the plates with the variation of the span length l_y and dead loads $\tilde{p}$ under a constant live load $\bar{p}$. In these figures, the vertical axis shows the ratios $\bar{w}_{\max}/\bar{w}_{0\max}$ and $\bar{M}_{\max}/\bar{M}_{0\max}$, and the horizontal axis live-load-to-dead-load ratio $\bar{p}/\tilde{p}$, where $\bar{w}_{\max}$ and $\bar{M}_{\max}$ indicate the respective maximum values of deflections $\bar{w}$ and stress couples $\bar{M}$ due to the live load $\bar{p}$ in each plate, respectively, and are obtained from the advanced governing equation including the effect of dead loads, Eq. (11.24). On the other hand, $\bar{w}_{0\max}$ and $\bar{M}_{0\max}$ are the maximum values corresponding to $\bar{w}_{\max}$ and $\bar{M}_{\max}$, respectively, and are obtained from the governing equation excluding the effect of dead loads, Eq. (11.12), in which $\bar{w}_0$ is easily obtained by changing $\tilde{w}$ and $\tilde{p}$ in Eq. (11.12) with $\bar{w}_0$ and $\bar{p}$, respectively. In these figures, when the value of the vertical axis is less than 1, it represents that the effect of the dead loads exists. If the effect of dead loads does not exist, the values of these ratios are always equal to 1. When dead loads $\tilde{p}$ are large in comparison with the constant live loads $\bar{p}$, i.e., $\bar{p}/\tilde{p} < 1$, the ratios $\bar{w}_{\max}/\bar{w}_{0\max}$ and $\bar{M}_{\max}/\bar{M}_{0\max}$ are smaller than 1. There is little difference between the ratios $\bar{w}_{\max}/\bar{w}_{0\max}$ and $\bar{M}_{\max}/\bar{M}_{0\max}$ in the numerical results, but this difference can be neglected in practice. This fact is also explained in the subsequent approximate solution.

The scale of horizontal axis in Fig. 11.4 differs from Figs. 11.2, 11.3, and 11.5, because the simply supported plate is greatly affected by the dimensions of the plate. Arranging the reductions from the ratios $\bar{w}_{\max}/\bar{w}_{0\max}$ and $\bar{M}_{\max}/\bar{M}_{0\max}$ obtained

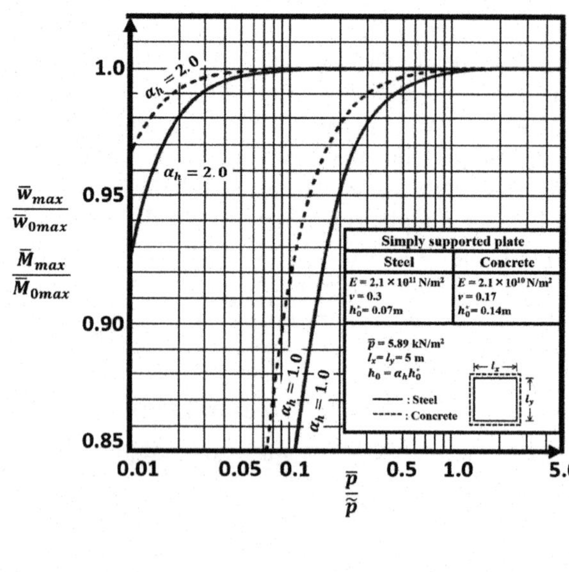

Fig. 11.2 Relationship between $\frac{\bar{w}_{\max}}{\bar{w}_{0\max}}$, $\frac{\bar{M}_{\max}}{\bar{M}_{0\max}}$ and $\frac{\bar{p}}{\tilde{p}}$ for a simply supported plate [24]

Fig. 11.3 Relationship between $\frac{\bar{w}_{\max}}{\bar{w}_{0\max}}$, $\frac{\bar{M}_{\max}}{\bar{M}_{0\max}}$ and $\frac{\bar{p}}{\tilde{p}}$ for a clamped plate [24]

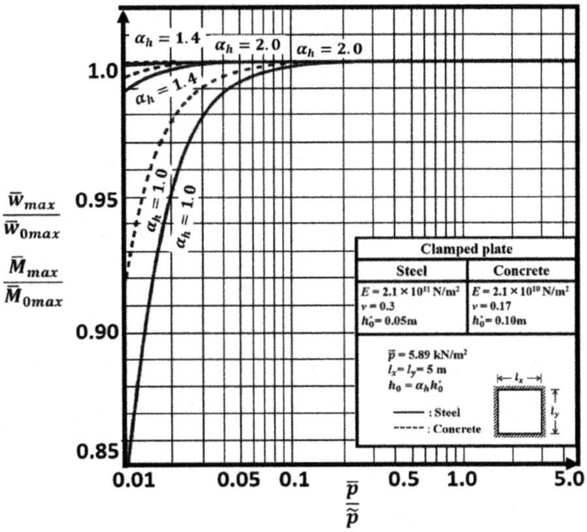

Fig. 11.4 Relationship between $\frac{\bar{w}_{\max}}{\bar{w}_{0\max}}$, $\frac{\bar{M}_{\max}}{\bar{M}_{0\max}}$ and $\frac{\bar{p}}{\tilde{p}}$ for a simply supported plate [24]

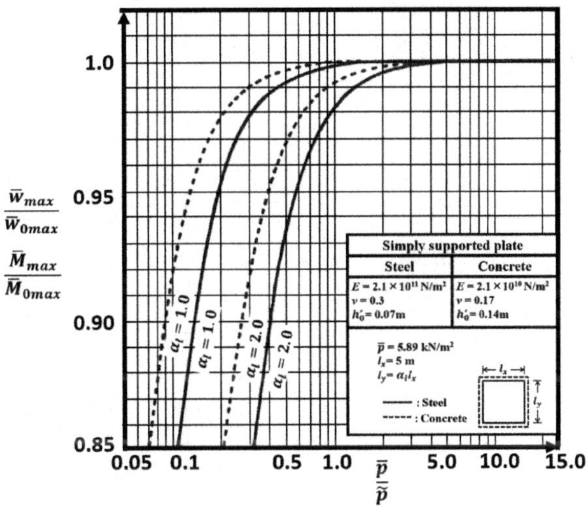

in Figs. 11.2 through 11.4, the effect of dead loads on simply supported plates is obtained as shown in Fig. 11.6.

These results demonstrate that an increase in dead load under a constant live load decreases the deflections and stress couples due to the live load and that this effect is large for a thin plate or a plate with a large span length. Thus, the effect of dead loads in plates is evident.

The above results represent that the effect of dead loads is large for a plate designed to have the lowest sectional efficiency. A plate designed with the lowest thickness essentially has a lower load-carrying capacity than a thick plate, and in this case, the

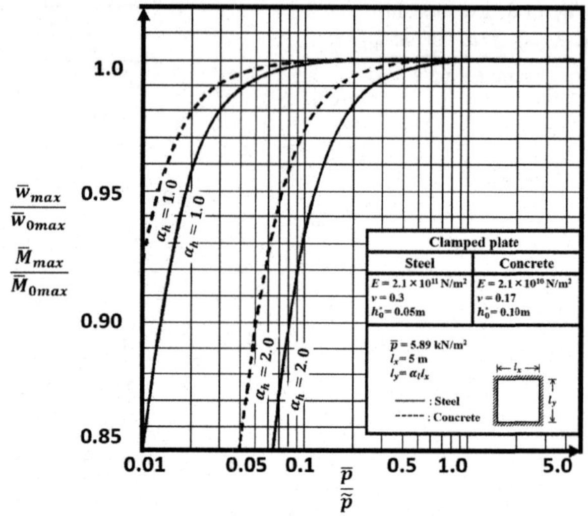

Fig. 11.5 Relationship between $\frac{\bar{w}_{\max}}{\bar{w}_{0\max}}$, $\frac{\bar{M}_{\max}}{\bar{M}_{0\max}}$ and $\frac{\bar{p}}{\tilde{p}}$ for a clamped plate [24]

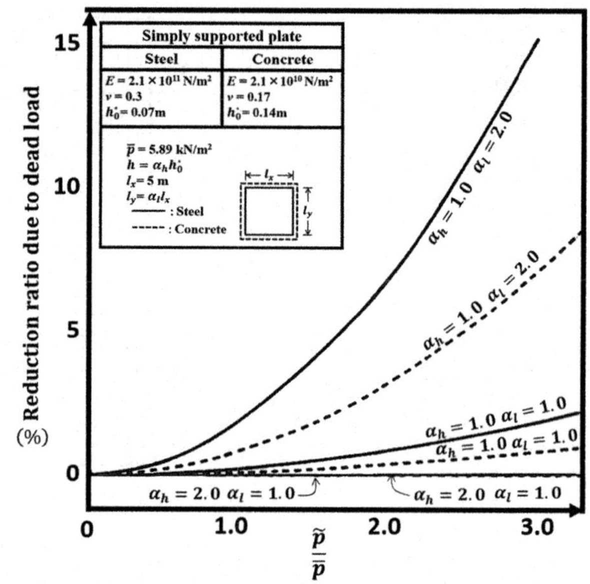

Fig. 11.6 Relationship between reduction ratio due to a dead load and $\tilde{p}/\bar{p}$ for a simply supported plate [24]

dead load is reduced as much as possible. Since this reduction in dead loads results in a reduction in the effect of dead loads, the plate becomes weaker for live loads. Present trends in structural design are based on a system in which structures are designed in terms of the sum of dead loads and live loads. Since live loads acting on structures are generally stipulated in the structural design code and are unchangeable in structural design, structural engineers are concerned about reducing the dead load. As a result, almost all new structures are being designed as light structures. Although the same degree of safety is desired for all structures, lightweight structures are deficient in terms of safety because the effect of dead loads in lightweight structures is smaller than that in heavy structures. In order to reflect this effect in structural design, it is necessary to reflect the reduction effect due to the existence of the dead loads on the live load, because the magnitude of live load is predetermined based on the intended used of the structure and independents of the existing dead loads. This opinion does not suggest increasing the dead load of structures. The author proposes that rather, in order to avoid the collapse of the plate due to live loads, the safety factor for lightweight structures should be increased to coincide with the safety factor for heavy structures.

11.5 Approximate Solution for Static Problems

In the preceding section, the effect of dead loads on static live loads has been demonstrated numerically using the Galerkin method. In order to clarify parameters governing the effect of dead loads, we are necessary to solve analytically the governing equation including the effect of dead loads for the static problems. However, because of the difficulty in solving this equation analytically, a closed-form solution is proposed by making some assumptions. Simply supported rectangular uniform plates are considered, because based on the preceding numerical results the effect of dead loads is remarkable in simply supported plates.

From Eq. (11.17), the current equilibrium equation, including the effect of dead loads, can be rewritten as follows:

$$\bar{w}_{,xxxx} + 2\bar{w}_{,xxyy} + \bar{w}_{,yyyy} = \frac{1}{D_0}(\bar{p} - D_0 R) \tag{11.26}$$

where R includes the effect of dead loads and is a function of $\tilde{w}\,(x,\ y)$ and $\bar{w}\,(x,\ y)$, as defined by

$$\begin{aligned} R = &-\frac{6}{h_0^2}[2\tilde{w}_{,x}\tilde{w}_{,xx}\bar{w}_{,x} + (\tilde{w}_{,x})^2\bar{w}_{,xx} + 2\tilde{w}_{,y}\tilde{w}_{,yy}\bar{w}_{,yy} + (\tilde{w}_{,y})^2\bar{w}_{,yy}] \\ &- \frac{6\nu}{h_0^2}[2\tilde{w}_{,y}\tilde{w}_{,xy}\bar{w}_{,x} + (\tilde{w}_{,y})^2\bar{w}_{,xx} + 2\tilde{w}_{,x}\tilde{w}_{,xy}\bar{w}_{,y} + (\tilde{w}_{,x})^2\bar{w}_{,yy}] \\ &- \frac{6(1-\nu)}{h_0^2}[(\tilde{w}_{,x}\tilde{w}_{,y}\bar{w}_{,y})_{,x} + (\tilde{w}_{,x}\tilde{w}_{,y}\bar{w}_{,x})_{,y}] \end{aligned} \tag{11.27}$$

On the other hand, the general governing equation excluding the effect of dead loads can be written as follows:

$$\bar{w}_{0,xxxx} + 2\bar{w}_{0,xxyy} + \bar{w}_{0,yyyy} = \frac{\bar{p}}{D_0} \tag{11.28}$$

where $\bar{w}_0$ represents the displacements due to live loads excluding the effect of dead loads. The subscript, 0, of the variable means to exclude the effect of dead loads. Here, the displacements $\bar{w}_0$ for a simply supported plate are obtained using Navier's solution as follows:

$$\bar{w}_0 = \sum_{m=1} \sum_{n=1} \bar{w}_{0mn} \, f_{mn} \tag{11.29}$$

where f_{mn} is given in Eq. (11.20) and $\bar{w}_{0mn}$ is obtained as follows:

$$\bar{w}_{0mn} = \frac{\bar{p}_{mn}}{D_0 \pi^4 \left[\left(\frac{m}{l_x} \right)^2 + \left(\frac{n}{l_y} \right)^2 \right]^2} \tag{11.30}$$

where the coefficients $\bar{p}_{mn}$ of the double Fourier expansion of live loads $\bar{p}$ are determined from Szilard [31]

Equation (11.26) cannot be solved analytically because R is a function of known displacements $\tilde{w}$ and unknown displacements $\bar{w}$. So, we assume that the unknown displacements $\bar{w}$ in R given by Eq. (11.27) are related to the displacements $\bar{w}_0$ excluding the effect of dead loads as follows:

$$\bar{w}(x, y) = \beta(x, y) \, \bar{w}_0(x, y) \tag{11.31}$$

where $\beta\,(x, y)$ is a coefficient indicating the effect of dead loads and is a function of x and y. It had been clarified from the numerical results in the preceding section that, for practical load ratios $\bar{p}/\tilde{p}$, the effect of dead loads has a reduction ratio $\bar{w}_{\max}/\bar{w}_{0\max}$ of between 1 and 0.85. $\beta\,(x, y)$ always takes a value of less than 1 and has an influence on the accuracy of the approximate solution proposed here. Then, based on both the preceding numerical results and the trial-and-error results for the proposed solution, the coefficient $\beta\,(x,\ y)$ for uniform live loads is assumed to be

$$\beta(x, y) = \beta_0 \sin\left(\frac{3\pi x}{l_x}\right) \sin\left(\frac{3\pi y}{l_y}\right) \tag{11.32}$$

where β_0 is the effect of dead loads on the displacements $\bar{w}$ at the midpoint ($x = l_x/2$ and $y = l_y/2$) of the plate and is assumed to be approximately 1. Hence, $\beta\,(x,\ y)$ becomes known function, so that $\bar{w}$ in R may be translated into known displacement $\bar{w}_0$. Thus, $R\,(\bar{w},\ \tilde{w})$, which is a function of both unknown displacements $\bar{w}$ and known displacement $\tilde{w}$, may be changed to $R\,(\bar{w}_0,\ \tilde{w})$, which is a function of all known displacements $\bar{w}_0$ and $\tilde{w}$. Consequently, R reduces to a function composed of all known displacements. Therefore, Eq. (11.26) is reduced to an analytically solvable form. Then, the right-hand side of Eq. (11.26) is redefined as

$$\bar{P} = \bar{p} - D_0 R \tag{11.33}$$

In order to find out parameters governing the effect of dead loads in nondimensioned expression, the quantities $\bar{P}$, $\bar{p}$, and R are expanded into a double Fourier series as follows:

$$\bar{P} = \sum_{m=1}\sum_{n=1} \bar{P}_{mn} f_{mn} \tag{11.34}$$

$$\bar{p} = \sum_{m=1}\sum_{n=1} \bar{p}_{mn} f_{mn} \tag{11.35}$$

$$R = \sum_{m=1}\sum_{n=1} R_{mn} f_{mn} \tag{11.36}$$

Substituting Eqs. (11.34) to (11.36) into Eq. (11.33), the Fourier coefficients $\bar{P}_{mn}$ can be written as

$$\bar{P}_{mn} = \bar{p}_{mn}\left[1 - \frac{D_0 R_{mn}}{\bar{p}_{mn}}\right] \tag{11.37}$$

The second term in the right-hand side of the above equation represents a reduction in the Fourier coefficients $\bar{p}_{mn}$ of live loads $\bar{p}$ due to the existence of dead loads. If the coefficients R_{mn} are known, then the coefficients $\bar{P}_{mn}$ become known and the displacements $\bar{w}_{mn}$ can be easily obtained by replacing $\bar{p}_{mn}$ in Eq. (11.30) with $\bar{P}_{mn}$.

The Fourier coefficients R_{mn} can be given by

$$R_{mn} = \frac{4}{l_x l_y} \int_0^{l_x} \int_0^{l_y} R f_{mn} \mathrm{d}x\mathrm{d}y \tag{11.38}$$

where R is a function of the known displacements $\bar{w}_0$ and $\tilde{w}$. Then, expressing the ratio of dead loads $\tilde{p}$ to live loads $\bar{p}$ as

$$\alpha = \frac{\bar{p}}{\tilde{p}} \tag{11.39}$$

the displacements $\tilde{w}$ produced by uniform dead loads $\tilde{p}$ become $1/\alpha$ times the displacements $\bar{w}_0$ produced by uniform live loads $\bar{p}$. Substituting the span ratio α_l ($=l_y/l_x$) and the expression

$$\bar{p}_{mn} = \bar{p}\bar{p}^{*}_{mn} \tag{11.40}$$

in Eq. (11.30), the coefficients $\bar{w}_{0mn}$ can be expressed in a nondimensional form as follows:

$$\bar{w}_{0mn} = \frac{\bar{p}(l_x)^4}{D_0}\bar{w}^{*}_{0mn} \tag{11.41}$$

where the superscript * denotes a nondimensioned value, and $\bar{w}^*_{0mn}$ and $\bar{p}^*_{mn}$ are defined as follows:

$$\bar{w}^*_{0mn} = \frac{\bar{p}^*_{mn}}{\pi^4\left[m^2 + \left(\frac{n}{\alpha_l}\right)^2\right]^2} \tag{11.42}$$

$$\bar{p}^*_{mn} = \frac{16}{\pi^2 mn} \quad (\text{for}\, m, n = 1, 3, 5, \ldots) \tag{11.43}$$

Applying these relations into R in Eq. (11.38), the Fourier coefficients R_{mn} can be written as

$$R_{mn} = \frac{\beta_0 (l_x)^8 \bar{p}^3}{h_0^2 D_0^3 \alpha^2} g_{mn} \tag{11.44}$$

where g_{mn} is a nondimensioned quantity, as given by Eq. (11.111) in Appendix, and depends only on the span ratio $\alpha_l = l_y / l_x$.

Since the Fourier coefficients R_{mn} have been given, the coefficients $\bar{P}_{mn}$ are known. Hence, the displacements $\bar{w}$ due to live loads, taking into consideration the effect of dead loads, yield

$$\bar{w} = \sum_{m=1} \sum_{n=1} \bar{w}_{mn} f_{mn} \tag{11.45}$$

where $\bar{w}_{mn}$ are

$$\bar{w}_{mn} = \frac{(l_x)^4 \bar{P}_{mn}}{D_0 \pi^4 \left[m^2 + \left(\frac{n}{\alpha_l}\right)^2\right]^2} \tag{11.46}$$

Substituting Eqs. (11.37) and (11.44) into Eq. (11.46), the displacements $\bar{w}$ can be obtained as

$$\bar{w} = [1 - k\alpha_w] \bar{w}^*_0 \frac{\bar{p}(l_x)^4}{D_0} \tag{11.47}$$

where k and α_w are defined as

$$k = \beta_0 \left[\frac{(l_x)^4 \tilde{p}}{h_0 D_0}\right]^2 \tag{11.48}$$

$$\alpha_w = \frac{\bar{w}^*_1}{\bar{w}^*_0} \tag{11.49}$$

and

$$\bar{w}_0^* = \sum_{m=1} \sum_{n=1} \bar{w}_{0mn}^* f_{mn}; \bar{w}_1^* = \sum_{m=1} \sum_{n=1} \bar{w}_{1mn}^* f_{mn} \tag{11.50}$$

Here, $\bar{w}^*_{1mn}$ is defined as

$$\bar{w}_{1mn}^* = \frac{g_{mn}}{\pi^4 \left[m^2 + \left(\frac{n}{\alpha_l} \right)^2 \right]^2} \tag{11.51}$$

The nondimensioned parameter α_w depends on only the span ratio α_l and can be evaluated beforehand.

Similarly, the stress couples $\bar{M}_x$, $\bar{M}_y$, and $\bar{M}_{xy}$ due to live loads can be given as

$$M_x = [1 - k\alpha_{Mx}] \bar{M}_{x0}^* \bar{p} l_x^2 \tag{11.52}$$

$$\bar{M}_y = [1 - k\alpha_{My}] \bar{M}_{y0}^* \bar{p} l_x^2 \tag{11.53}$$

$$\bar{M}_{xy} = [1 - k\alpha_{Mxy}] \bar{M}_{xy0}^* \bar{p} l_x^2 \tag{11.54}$$

where

$$\alpha_{Mx} = \frac{\bar{M}_{x1}^*}{\bar{M}_{x0}^*}; \quad \alpha_{My} = \frac{\bar{M}_{y1}^*}{\bar{M}_{y0}^*}; \quad \alpha_{Mxy} = \frac{\bar{M}_{xy1}^*}{\bar{M}_{xy0}^*} \tag{11.55}$$

and

$$\left.\begin{aligned}
\bar{M}_{x0}^* &= \sum_{m=1} \sum_{n=1} \left[(m\pi)^2 + \nu \left(\frac{n\pi}{\alpha_l} \right)^2 \right] \bar{w}_{0mn}^* f_{mn} \\
\bar{M}_{x1}^* &= \sum_{m=1} \sum_{n=1} \left[(m\pi)^2 + \nu \left(\frac{n\pi}{\alpha_l} \right)^2 \right] \bar{w}_{1mn}^* f_{mn} \\
\bar{M}_{y0}^* &= \sum_{m=1} \sum_{n=1} \left[\left(\frac{n\pi}{\alpha_l} \right)^2 + \nu (m\pi)^2 \right] \bar{w}_{0mn}^* f_{mn} \\
\bar{M}_{y1}^* &= \sum_{m=1} \sum_{n=1} \left[\left(\frac{n\pi}{\alpha_l} \right)^2 + \nu (m\pi)^2 \right] \bar{w}_{1mn}^* f_{mn} \\
\bar{M}_{xy0}^* &= -(1-\nu) \sum_{m=1} \sum_{n=1} \frac{mn\pi^2}{\alpha_l} \bar{w}_{0mn}^* \cos\left(\frac{m\pi x}{l_x} \right) \cos\left(\frac{n\pi y}{l_y} \right) \\
\bar{M}_{xy1}^* &= -(1-\nu) \sum_{m=1} \sum_{n=1} \frac{mn\pi^2}{\alpha_l} \bar{w}_{1mn}^* \cos\left(\frac{m\pi x}{l_x} \right) \cos\left(\frac{n\pi y}{l_y} \right)
\end{aligned}\right\} \tag{11.56}$$

Similarly, the transverse shearing forces $\bar{Q}_x$ and $\bar{Q}_y$ due to live loads can be obtained as

$$\bar{Q}_x = [1 - k\alpha_{Qx}] \bar{Q}_{x0}^* \bar{p} l_x; \bar{Q}_y = [1 - k\alpha_{Qy}] \bar{Q}_{y0}^* \bar{p} l_x \tag{11.57}$$

where

$$\alpha_{Qx} = \frac{\bar{Q}^*_{x1}}{\bar{Q}^*_{x0}}; \quad \alpha_{Qy} = \frac{\bar{Q}^*_{y1}}{\bar{Q}^*_{y0}} \tag{11.58}$$

and

$$\left.\begin{aligned}
\bar{Q}^*_{x0} &= \sum_{m=1}\sum_{n=1}\left[(m\pi)^3 + m\pi\left(\tfrac{n\pi}{\alpha_l}\right)^2\right]\bar{w}^*_{0mn}\cos\left(\tfrac{m\pi x}{l_x}\right)\sin\left(\tfrac{n\pi y}{l_y}\right)\\
\bar{Q}^*_{x1} &= \sum_{m=1}\sum_{n=1}\left[(m\pi)^3 + m\pi\left(\tfrac{n\pi}{\alpha_l}\right)^2\right]\bar{w}^*_{1mn}\cos\left(\tfrac{m\pi x}{l_x}\right)\sin\left(\tfrac{n\pi y}{l_y}\right)\\
\bar{Q}^*_{y0} &= \sum_{m=1}\sum_{n=1}\left[(m\pi)^2\left(\tfrac{n\pi}{\alpha_l}\right) + \left(\tfrac{n\pi}{\alpha_l}\right)^3\right]\bar{w}^*_{0mn}\sin\left(\tfrac{m\pi x}{l_x}\right)\cos\left(\tfrac{n\pi y}{l_y}\right)\\
\bar{Q}^*_{y1} &= \sum_{m=1}\sum_{n=1}\left[(m\pi)^2\left(\tfrac{n\pi}{\alpha_l}\right) + \left(\tfrac{n\pi}{\alpha_l}\right)^3\right]\bar{w}^*_{1mn}\sin\left(\tfrac{m\pi x}{l_x}\right)\cos\left(\tfrac{n\pi y}{l_y}\right)
\end{aligned}\right\} \tag{11.59}$$

In the plate theory, the deflection $\bar{w}_0$, stress couples $\bar{M}_{x0}$, $\bar{M}_{y0}$, and $\bar{M}_{xy0}$, and transverse shearing forces $\bar{Q}_{x0}$ and $\bar{Q}_{y0}$, which exclude the reduction effect of dead loads, are related as follows:

$$\left.\begin{aligned}
\bar{w}_0 &= \bar{w}^*_0\tfrac{\bar{p}l_x^4}{D_0}\\
\bar{M}_{x0} = \bar{M}^*_{x0}\bar{p}l_x^2;\ \bar{M}_{y0} &= \bar{M}^*_{y0}\bar{p}l_x^2;\ \bar{M}_{xy0} = \bar{M}^*_{xy0}\bar{p}l_x^2\\
\bar{Q}_{x0} = \bar{Q}^*_{x0}\bar{p}l_x;\ \bar{Q}_{y0} &= \bar{Q}^*_{y0}\bar{p}l_x;\ \bar{Q}_{xy0} = \bar{Q}^*_{xy0}\bar{p}l_x
\end{aligned}\right\} \tag{11.60}$$

Thus, applying Eq. (11.60) to Eqs. (11.47), (11.52) through (11.54), and (11.57), the reduction ratios due to the dead loads $\tilde{p}$ on the action of live loads $\bar{p}$ are expressed as follows:

$$\frac{\bar{w}}{\bar{w}_0} = 1 - k\alpha_w \tag{11.61}$$

$$\frac{\bar{M}_x}{\bar{M}_{x0}} = 1 - k\alpha_{Mx} \tag{11.62}$$

$$\frac{\bar{M}_y}{\bar{M}_{y0}} = 1 - k\alpha_{My} \tag{11.63}$$

$$\frac{\bar{M}_{xy}}{\bar{M}_{xy0}} = 1 - k\alpha_{Mxy} \tag{11.64}$$

$$\frac{\bar{Q}_x}{\bar{Q}_{x0}} = 1 - k\alpha_{Qx} \tag{11.65}$$

Table 11.1 Values of α_w, α_{Mx}, α_{My}, α_{Mxy}, α_{Qx}, and α_{Qy} ($v = 0.3$)

$\alpha_l = l_y/l_x$	α_w	α_{Mx}	α_{My}	α_{Mxy}	α_{Qx}	α_{Qy}
1.0	0.280×10^{-4}	0.295×10^{-4}	0.295×10^{-4}	0.263×10^{-4}	0.538×10^{-4}	0.538×10^{-4}
1.1	0.403×10^{-4}	0.445×10^{-4}	0.422×10^{-4}	0.376×10^{-4}	0.846×10^{-4}	0.703×10^{-4}
1.2	0.547×10^{-4}	0.624×10^{-4}	0.618×10^{-4}	0.499×10^{-4}	0.123×10^{-3}	0.864×10^{-4}
1.3	0.708×10^{-4}	0.829×10^{-4}	0.861×10^{-4}	0.624×10^{-4}	0.170×10^{-3}	0.101×10^{-3}
1.4	0.881×10^{-4}	0.105×10^{-3}	0.115×10^{-3}	0.742×10^{-4}	0.222×10^{-3}	0.114×10^{-3}
1.5	0.106×10^{-3}	0.130×10^{-3}	0.148×10^{-3}	0.867×10^{-4}	0.280×10^{-3}	0.125×10^{-3}
1.6	0.125×10^{-3}	0.155×10^{-3}	0.185×10^{-3}	0.105×10^{-3}	0.341×10^{-3}	0.155×10^{-3}
1.7	0.144×10^{-3}	0.180×10^{-3}	0.226×10^{-3}	0.124×10^{-3}	0.405×10^{-3}	0.188×10^{-3}
1.8	0.163×10^{-3}	0.206×10^{-3}	0.270×10^{-3}	0.144×10^{-3}	0.471×10^{-3}	0.221×10^{-3}
1.9	0.183×10^{-3}	0.232×10^{-3}	0.315×10^{-3}	0.165×10^{-3}	0.537×10^{-3}	0.255×10^{-3}
2.0	0.202×10^{-3}	0.258×10^{-3}	0.363×10^{-3}	0.186×10^{-3}	0.602×10^{-3}	0.289×10^{-3}
3.0	0.379×10^{-3}	0.475×10^{-3}	0.763×10^{-3}	0.436×10^{-3}	0.114×10^{-2}	0.570×10^{-3}
4.0	0.541×10^{-3}	0.626×10^{-3}	0.988×10^{-3}	0.669×10^{-3}	0.145×10^{-2}	0.724×10^{-3}
5.0	0.690×10^{-3}	0.736×10^{-3}	0.110×10^{-2}	0.834×10^{-3}	0.163×10^{-2}	0.799×10^{-3}

$$\frac{\bar{Q}_y}{\bar{Q}_{y0}} = 1 - k\alpha_{Qy} \tag{11.66}$$

Thus, considering of the effect of dead loads, it is clarified that the reductions in the displacements, stress couples, and transverse shear forces produced by live loads always depend on k in the second terms of the right-hand sides of the above equations. Then, since the nondimensioned coefficients α_w, α_{Mx}, α_{My}, α_{Mxy}, α_{Qx}, and α_{Qy} depend only on the span ratio $\alpha_l(=l_y/l_x)$, they can be given by Tables 11.1 and 11.2 for $v = 0.3$ and 0.17, respectively. The effect of dead loads is obtained from the product of k and the values given in Tables 11.1 and 11.2. The product is small, as is the effect itself. Consequently, the slight variation in the values shown in Tables 11.1 and 11.2 may be disregarded in practical use. This explains the phenomenon, whereby there is little difference between the displacements and stress couples due to the effect of dead loads, as indicated by the preceding numerical results.

Table 11.2 Values of α_w, α_{Mx}, α_{My}, α_{Mxy}, α_{Qx}, and α_{Qy} ($\nu = 0.17$)

$\alpha_l = l_y/l_x$	α_w	α_{Mx}	α_{My}	α_{Mxy}	α_{Qx}	α_{Qy}
1.0	0.280×10^{-4}	0.299×10^{-4}	0.299×10^{-4}	0.263×10^{-4}	0.538×10^{-4}	0.538×10^{-4}
1.1	0.403×10^{-4}	0.448×10^{-4}	0.418×10^{-4}	0.376×10^{-4}	0.846×10^{-4}	0.703×10^{-4}
1.2	0.547×10^{-4}	0.625×10^{-4}	0.617×10^{-4}	0.499×10^{-4}	0.123×10^{-3}	0.864×10^{-4}
1.3	0.708×10^{-4}	0.826×10^{-4}	0.868×10^{-4}	0.624×10^{-4}	0.170×10^{-3}	0.101×10^{-3}
1.4	0.881×10^{-4}	0.104×10^{-3}	0.117×10^{-3}	0.742×10^{-4}	0.222×10^{-3}	0.114×10^{-3}
1.5	0.106×10^{-3}	0.128×10^{-3}	0.153×10^{-3}	0.867×10^{-4}	0.280×10^{-3}	0.125×10^{-3}
1.6	0.125×10^{-3}	0.152×10^{-3}	0.195×10^{-3}	0.105×10^{-3}	0.341×10^{-3}	0.155×10^{-3}
1.7	0.144×10^{-3}	0.177×10^{-3}	0.243×10^{-3}	0.124×10^{-3}	0.405×10^{-3}	0.188×10^{-3}
1.8	0.163×10^{-3}	0.201×10^{-3}	0.298×10^{-3}	0.144×10^{-3}	0.471×10^{-3}	0.221×10^{-3}
1.9	0.183×10^{-3}	0.226×10^{-3}	0.349×10^{-3}	0.165×10^{-3}	0.537×10^{-3}	0.255×10^{-3}
2.0	0.202×10^{-3}	0.251×10^{-3}	0.397×10^{-3}	0.186×10^{-3}	0.602×10^{-3}	0.289×10^{-3}
3.0	0.379×10^{-3}	0.458×10^{-3}	0.845×10^{-3}	0.436×10^{-3}	0.114×10^{-2}	0.570×10^{-3}
4.0	0.541×10^{-3}	0.606×10^{-3}	0.114×10^{-2}	0.669×10^{-3}	0.145×10^{-2}	0.724×10^{-3}
5.0	0.690×10^{-3}	0.717×10^{-3}	0.130×10^{-2}	0.834×10^{-3}	0.163×10^{-2}	0.799×10^{-3}

Subsequently, the exactness of the approximate solution proposed here is examined by comparing the results of the proposed solution with the above-described numerical results obtained by the Galerkin method, in which the coefficient β_0 is assumed to be 1. All of the results showed excellent agreement, except in the case of $\alpha_l = 2$ in Fig. 11.4, for simply supported plates. However, this difference is negligible for practical use and shows error toward the side of safety. The exactness of the proposed solution has been proven. Thus, the effect of dead loads is governed by the value of k given in Eq. (11.48). Namely, the effect is proportional to the eighth power of the span length l_x and the second power of the dead load $\tilde{p}$ and inversely proportional to the second power of the thickness h and of the bending rigidity of the plate, D_0.

Then, the factor R obtained by substituting Eq. (11.44) into Eq. (11.36) decreases live loads $\bar{p}$, as estimated from the right-hand side of Eq. (11.26). Consequently, this reduction can be explained by the effects of dead loads for the displacements, stress couples, and transverse shear forces caused by live loads. Since heavy plates have a larger value of R than lightweight plates, the effect of dead loads is greater for heavy plates than for lightweight plates. As such, in order to achieve parity in safety, the safety factor for lightweight plates must be increased by considering the effect of dead loads.

Next, the reduction effect due to the dead loads on the total maximum deflections of simply supported plates is accounted for in the relation $(\bar{w} + \tilde{w})_{\max}/(\bar{w}_0 + \tilde{w})_{\max}$. From the relation $\bar{w}_0/\tilde{w} = \bar{p}/\tilde{p}$ (=α) and Eq. (11.47), we have

$$\frac{(\bar{w} + \tilde{w})_{\max}}{(\bar{w}_0 + \tilde{w})_{\max}} = 1 - \frac{\alpha k \alpha_w}{1 + \alpha} \tag{11.67}$$

Hence, the reduction effect of dead loads on the total maximum deflections of the plate subjected to the dead load $\tilde{p}$ and live load $\bar{p}$ is governed by the factors α (=$\bar{p}/\tilde{p}$) and k. Similar relationships for the stress couples $\bar{M}_x$, $\bar{M}_y$, and $\bar{M}_{xy}$, and the transverse shear forces $\bar{Q}_x$ and $\bar{Q}_y$ can be obtained by replacing α_w with α_{Mx}, α_{My}, α_{Mxy}, α_{Qx}, and α_{Qy} respectively.

11.6 How to Reflect the Effect of Dead Loads on Static Live Loads

We present how to apply the effect of dead loads to the modification of static live loads. We consider an equivalent steel plate with simply supported ends, span lengths $l_x = l_y$ =6 m, a thickness h_0 =0.05 m, Young's modulus $E = 2.1 \times 10^{11}$ N/m^2, and Poisson's ratio ν =0.3. The flexural rigidity D_0 =240 × 10^4 N m.

Step 1: Evaluate the value of k from Eq. (11.48), in which β_0 is assumed to be 1, to obtain

$$k = \left[\frac{(l_x)^4 \tilde{p}}{h_0 D_0}\right]^2 = 1 \times \left[\frac{6^4}{0.05 \times 240 \times 10^4}\right]^2 \tilde{p}^2 = 0.0001166\ \tilde{p}^2 \tag{11.68}$$

where $\tilde{p}$ (N/m^2) is a uniform dead load in the current plate. Also, the value of α_w for $\alpha_l = l_y/l_x$ =1 is 0.280 × 10^{-4} obtained from Table 11.1.

Step 2: From Eq. (11.61), the reduction ratio of the maximum deflections produced by live loads takes the following value:

$$\frac{\bar{w}}{\bar{w}_0} = 1 - k\alpha_w = 1 - 0.0001166\ \tilde{p}^2 \times 0.280 \times 10^{-4} = 1 - 3.264 \times 10^{-9} \tilde{p}^2 \tag{11.69}$$

Step 3: Since this reduction effect is influenced by the magnitude of dead loads $\tilde{p}$ (N/m^2), the action of live loads is reduced greater on heavy plates than on lightweight plates. The dead loads of safe plates are denoted by $\tilde{p}_0$. Although this value cannot be determined unconditionally, it will be decided from the results obtained from safe buildings in past designs and experimental results. The reduction ratios of the maximum deflections produced by live loads for safe plates are given by

$$\frac{\bar{w}}{\bar{w}_0} = 1 - 3.264 \times 10^{-9} \tilde{p}_0^2 \tag{11.70}$$

Step 4: Since the reduction effect due to dead loads also contributes to the safety of the structure, the magnitude of live loads acting on a plate with dead loads $\tilde{p}$ must increase corresponding to the reduction ratio. The incremental ratio for safety is given by

$$\text{incremental ratio for safety} = \frac{1 - 3.264 \times 10^{-9} \tilde{p}^2}{1 - 3.264 \times 10^{-9} \tilde{p}_0^2} \tag{11.71}$$

For example, assuming that $\tilde{p}_0 = 3922\ \mathrm{N/m^2}$ and $\tilde{p} = 196\ \mathrm{N/m^2}$, the live loads $\bar{p}$ acting on the current plate must be increased by 1.053 times from Eq. (11.71).
Step 5: The effect of dead loads on the total deflection of the current static plate can be obtained from Eq. (11.67) for arbitrary live loads.

Thus, it is clarified from Eq. (11.48) that the increase of static live loads in order to prevent the collapse of structures due to live loads (snow loads) depends on the span length, the bending rigidity, the thickness of the plates used, and the standard safety value $\tilde{p}_0$ for dead loads. This standard safety value of dead loads is determined experimentally or through the investigation of collapsed structures, as mentioned above. However, for practical use, it is recommended that dead loads of reinforced concrete plates are considered as the standard value of dead loads, rather than the dead loads used commonly for plates in steel structures. This method will have the same effect when used on live loads for steel plates and reinforced concrete plates, except for different treatments of the redundancy of structural materials used to ensure safety.

11.7 Transverse Free Vibration Based on the Galerkin Method

We examine the effect of dead loads for transverse free vibration of plates. The advanced equation of free vibrations including the effect of dead loads is obtained from Eq. (11.14)

$$\frac{m_0 \ddot{\bar{w}}}{D_0} + \nabla^4 \bar{w} + R(\bar{w}, \tilde{w}) = 0 \tag{11.72}$$

Employing the method of separation of variables, we have

$$\bar{w}(x, y, t) = \bar{W}(x, y)\Phi(t) \tag{11.73}$$

in which $\bar{W}(x, y)$ is a deflection function depending on x and y and $\Phi(t)$ is a deflection function depending on t. The term $R(\tilde{w}, \tilde{w})$ indicating the effect of dead loads is a function of x, y, and t, as given in Eq. (11.16). It can also be expressed as in Eq. (11.74) by the method of separation of variables, because $\tilde{w}(x, y)$ due to dead loads are independent of time t. Hence,

$$R(x, y, t) = \bar{R}(x, y)\Phi(t) \tag{11.74}$$

in which $\bar{R}(x, y)$ is a function of x and y which is defined as

$$\begin{aligned}\bar{R}(x, y) = &-\frac{6}{h_0^2}[2\tilde{w}_{,x}\tilde{w}_{,xx}\bar{W}_{,x} + (\tilde{w}_{,x})^2\bar{W}_{,xx} + 2\tilde{w}_{,y}\tilde{w}_{,yy}\bar{W}_{,y} + (\tilde{w}_{,y})^2\bar{W}_{,yy}] \\ &- \frac{6\upsilon}{h_0^2}[2\tilde{w}_{,y}\tilde{w}_{,xy}\bar{W}_{,x} + (\tilde{w}_{,y})^2\bar{W}_{,xx} + 2\tilde{w}_{,x}\tilde{w}_{,xy}\bar{W}_{,y} + (\tilde{w}_{,x})^2\bar{W}_{,yy}] \\ &- \frac{6(1-\upsilon)}{h_0^2}[(\tilde{w}_{,x}\tilde{w}_{,y}\bar{W}_{,y})_{,x} + (\tilde{w}_{,x}\tilde{w}_{,y}\bar{W}_{,x})_{,y}]\end{aligned} \tag{11.75}$$

Substituting Eqs. (11.73) and (11.74) into Eq. (11.72), we have the two equations for $\bar{W}(x, y)$ and $\Phi(t)$

$$\nabla^4\bar{W} - k^4\bar{W} + \bar{R}(x, y) = 0 \tag{11.76}$$

and

$$\ddot{\Phi} + \omega^2\Phi = 0 \tag{11.77}$$

in which ω^2 is a constant and k is defined as

$$k^2 = \omega\sqrt{\frac{m_0}{D_0}} \tag{11.78}$$

$\bar{W}(x, y)$ is expressed in power series expansions as follows:

$$\bar{W}(x, y) = \sum_{m=1}^{\infty}\sum_{n=1}^{\infty}\bar{W}_{mn}f_{mn}(x, y) \tag{11.79}$$

in which $\bar{W}_{mn}$ are unknown coefficients. Similarly, the static displacements $\tilde{w}(x, y)$ are also expressed in power series expansions as follows:

$$\tilde{w}(x, y) = \sum_{m=1}^{\infty}\sum_{n=1}^{\infty}\tilde{w}_{mn}f_{mn}(x, y) \tag{11.80}$$

in which the coefficients $\tilde{w}_{mn}$ are previously determined. Substituting Eqs. (11.79) and (11.80) into Eq. (11.76), the Galerkin equation becomes

$$\delta \bar{W}_{mn} : \sum_{\bar{m}=1} \sum_{\bar{n}=1} \bar{W}_{\bar{m}\bar{n}} [A_{mn\bar{m}\bar{n}} - k^4 B_{mn\bar{m}\bar{n}}] = 0 \tag{11.81}$$

in which the coefficients $A_{mn\bar{m}\bar{n}}$ and $B_{mn\bar{m}\bar{n}}$ are defined as

$$\begin{aligned} A_{mn\bar{m}\bar{n}} = \int_0^{l_x} \int_0^{l_y} \Big(& f_{\bar{m}\bar{n},xxxx} + 2 f_{\bar{m}\bar{n},xxyy} + f_{\bar{m}\bar{n},yyyy} \\ & - \sum_{\tilde{m}=1} \sum_{\tilde{n}=1} \sum_{\hat{m}=1} \sum_{\hat{n}=1} \tilde{w}_{\tilde{m}\tilde{n}} \tilde{w}_{\hat{m}\hat{n}} \frac{6}{h_0^2} \Big\{ 2 f_{\tilde{m}\tilde{n},x} f_{\hat{m}\hat{n},xx} f_{\bar{m}\bar{n},x} + f_{\tilde{m}\tilde{n},x} f_{\hat{m}\hat{n},x} f_{\bar{m}\bar{n},xx} \\ & + 2 f_{\tilde{m}\tilde{n},y} f_{\hat{m}\hat{n},yy} f_{\bar{m}\bar{n},y} + f_{\tilde{m}\tilde{n},y} f_{\hat{m}\hat{n},y} f_{\bar{m}\bar{n},yy} + \upsilon (2 f_{\tilde{m}\tilde{n},y} f_{\hat{m}\hat{n},xy} f_{\bar{m}\bar{n},y} \\ & + f_{\tilde{m}\tilde{n},y} f_{\hat{m}\hat{n},y} f_{\bar{m}\bar{n},xx} + 2 f_{\tilde{m}\tilde{n},x} f_{\hat{m}\hat{n},xy} f_{\bar{m}\bar{n},y} + f_{\tilde{m}\tilde{n},x} f_{\hat{m}\hat{n},x} f_{\bar{m}\bar{n},yy}) \\ & + (1 - \upsilon) [(f_{\tilde{m}\tilde{n},x} f_{\hat{m}\hat{n},y} f_{\bar{m}\bar{n},y})_{,x} + (f_{\tilde{m}\tilde{n},x} f_{\hat{m}\hat{n},y} f_{\bar{m}\bar{n},x})_{,y}] \Big\} \Big) f_{mn} \mathrm{d}x \, \mathrm{d}y \end{aligned} \tag{11.82}$$

$$B_{mn\bar{m}\bar{n}} = \int_0^{l_x} \int_0^{l_y} f_{\bar{m}\bar{n}} f_{mn} dx \, dy \tag{11.83}$$

Equations (11.81) are a system of linear, homogeneous, simultaneous algebraic equations with respect to the unknown displacement coefficients $\bar{W}_{\bar{m}\bar{n}}$. Solving Eq. (11.81) as eigenvalue problems for *k*, the *m*, *n*-th natural frequencies ω_{mn} are determined by employing the *m*, *n*-th value of *k* into Eq. (11.78).

11.8 Closed-Form Solution for Transverse Free Vibrations

In the preceding section, we have discussed to solve the transverse free vibration, which includes the effect of dead loads, by means of computational method based on the Galerkin method. For practical use, the closed-form solution is benefit instead of the computational method. So, we present the closed-form solution including the effect of dead loads in the transverse free vibration of simply supported rectangular plates.

When the effect of dead loads is small, the natural function $\bar{W}$ in $\bar{R}(x, y)$ given in Eq. (11.75) may be assumed as

$$\bar{W}(x, y) = \beta \bar{W}_0(x, y) \tag{11.84}$$

in which β is the ratio of reduction of the dynamic deflections produced by live loads, occurring from the effect of dead loads, and $\bar{W}_0$ is the natural function excluding the effect of dead loads, which is given by

$$\nabla^4 \bar{W}_0 - k^4 \bar{W}_0 = 0 \tag{11.85}$$

Assuming that the ratio β is a known constant, $\bar{R}(x, y)$ may change from a function $\bar{R}(\bar{w}, \tilde{w})$, which is composed of both the unknown displacement coefficient $\bar{W}$ and the known displacement $\tilde{w}$, to a function $\bar{R}(\bar{w}_0, \tilde{w})$, which is composed of all known displacements $\bar{W}_0$ and $\tilde{w}$. Then, $\bar{R}(x, y)$ can be expressed in a double Fourier series given by

$$\bar{R}(x, y) = \sum_{m=1} \sum_{n=1} \bar{R}_{mn} f_{mn}(x, y) \tag{11.86}$$

in which $\bar{R}_{mn}$ are Fourier coefficients and f_{mn} are natural functions given by

$$f_{mn} = \sin\left(\frac{m\pi x}{l_x}\right) \sin\left(\frac{n\pi y}{l_y}\right) \tag{11.87}$$

Assuming the following expression for $\bar{W}(x, y)$ in Eq. (11.76)

$$\bar{W}(x, y) = \sum_{m=1} \sum_{n=1} f_{mn} \tag{11.88}$$

and substituting Eqs. (11.86) and (11.88) into Eq. (11.76), k^4_{mn} for selected integers m and n can be obtained as

$$k^4_{mn} = k^4_{0mn} + \bar{R}_{mn} \tag{11.89}$$

in which k^4_{0mn}, which excludes the effect of dead loads, can be given by Eq. (11.85) as

$$k^4_{0mn} = \left[\left(\frac{m\pi}{l_x}\right)^2 + \left(\frac{n\pi}{l_y}\right)^2\right]^2 \tag{11.90}$$

Applying Eq. (11.89) into Eq. (11.78), the natural frequencies ω_{mn} including the effect of dead loads can be obtained by

$$\omega_{mn} = \omega_{0mn} \sqrt{1 + \frac{\bar{R}_{mn}}{k^4_{0mn}}} \tag{11.91}$$

in which ω_{0mn} are the well-known natural frequencies of simply supported plates excluding the effect of dead loads and are defined as

$$\omega_{0mn} = \left(\frac{\pi}{l_x}\right)^2 \left[m^2 + \left(\frac{n}{\alpha_l}\right)^2\right] \sqrt{\frac{D_0}{m_0}} \tag{11.92}$$

in which $\alpha_l = l_y/l_x$ is the aspect ratio.

On the other hand, the Fourier coefficients $\bar{R}_{mn}$ are given from Eq. (11.86) as

$$\bar{R}_{mn} = \frac{4}{l_x l_y} \int_0^{l_x} \int_0^{l_y} \bar{R}(x, y) f_{mn} \mathrm{d}x\,\mathrm{d}y \tag{11.93}$$

Static displacements $\tilde{w}$ included in $\bar{R}(x, y)$ can be written as

$$\tilde{w} = \sum_{m=1} \sum_{n=1} \tilde{w}_{mn} f_{mn} \tag{11.94}$$

for simply supported uniform plates, where the Fourier coefficients $\tilde{w}_{mn}$ for uniform dead loads $\tilde{p}$ are expressed as

$$\tilde{w}_{mn} = \tilde{w}^*_{mn} \frac{\tilde{p} l_x^4}{D_0} \tag{11.95}$$

in which $\tilde{w}^*_{mn}$ are the nondimensioned values of $\tilde{w}_{mn}$. $\tilde{w}^*_{mn}$ are given from Navier's double Fourier solution, for example in Szilard [30]. Then, substituting Eqs. (11.75), (11.84), (11.88), and (11.94) into Eq. (11.93), the coefficients $\bar{R}_{mn}$ can be obtained as

$$\bar{R}_{mn} = \beta \left(\frac{l_x^2 \tilde{p}}{h_0 D_0} \right)^2 g_{mn} \tag{11.96}$$

where g_{mn} are nondimensioned quantities depending on only the aspect span ratio α_l. Substituting Eq. (11.96) into (11.91), the approximate equation for the natural frequencies is given in the closed form

$$\omega_{mn} = \omega_{0mn} \sqrt{1 + k_\omega G_{mn}} \tag{11.97}$$

where the notation k_ω is

$$k_\omega = \beta \left[\frac{(l_x)^4 \tilde{p}}{h_0 D_0} \right]^2 \tag{11.98}$$

The nondimensioned coefficients G_{mn} are given in Table 11.3. The computational result shows that the values of G_{mn} are little influenced by the values of Poisson's ratios ($\nu = 0.3$ and 0.17 are used).

Natural frequencies ω_{mn} including the effect of dead loads have been presented in closed form. The result neglecting the second term in the root of Eq. (11.97) agrees with the well-known natural frequencies excluding the effect of dead loads. Subsequently, the effectiveness of Eq. (11.97) is examined by comparing with results obtained from Eq. (11.81), in which the ratio β is assumed to be 1 for simplicity. The results are in good agreements in all cases, as shown in Fig. 11.7. The difference

Table 11.3 Values of G_{mn} (v = 0.3 or v = 0.17) [23]

$\alpha_l = l_y/l_x$	G_{11} first mode	G_{12} second mode	G_{22} third mode
1.0	0.3311×10^{-4}	0.8331×10^{-5}	0.2644×10^{-5}
1.1	0.4826×10^{-4}	0.1166×10^{-4}	0.3873×10^{-5}
1.2	0.6746×10^{-4}	0.1566×10^{-4}	0.5482×10^{-5}
1.3	0.9085×10^{-4}	0.2043×10^{-4}	0.7510×10^{-5}
1.4	0.1183×10^{-3}	0.2617×10^{-4}	0.9970×10^{-5}
1.5	0.1493×10^{-3}	0.3304×10^{-4}	0.1285×10^{-4}
1.6	0.1835×10^{-3}	0.4122×10^{-4}	0.1612×10^{-4}
1.7	0.2200×10^{-3}	0.5085×10^{-4}	0.1975×10^{-4}
1.8	0.2584×10^{-3}	0.6203×10^{-4}	0.2367×10^{-4}
1.9	0.2979×10^{-3}	0.7480×10^{-4}	0.2785×10^{-4}
2.0	0.3380×10^{-3}	0.8919×10^{-4}	0.3224×10^{-4}
3.0	0.7064×10^{-3}	0.3038×10^{-3}	0.8035×10^{-4}
4.0	0.9695×10^{-3}	0.5690×10^{-3}	0.1248×10^{-3}
5.0	0.1154×10^{-2}	0.8123×10^{-3}	0.1609×10^{-3}

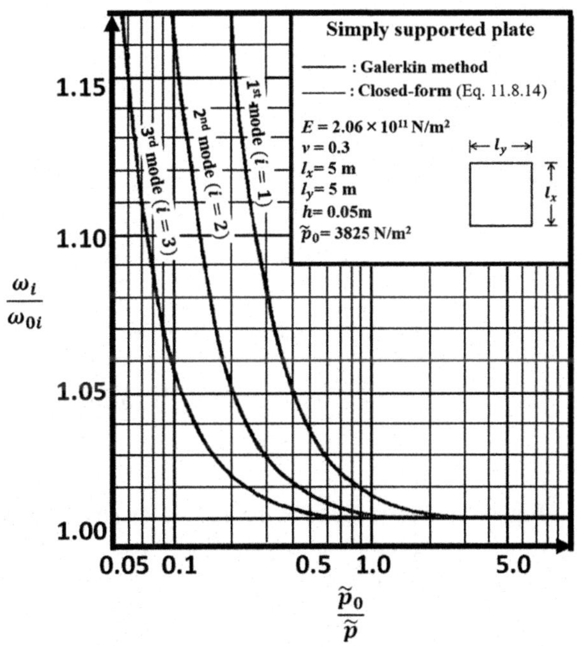

Fig. 11.7 Relationships between closed-form solution and Galerkin method on natural frequencies [23]

using Eq. (11.97), (approximate equation), and Eq. (11.81), (rigorous equation), is imperceptible. In Fig. 11.7, $\tilde{p}_0$ denotes a reference value of dead loads.

11.9 Dynamic Analyses Based on the Galerkin Method

We examine the effect of dead loads on the dynamic problem of plates. For simplicity, let us consider the dynamic solution for simply supported plates. To solve Eq. (11.14), employing the method of separation of variables, we have

$$\bar{w}(x, y, t) = \sum_{m=1}\sum_{n=1} \Phi_{mn}(t) f_{mn}(x, y) \tag{11.99}$$

in which Φ_{mn} are the unknown displacement coefficients with respect to time t.

First, we consider the analytical method for the advanced equation of motion of plates including the effect of dead loads by means of the Galerkin method. Applying Eq. (11.99) to Eq. (11.14), the Galerkin equation may be written as

$$\begin{aligned} \delta\Phi_{mn}:\ \sum_{\bar{m}=1}\sum_{\bar{n}=1} \Phi_{\bar{m}\bar{n}}[A^*_{mn\bar{m}\bar{n}}] &= -\sum_{\bar{m}=1}\sum_{\bar{n}=1} \ddot{\Phi}_{\bar{m}\bar{n}} \frac{m_0}{D_0} \int_0^{l_x}\int_0^{l_y} f_{mn} f_{\bar{m}\bar{n}} \mathrm{d}x\,\mathrm{d}y \\ &+ \int_0^{l_x}\int_0^{l_y} \frac{\bar{p}(x, y, t)}{D_0} f_{mn} \mathrm{d}x\,\mathrm{d}y \end{aligned} \tag{11.100}$$

in which

$$\begin{aligned} A^*_{mn\bar{m}\bar{n}} = \int_0^{l_x}\int_0^{l_y} & f_{mn}\Big\{\nabla^4 f_{\bar{m}\bar{n}} \\ & - \frac{6}{h_0^2}[2\tilde{w}_{,x}\tilde{w}_{,xx} f_{\bar{m}\bar{n},x} + (\tilde{w}_{,x})^2 f_{\bar{m}\bar{n},xx} + 2\tilde{w}_{,y}\tilde{w}_{,yy} f_{\bar{m}\bar{n},y} + (\tilde{w}_{,y})^2 f_{\bar{m}\bar{n},yy}] \\ & - \frac{6\nu}{h_0^2}[2\tilde{w}_{,y}\tilde{w}_{,xy} f_{\bar{m}\bar{n},x} + (\tilde{w}_{,y})^2 f_{\bar{m}\bar{n},xx} + 2\tilde{w}_{,x}\tilde{w}_{,xy} f_{\bar{m}\bar{n},y} + (\tilde{w}_{,x})^2 f_{\bar{m}\bar{n},yy}] \\ & - \frac{6(1-\nu)}{h_0^2}[(\tilde{w}_{,x}\tilde{w}_{,y})_{,x} f_{\bar{m}\bar{n},y} + (\tilde{w}_{,x}\tilde{w}_{,y})_{,y} f_{\bar{m}\bar{n},x} + 2(\tilde{w}_{,x}\tilde{w}_{,y}) f_{\bar{m}\bar{n},xy}]\Big\}\mathrm{d}x\,\mathrm{d}y \end{aligned} \tag{11.101}$$

Equation (11.100) can be solved by means of step-by-step integration methods, in which Eq. (11.100) is translated to linear simultaneous algebraic equations with respect to the unknown incremental displacement coefficients $\Delta\bar{w}_{mn}$.

11.10 Dynamic Analyses Based on the Closed-Form Approximate Solution

Next, we consider the closed-form solution to the dynamic problem including the effect of dead loads. $R(x, y, t)$, which is a function of static deflections $\tilde{w}$ and dynamic deflections $\bar{w}$, similar to Eq. (11.99), is expressed by separation of variables as

$$R(x, y, t) = \sum_{m=1}\sum_{n=1} \Phi_{mn}(t) R_{mn}(x, y) \tag{11.102}$$

Substituting Eqs. (11.99) and (11.102) into Eq. (11.14) yields

$$\sum_{m=1}\sum_{n=1}\left\{\frac{m_0}{D_0}\ddot{\Phi}_{mn}(t) f_{mn} + \Phi_{mn}(t)\left[\nabla^4 f_{mn} + R_{mn}(x, y)\right]\right\} = \frac{\bar{p}(x, y, t)}{D_0} \tag{11.103}$$

Applying Eqs. (11.76) and (11.78) to Eq. (11.103) and noticing that f_{mn} are natural functions of the plate, we have

$$\sum_{m=1}\sum_{n=1}\left[\ddot{\Phi}_{mn}(t) + \Phi_{mn}(t)\omega_{mn}^2\right] f_{mn} = \frac{\bar{p}(x, y, t)}{m_0} \tag{11.104}$$

Multiplying both sides of the above equation by f_{mn} and integrating from 0 to l_x with respect to x and from 0 to l_y with respect to y, Eq. (11.104) becomes

$$\ddot{\Phi}_{mn}(t) + \omega_{mn}^2 \Phi_{mn}(t) = \frac{1}{m_0 a} Q_{mn}(t) \tag{11.105}$$

where $Q_{mn}(t)$ and a are defined as

$$Q_{mn}(t) = \int_0^{l_x}\int_0^{l_y} \bar{p}(x, y, t) f_{mn} \mathrm{d}x\, \mathrm{d}y \tag{11.106}$$

$$a = \int_0^{l_x}\int_0^{l_y} f_{mn} f_{\bar{m}\bar{n}} \mathrm{d}x\, \mathrm{d}y = \frac{l_x l_y}{4} \tag{11.107}$$

The general solution of Eq. (11.105) is given as

$$\Phi_{mn}(t) = A_{mn}\cos(\omega_{mn} t) + B_{mn}\sin(\omega_{mn} t) + \frac{1}{m_0 a \omega_{mn}} \int_0^t Q_{mn}(\tau)\sin\left[\omega_{mn}(t - \tau)\right] d\tau \tag{11.108}$$

where the coefficients A_{mn} and B_{mn} are constants determined from initial conditions. Thus, we can obtain the closed-form approximate solution for the dynamic problem of plates including the effect of dead loads.

Next, we propose the closed-form solution for plates subject to the following harmonic dynamic live loads using Eq. (11.108).

$$\bar{p}(x, y, t) = \bar{p}_0 \sin(\omega_p t) \tag{11.109}$$

in which ω_p is a frequency of external load and $\bar{p}_0$ indicates a constant intensity of the live loads. Equation (11.108) reduces to

$$\Phi_{mn}(t) = \frac{16\bar{p}_0}{\pi^2 m_0 mn\omega_{mn}} \frac{\omega_{mn}\sin(\omega_p t) - \omega_p \sin(\omega_{mn} t)}{(\omega_{mn}^2 - \omega_p^2)} \tag{11.110}$$

The dynamic deflections $\bar{w}$ produced by the dynamic live loads are given by applying Eq. (11.110) to Eq. (11.99).

Thus, the dynamic deflections $\bar{w}$ including the effect of dead loads have been presented. The expression for the dynamic deflections $\bar{w}_0$ excluding the effect of dead loads is given by replacing ω_{mn} in Eq. (11.110) with ω_{0mn}, and the result agrees with the previous result. The effect of dead loads on dynamic deflections of simply supported uniform plates is governed by k_ω given by Eq. (11.98), which determines practically the natural frequencies including the effect of dead loads. It is clarified that the physical factors governing the effect of dead loads are proportional to the eighth power of the span length l_x and the second power of the dead load $\tilde{p}$, and inversely proportional to the second power of the thickness h_0 and of the bending rigidity of the plate D_0.

Once dynamic displacements $\bar{w}(x, y, t)$ due to live loads are given, the stress couples $\bar{M}_x, \bar{M}_y$, and $\bar{M}_{xy}$ due to live loads are obtained from the well-known relations $\bar{M}_x = -D_0(\bar{w}_{,xx} + v\bar{w}_{,yy})$, and etc.

11.11 Numerical Results to Dynamic Live Loads

We examine the exactness of the closed-form solution proposed here for simply supported plates from numerical calculations. Plates are assumed to be an equivalent solid steel plate given from a steel structure. Data used in the numerical computation are as follows: span lengths $l_x = l_y = 5$ m; a reference thickness $h_0^* = 0.07$ m; uniformly distributed live load $\bar{p} = 5884$ N/m^2; Young's modulus $E = 2.06 \times 10^{11}$ N/m^2; Poisson's ratio $v = 0.3$; and $\omega_p = \omega_{011}/3$, in which ω_{011} is the first natural frequency of the plate both subjected to only dead loads $\tilde{p}$ and excluding the effect of dead loads.

Figure 11.8 shows the time history of dynamic deflection $\bar{w}$ at the midpoint of the plate with the ratio $\bar{p}/\tilde{p} = 0.2$. The solid line represents results excluding the effect of dead loads. The broken line denotes results obtained using the proposed approximate solution including the effect of dead loads, given in Eq. (11.110). Circular notation denotes results obtained from Eq. (11.100), including the effect of dead loads. The proposed approximate solution shows excellent agreement with the results obtained using Eq. (11.100). The numerical result shows that dynamic deflections $\bar{w}$ due to dynamic live loads are decreased, considering the effect of dead loads. Similarly, Fig. 11.9 describes the dynamic deflections $\bar{w}$ at the midpoint of the plate with the ratio $\bar{p}/\tilde{p} = 0.5$. The results obtained using the proposed approximate solution agree with the results obtained using Eq. (11.110). The effect of dead loads on dynamic deflection $\bar{w}$ for the current plate subjected to dead loads given by the ratio $\bar{p}/\tilde{p} = 0.5$ has little effect. Figure 11.10 shows the time history of the dynamic deflection $\bar{w}$ at the midpoint of the current plate with the ratio $\bar{p}/\tilde{p} = 1$. It is seen that the effect of dead loads in this case is negligible.

Next, Figs. 11.11 and 11.12 demonstrate the effect of dead loads on the maximum value of dynamic deflections at the midpoint of the plate, in which only the values of the uniform dead loads $\tilde{p}$ vary under the constant live loads $\bar{p} = 5884$ N/m^2. In these figures, the ratio $\bar{w}_{max}/\bar{w}_{0max}$ implies that the maximum value of dynamic deflections due to the constant live load decreases owing to the effect of dead loads, in which $\bar{w}_{0max}$ is the maximum value of dynamic deflections excluding the effect of dead

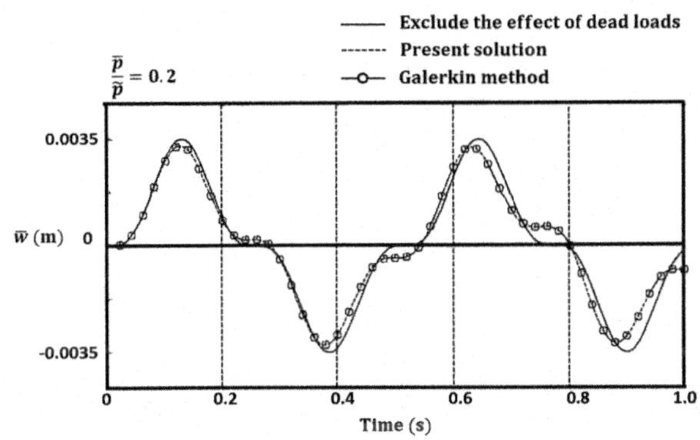

Fig. 11.8 Dynamic deflection $\bar{w}$ of simply supported plates ($\bar{p}/\tilde{p} = 0.2$) [23]

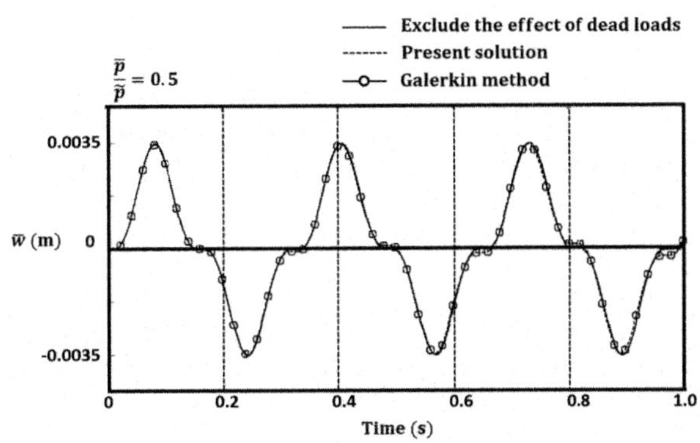

Fig. 11.9 Dynamic deflection $\bar{w}$ of simply supported plates ($\bar{p}/\tilde{p} = 0.5$) [23]

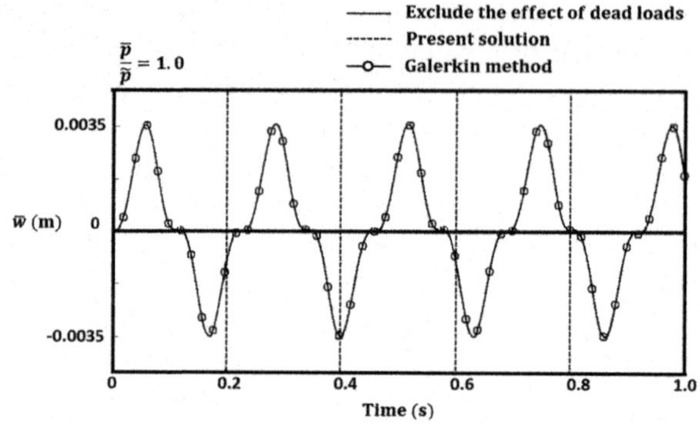

Fig. 11.10 Dynamic deflection $\bar{w}$ of simply supported plates ($\bar{p}/\tilde{p} = 1$) [23]

loads. For example, in Fig. 11.11, the maximum dynamic deflections $\bar{w}_{\max}$ of the plate whose thickness ratio $\alpha_h(=h_0/h_0^*) = 1$ decreases 0.989 times $\bar{w}_{0\max}$ when the dead loads $\tilde{p}$ are twice the live loads $\bar{p}$, namely $\bar{p}/\tilde{p} = 0.5$.

When the thickness ratio α_h is 2, the effect of dead loads becomes small. Figure 11.12 shows the effect of dead loads on the maximum dynamic deflections for plates with aspect ratios $\alpha_l(=l_y/l_x) = 1$ and 2. The effect of dead loads for plates subjected to the same live load is remarkable as dead loads become large. Although at $\alpha_l = 2$ there is a little difference between the closed-form solution indicated by a broken line and the rigorous method using Eq. (11.100) indicated by a solid line, the difference is negligible for practical use.

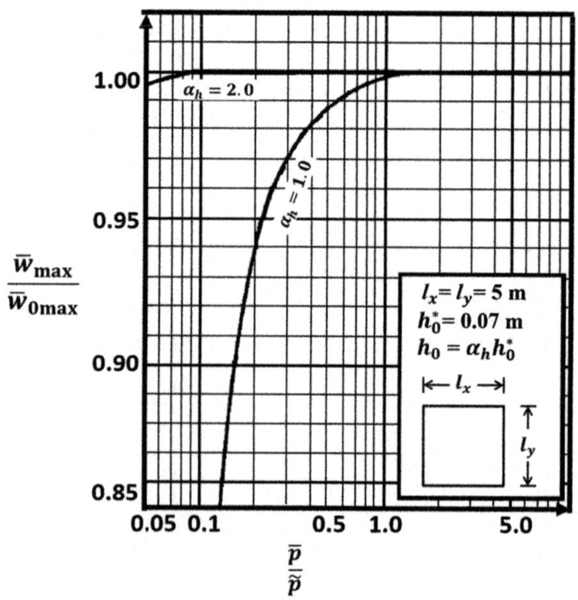

Fig. 11.11 Relationships between $\bar{w}_{\max}/\bar{w}_{0\max}$ and $\bar{p}/\tilde{p}$ for $\alpha_h = 1$ and 2 [23]

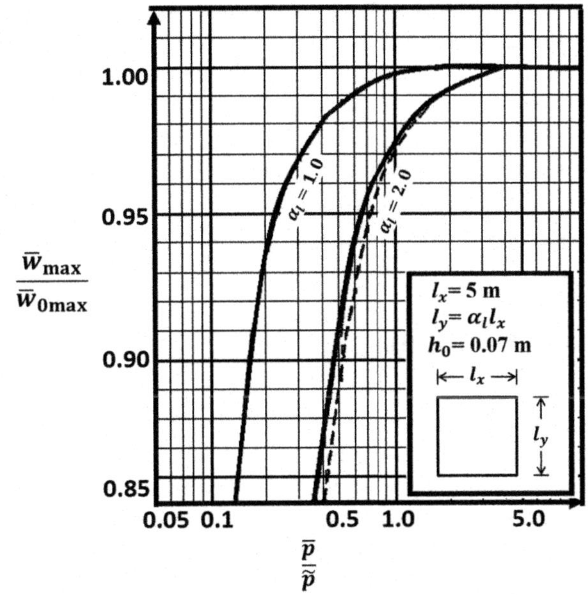

Fig. 11.12 Relationships between $\bar{w}_{max}/\bar{w}_{0max}$ and $\bar{p}/\tilde{p}$ for $\alpha_l = 1$ and 2 [23]

The numerical results suggest that dynamic deflections due to live loads for plates with heavyweight dead loads are minimized by considering the effect of dead loads. When plates are subjected to the same live load, the influence due to the live load is larger in lightweight steel structures than in heavyweight reinforced concrete structures. In order to safely design structures of all kinds, it is suggested that design live loads acting on steel structures should be increased. It also requires further examination from the point of limit state to apply this suggestion to structures. However, under the present condition in which most structures are designed by elastic design method, this suggestion presents an important matter to dynamic plates.

11.12 Method Reflected the Effect of Dead Loads in Dynamic Problems

We propose how to apply the effect of dead loads to the dynamic live loads derived from design codes. The live loads $\bar{p}$ are assumed to be harmonic given in Eq. (11.109). This procedure is presented as follows:

Step 1: Determine from Eqs. (11.110) and (11.99) the maximum value $\bar{w}_{max}(\tilde{p},\bar{p})$ of $\bar{w}$ for current plates subjected to dead loads $\tilde{p}$ and live loads $\bar{p}$, in which the dead loads are determined by designers.

Step 2: Denote a standard dead load, in which the current plate is considered to be safe, by $\tilde{p}_0$. This standard value is previously determined by experiments or

investigations of collapsed structures. It is recommended practically that the standard value of dead loads uses the value of dead loads for reinforced concrete plates instead of those used in steel plates. Thus, the maximum value $\bar{w}_{\max}(\tilde{p}_0,\bar{p})$ of $\bar{w}(x, y, t)$ for the current plates subjected to both the standard dead loads $\tilde{p}_0$ and the live loads $\bar{p}$ is obtained from Eqs. (11.110) and (11.99).
Step 3: In order to have the same safety as reinforced concrete structures subject to the live loads $\bar{p}$, the live load in steel structures should be raised according to the ratio $\bar{w}_{\max}(\tilde{p},\bar{p})$ / $\bar{w}_{\max}(\tilde{p}_0,\bar{p})$.

11.13 Conclusions

The phenomenon whereby the deflection, stress couples, and shear forces produced by static and dynamic live loads decrease due to the existence of the dead loads on elastic plates has been confirmed by means of the advanced governing equation for plates that includes the effect of dead loads.

For static plates, key factors dominating the effect of dead loads in simply supported plates have been clearly shown in a closed-form solution. The effect due to dead loads is demonstrated to be proportional to the eighth power of the span length l_x and the second power of the dead load $\tilde{P}$, and inversely proportional to the second power of the thickness h and bending stiffness D_0. Finally, a method reflecting the effect of dead loads on live loads has been proposed for practical use. The effect of dead loads is demonstrated to be an important consideration with respect to the safety of simply supported and lightweight plates.

For dynamic plates, the closed-form approximate solutions for the free and forced vibration problems have been presented. The effect of dead loads on dynamic plates of simply supported uniform plates is clarified to be practically governed by k_ω given by Eq. (11.98), in which k_ω is a function of the dead loads, span length, thickness, and bending rigidity of the plates. Finally, it has been shown how to apply the effect of dead loads to the modification of dynamic live loads. The idea proposed here will be an important consideration in the design of dynamic plates.

Appendix: Values of g_{mn}

The notation g_{mn} in Eq. (11.44) is defined as follows:

$$g_{mn} = 24\pi^4 \sum_{\bar{m}=1} \sum_{\bar{n}=1} \sum_{\tilde{m}=1} \sum_{\tilde{n}=1} \sum_{\hat{m}=1} \sum_{\hat{n}=1} \frac{\bar{p}^*_{\bar{m}\,\bar{n}}}{\pi^4\left[\bar{m}^2 + \left(\frac{\bar{n}}{\alpha_l}\right)^2\right]^2}$$

$$\frac{\bar{p}^*_{\hat{m}\,\hat{n}}}{\pi^4\left[\hat{m}^2 + \left(\frac{\hat{n}}{\alpha_l}\right)^2\right]^2} \frac{\bar{p}^*_{\tilde{m}\,\tilde{n}}}{\pi^4\left[\tilde{m}^2 + \left(\frac{\tilde{n}}{\alpha_l}\right)^2\right]^2} k_0(\bar{m}, \bar{n}; \tilde{m}, \tilde{n}; \hat{m}, \hat{n}; m, n) \quad (11.111)$$

where $k_0(\bar{m}, \bar{n}; \tilde{m}, \tilde{n}; \hat{m}, \hat{n}; m, n)$ takes the value

$$\begin{aligned} k_0(\bar{m}, \bar{n}; \tilde{m}, \tilde{n}; \hat{m}, \hat{n}; m, n) &= 2\bar{m}\tilde{m}\hat{m}^2 F_{sc}(\hat{m}, m, \tilde{m}, \bar{m}) F_{ss}(\bar{n}, \tilde{n}, \hat{n}, n) \\ &+ \bar{m}^2 \tilde{n}\hat{m} F_{sc}(\bar{m}, m, \tilde{m}, \hat{m}) F_{ss}(\bar{n}, \tilde{n}, \hat{n}, n) + \frac{2\bar{n}\tilde{n}\hat{n}^2}{\alpha_l^4} F_{ss}(\bar{m}, \tilde{m}, \hat{m}, m) F_{sc}(\hat{n}, n, \bar{n}, \tilde{n}) \\ &+ \frac{\bar{n}^2\tilde{n}\hat{n}}{\alpha_l^4} F_{ss}(\bar{m}, \tilde{m}, \hat{m}, m) F_{sc}(\tilde{n}, n, \bar{n}, \hat{n}) + \frac{\nu}{\alpha_l^2}[-2\bar{m}\tilde{n}\hat{m}\hat{n} F_{sc}(\tilde{m}, m, \bar{m}, \hat{m}) F_{sc}(\bar{n}, n, \tilde{n}, \hat{n}) \\ &+ \bar{m}^2 \tilde{n}\hat{n} F_{ss}(\bar{m}, \tilde{m}, \hat{m}, m) F_{sc}(\bar{n}, n, \tilde{n}, \hat{n}) - 2\bar{n}\tilde{m}\hat{m}\hat{n} F_{sc}(\bar{m}, m, \tilde{m}, \hat{m}) F_{sc}(\tilde{n}, n, \bar{n}, \hat{n}) \\ &+ \bar{n}^2 \tilde{m}\hat{m} F_{sc}(\bar{m}, m, \tilde{m}, \hat{m}) F_{ss}(\tilde{n}, \hat{n}, \bar{n}, n)] + \frac{(1-\nu)}{\alpha_l^2}[\bar{n}\bar{m}^2 \hat{n} F_{ss}(\bar{m}, \tilde{m}, \hat{m}, m) F_{sc}(\tilde{n}, n, \bar{n}, \hat{n}) \\ &- \bar{n}\tilde{m}\hat{m}\hat{n} F_{sc}(\bar{m}, m, \tilde{m}, \hat{m}) F_{sc}(\tilde{n}, n, \bar{n}, \hat{n}) - 2\bar{m}\bar{n}\hat{m}\hat{n} F_{sc}(\hat{m}, m, \bar{m}, \tilde{m}) F_{sc}(\tilde{n}, n, \bar{n}, \hat{n}) \\ &- \bar{m}\tilde{m}\bar{n}\hat{n} F_{sc}(\hat{m}, m, \bar{m}, \tilde{m}) F_{sc}(\bar{n}, n, \tilde{n}, \hat{n}) + \bar{m}\tilde{m}\hat{n}^2 F_{sc}(\hat{m}, m, \bar{m}, \tilde{m}) F_{ss}(\tilde{n}, \hat{n}, \bar{n}, n)] \end{aligned} \quad (11.112)$$

Here, F_{ss} and F_{sc} indicate integrals given by

$$F_{ss}(m_1, m_2, m_3, m_4) = \int_0^1 \sin(3\pi\xi)\sin(m_1\pi\xi)\sin(m_2\pi\xi)\sin(m_3\pi\xi)\sin(m_4\pi\xi)d\xi \quad (11.113)$$

$$F_{sc}(m_1, m_2, m_3, m_4) = \int_0^1 \sin(3\pi\xi)\sin(m_1\pi\xi)\sin(m_2\pi\xi)\cos(m_3\pi\xi)\cos(m_4\pi\xi)d\xi \quad (11.114)$$

References

1. Timoshenko, S., Woinowsky-Krieger, S.: Theory of Plates and Shells. McGraw-Hill Book Company, New York (1959)
2. Volterra, E., Zachmanoglou, E.C.: Dynamics of Vibrations. Charles, E., Merrill Books, Columbus, Ohio (1965)
3. Shimpi, R.P., Patel, H.G., Arya, H.: New first-order shear deformation plate theories. J. Appl. Mech-Trans. ASME **74**(3), 523–533 (2007)
4. Wu, J.H., Liu, A.Q., Chen, H.L.: Exact solutions for free-vibration analysis of rectangular plates using bessel functions. J. Appl. Mech-Trans. ASME **74**(6), 1247–1251 (2007)
5. Boscolo, M., Banerjee, J.R.: Dynamic stiffness elements and their applications for plates using first order shear deformation theory. Comput. Struct. **89**(3–4), 395–410 (2011)
6. Fang, X., Hu, C., Huang, W.: Dynamic stress concentration of a circular cutout buried in semi-infinite plates subjected to flexural waves. J. Appl. Mech-Trans. ASME **74**(2), 382–387 (2007)

7. Lee, W., Chen, J.T.: Free vibration analysis of a circular pate with multiple circular holes by using indirect BIEM and addition theorem. J. Appl. Mech. ASME **78**(1) (2011)
8. Tanveer, M., Singh, A.V.: Linear and nonlinear dynamic responses of various shaped laminated composite plates. J. Comput. Nonlinear Dyn **4**(4) (2009)
9. Rao, G.V., Saheb, K.M.: Simple formula to study the large amplitude free vibrations of beams and plates. J. Appl. Mech. **75**(1), 014505-1-4 (2008)
10. Takabatake, H.: Static analyses of elastic plates with voids. Int. J. Solids Struct. **28**(2), 179–196 (1991)
11. Takabatake, H.: Dynamic analyses of elastic plates with voids. Int. J. Solids Struct. **28**(7), 879–895 (1991)
12. Takabatake, H., Imaizumi, T., Okatomi, K.: Simplified analysis of rectangular plates with stepped thickness—discussion. J. Struct. Eng. ASCE **122**(7), 839–840 (1996)
13. Takabatake, H., Yanagisawa, N., Kawano, T.: A simplified analysis of rectangular cellular plates. Int. J. Solids Struct. **33**(14), 2055–2074 (1996)
14. Takabatake, H., Morimoto, H., Fujiwara, T., Honma, T.: Simplified analysis of circular plates including voids. Comput. Struct. **58**(2), 263–275 (1996)
15. Takabatake, H., Kajiwara, K., Takesako, R.: A simplified analysis of circular cellular plates. Comput. Struct. **61**(5), 789–804 (1996)
16. Takabatake, H.: Dynamic analysis of rectangular with stepped thickness subjected to moving loads including additional mass. J. Sound Vib. **213**(5), 829–842 (1998)
17. Takabatake, H., Nagareda, Y.: A simplified analysis of elastic plates with edge beams. Comput. Struct. **70**(2), 129–139 (1999)
18. Takabatake, H.: Effects of dead loads in static beams. J. Struct. Eng. ASCE **116**(4), 1102–1120 (1990)
19. Takabatake, H.: Effect of dead loads on natural frequencies of beams. J. Struct. Eng. ASCE **117**(4), 1039–1052 (1991)
20. Takabatake, H.: Effects of dead loads on dynamic analyses of beams. Earthq. Struct. **1**(4), 411–425 (2010)
21. Zhou, S.J., Zhu, X.: Analysis of effect of dead loads on natural frequencies of beams using finite-element techniques. J. Struct. Eng. ASCE **122**(5), 512–516 (1996)
22. Kelly, J.M., Sackman, J.L., Javid, A.: The influence of preform on the modes of vibrating beams. Earthq. Eng. Struct. Dyn. **20**(12), 1145–1157 (1991)
23. Takabatake, H.: Effects of dead loads in dynamic plates. J. Struct. Eng. ASCE **118**(1), 34–51 (1992)
24. Takabatake, H.: Effects of dead loads on the static analysis of plates. Struct. Eng. Mech. **42**(6), 761–781 (2012)
25. Mostaghel, N., Fu, K.C., Yu, Q.: Shifting natural frequencies of plates through preforming. Earthq. Eng. Struct. Dyn. **24**(3), 411–418 (1995)
26. Yu, O., Mostaghel, N., Fu, K.C.: Effect of initial curvature on natural frequency of thin plate on hinge supports. J. Eng. Mech. ASCE **120**(4), 796–813 (1994)
27. Zhou, S.J.: Load-induced stiffness matrix of plates. Can. J. Civ. Eng. **29**(1), 181–184 (2002)
28. Durmaz, M., Daloğlu, A.T.: Frequency analysis of ground snow data and production of the snow load map using geographic information system for the eastern black sea region of turkey. J. Struct. Eng. ASCE **132**(7), 1166–1177 (2006)
29. Washizu, K.: Variational Methods in Elasticity and Plasticity, 3rd edn. Pergamon Press, New York, New York (1982)
30. Haroun, M.A.: Dynamic Analysis of Liquid Storage Tanks. Earthquake Engineering Research Laboratory, California Institute of Technology, EERL 80-04 (1980)
31. Szilard, R.: Theory and Analysis of Plates-Classical and Numerical Methods. Prentice-Hall, Englewood Cliffs, New Jersey (1974)

Part V
Effects of Dead Loads on Elastic Beams

Chapter 12
Effects of Dead Loads on Static and Free Vibration Problems of Beams

12.1 Introduction

In the preceding chapter, we have considered the effect of dead loads on the static and dynamic problems of rectangular elastic plates. This Chapter and Chap. 13 presents the effect of dead loads on the static and dynamic analyses of elastic beams.

The collapse of structures due to snow loads on roofs happens especially on steel structures. It is concluded in one report [1] that the collapse of structures caused by snow loads is due to a heavy snowfall exceeding the maximum depth for design. This consideration, however, does not sufficiently explain why the collapse of heavy concrete structures is a much rarer occurrence than for those structures composed of steel. This part considers that, in addition to the aforementioned overload of snow, the dead load of structures plays an integral part in the collapse of structures due to snow loads.

Most structures are designed by the elastic design method. The notable difference between reinforced concrete structures and steel structures is in the magnitude of the dead loads. The dead load of reinforced concrete structures is in some cases 10 times larger than that of steel structures. Structures with heavy dead loads are advantageous in both the reserve strength and the effect of dead loads. The former reserve strength is based on the following reason: For structures subject to the same live load, structures with heavy dead loads need a large cross section. This large cross section, which is determined by means of the elastic design method, has a larger reserve strength up to the collapse. This reserve strength prevents the structure from collapsing as an overload acts on the structure.

On the other hand, the later effect of dead loads decreases the displacements and stress couples produced by live loads. Beams like structural members using in building are always subjected to dead loads before live loads are applied. The inherent property of dead loads occurs conservative initial bending stresses. Since live loads always act on the deformed state produced by dead loads, the strain energy produced by live loads should include the work done by the conservative initial

H. Takabatake, *Simplified Analytical Methods of Elastic Plates*,
https://doi.org/10.1007/978-981-13-0086-8_12

bending stresses. Due to the interaction between dead loads and live loads, external disturbances such as displacements and stress couples produced by live loads are considered to be affected by dead load. However, since the effect of dead loads is unknown nature, in present design codes, live loads exclude the effect of dead loads. In order to prevent the collapse of structures by snow loads, an explanation of this effect is indispensable.

There are numerous studies concerning the static and dynamic problems of beams. However, a study considering the interaction between dead loads and live loads is not found. The new attention concerning this interaction is considered to become the important jumping-off point for an extension of elementary beam theory.

This chapter is to clarify the effect of dead loads on the static and transverse free vibration problems for elastic beams. Concerning the static problems is as follows: First, the advanced governing equation of beams, including the effect of dead loads, is proposed by means of Hamilton's principle. Second, the existence of the effect of dead loads for simply supported beams and clamped beams is proven from numerical calculations for static beams by means of the Galerkin method. Third, key physical factors governing the effect of dead loads on the static problem are represented from the closed-form approximate solution. The exactness of the proposed approximate solution is proven from numerical results. Forth, a method that reflects the effect of dead loads on live loads is proposed in an example.

Next, concerning the transverse free vibration problems is as follows: Fifth, the effect of dead loads on the natural frequencies of simply supported and clamped beams is examined from numerical computations using the Galerkin method. Last, the key physical factors governing the effect of dead loads on the natural frequencies of simply supported beams are extracted from a closed-form approximate solution of the frequency equation, including the effect of dead loads. The results are easily applicable to practical design. The methodology proposed in this chapter is mainly based on references [2, 3].

12.2 Advanced Governing Equations of Beams Including Effect of Dead Loads

We consider a uniform straight beam without initial imperfections used a Cartesian coordinate system, as shown in Fig. 12.1, in which the axis x passes through the centroidal axis of the beam, and the y- and x-axes are the principal axes of the beam. We also assume that external loads acting on the beam are only transverse loads and that axial forces are absent because of considering the bending problem of the beam. Denoting the transverse deflection due to a dead load $\tilde{p}$ per unit length acting on the beam by $\tilde{w}$, this deformed state is defined as the reference state of the beam. A live load $\bar{p}$ per unit length acts always on this reference state. The deflection $\bar{w}$ due to the live load is measured from the reference state. The deflections, $\tilde{w}$ and $\bar{w}$,

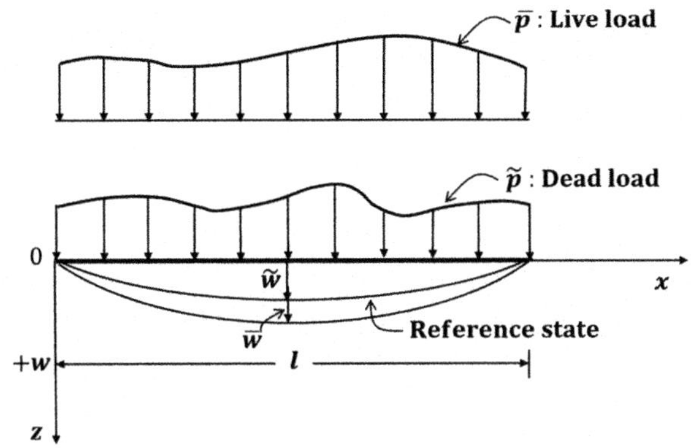

Fig. 12.1 Coordinates and load distributions of beam [2]

and transverse loads, $\tilde{p}$ and $\bar{p}$, are defined positive when they point in the positive direction of the coordinate axis z.

Assuming the Euler–Bernoulli hypothesis for the current beam, the strain–displacement relations can be expressed from Washizu [4] by

$$\varepsilon_x = -zw'' + \frac{1}{2}(w')^2 \quad \text{(nonlinear)} \tag{12.1}$$

$$\varepsilon_x = -zw'' \quad \text{(linear)} \tag{12.2}$$

in which ε_x is bending normal strain of the beam in the bending; w indicates deflection; and dashes indicate the differentiation with respect to x.

The equation of motion of beams including the effect of dead loads is proposed by means of the following Hamilton's principle in the similar method presented in the preceding chapters:

$$\delta I = \delta \int_{t_1}^{t_2} (T - U - V)\mathrm{d}t = 0 \tag{12.3}$$

in which T, U, and V indicate the kinetic energy, strain energy; and potential energy produced by external loads, respectively. δ denotes variational operator taken during the indicated time interval from t_1 to t_2.

The strain energy U can be written as

$$U = \bar{U} + \tilde{U} \tag{12.4}$$

in which $\bar{U}$ and $\tilde{U}$ are the strain energy produced by live load $\bar{p}$ and the dead load $\tilde{p}$, respectively, as given by

$$\bar{U} = \frac{1}{2} \iiint \bar{\sigma}_x \bar{\varepsilon}_x \mathrm{d}x\, \mathrm{d}y\, \mathrm{d}z \tag{12.5}$$

$$\tilde{U} = \iiint \tilde{\sigma}_x \bar{\varepsilon}_x \mathrm{d}x\, \mathrm{d}y\, \mathrm{d}z \tag{12.6}$$

where the normal stresses $\bar{\sigma}_x$ and $\tilde{\sigma}_x$ are the bending stresses produced by the live load $\bar{p}$ and the dead load $\tilde{p}$, respectively; and the normal strain $\bar{\varepsilon}_x$ is the strain produced by the live load. The stresses $\tilde{\sigma}_x$ produced by the dead load are inherent and conservative initial stresses. Now, supposing linear stress–strain relations, we have

$$\bar{\sigma}_x = E\bar{\varepsilon}_x \tag{12.7}$$

$$\tilde{\sigma}_x = E\tilde{\varepsilon}_x \tag{12.8}$$

in which E is Young's modulus; and $\tilde{\varepsilon}_x$ indicates the conservative normal strain produced by the dead loads. The strain energy $\bar{U}$ is calculated from the linear strain –displacement relation given in Eq. (12.2). On the other hand, the strain energy $\tilde{U}$ is calculated from the nonlinear strain–displacement relation given in Eq. (12.1) in order to consider the effect of dead loads. Since the displacements $\tilde{w}$ are previously known, the interaction effect between the dead loads and live loads is taken into account by the use of this different types of the strain–displacement relation for the strain energies $\bar{U}$ and $\tilde{U}$. It should be pointed out here that the effect of dead loads considered in this chapter is not the well-known stiffening effect of midplane stretching in beams, because the current beam is considered only in the bending state, in which the longitudinal displacement can be neglected in the strain–displacement relations. Hence, the strain energy U can be expressed as

$$U = \int_0^l \left[\frac{1}{2} EI(\bar{w}'')^2 + \frac{EA}{4} (\tilde{w}')^2 (\bar{w}')^2 + EI\, \tilde{w}'' \bar{w}'' \right] \mathrm{d}x \tag{12.9}$$

in which A is the cross-sectional area of beams; I is the principal moment of inertia; and l is the length of the beam.

Next, the potential energy produced by the external loads acting on the centroidal axis of the beam can be written as

$$V = -\int_0^l (\bar{p} + \tilde{p}) \bar{w} \mathrm{d}x + \int_0^l p_D \dot{\bar{w}} \mathrm{d}x \tag{12.10}$$

where p_D is the damping force per unit length and is defined by

$$p_D = c\, \dot{\bar{w}} \tag{12.11}$$

where c is damping coefficient per unit length (N/m/s/m).

Also, the kinetic energy T neglecting the effect of rotatory inertia of the beam is expressed as

$$T = \int_0^l \frac{\rho A}{2} (\dot{\bar{w}})^2 \mathrm{d}x \tag{12.12}$$

where ρ is the mass density of the material of the beam; and dot, $\cdot$, represents the differentiation with respect to time t.

Substituting Eqs. (12.9)–(12.12) into Eq. (12.3), we have

$$\begin{aligned} \delta I = & \int_{t_1}^{t_2} \int_0^l \{-\rho A \ddot{\bar{w}} - c\dot{\bar{w}} - (EI\bar{w}'')'' \\ & + \frac{1}{2}[EA(\tilde{w}')^2 \bar{w}']' - (EI\tilde{w}'')'' + (\bar{p} + \tilde{p})\Big\} \delta\bar{w}\, \mathrm{d}x\, \mathrm{d}t \\ & + \int_{t_1}^{t_2} \left\{ \left[(EI\bar{w}'' + EI\tilde{w}'')' - \frac{EA}{2}(\tilde{w}')^2 \bar{w}' \right] \delta\bar{w} \right\}_0^l \mathrm{d}t \\ & - \int_{t_1}^{t_2} \left[(EI\bar{w}'' + EI\tilde{w}'')\delta\bar{w}' \right]_0^l \mathrm{d}t = 0 \end{aligned} \tag{12.13}$$

Now, in the reference state, the following well-known equilibrium equation and boundary conditions for the beams subject to only dead loads $\tilde{p}$ must exist:

$$(EI\tilde{w}'')'' - \tilde{p} = 0 \tag{12.14}$$

$$\begin{aligned} &\tilde{w} = 0 \text{ or } \tilde{w}''' = 0 \quad \text{at} \quad x = 0 \text{ and } x = l \\ &\tilde{w}' = 0 \text{ or } \tilde{w}'' = 0 \quad \text{at} \quad x = 0 \text{ and } x = l \end{aligned} \tag{12.15}$$

In view of these expressions, from Eq. (12.13) the advanced governing equation of the beam subject to live loads $\bar{p}$, which includes the effect of dead loads, can be obtained

$$\rho A \ddot{\bar{w}} + c\dot{\bar{w}} + (EI\,\bar{w}'')'' - \frac{1}{2}[EA(\tilde{w}')^2 \bar{w}']' - \bar{p} = 0 \tag{12.16}$$

and the boundary conditions are

$$\bar{w} = 0 \text{ or } EI\,\bar{w}''' - \frac{EA}{2}(\tilde{w}')^2 \bar{w}' = 0 \text{ at} \quad x = 0 \text{ and } x = l \tag{12.17}$$

$$\bar{w}' = 0 \text{ or } \bar{w}'' = 0 \quad \text{at } x = 0 \text{ and } x = l \tag{12.18}$$

We must note that since the deflection $\tilde{w}$ due to dead loads is initially known, Eqs. (12.16)–(12.18) are linear expressions with respect to the unknown deflections $\bar{w}$ produced by live loads. The terms 1/2 $[EA\ (\tilde{w}')^2 \bar{w}']'$ in Eq. (12.16) and $(EA/2)(\tilde{w}')^2 \bar{w}'$ in Eq. (12.17) represent the effect of dead loads. The expressions neglecting these terms concerning the dead loads agree with the general well-known beam equations.

12.3 Numerical Results Using Galerkin Method for Static Problems

We examine the effect of dead loads on a uniform beam subjected to a static live load from numerical calculations used the Galerkin method. The current Galerkin equation can be expressed as

$$\int_0^l Q \cdot \delta\bar{w}\mathrm{d}x = 0 \tag{12.19}$$

in which Q represents the equilibrium equation given by ignoring the inertia and damping terms in Eq. (12.16) is defined as follows:

$$Q = \bar{w}'''' - \frac{1}{r^2}\tilde{w}'\tilde{w}''\bar{w}' - \frac{1}{2r^2}(\tilde{w}')^2\bar{w}'' - \frac{\bar{p}}{EI} = 0 \tag{12.20}$$

in which r is a radius of gyration of the beam.

Now, expressing the deflection $\bar{w}$ due to live loads by a power series expansion, we have

$$\bar{w} = \sum_{n=1} \bar{w}_n f_n \quad (n = 1, 2, 3, \ldots, \infty) \tag{12.21}$$

in which $\bar{w}_n$ are unknown displacement coefficients. The shape functions f_n must satisfy the prescribed boundary conditions of the beam. The following shape functions are used for simply supported beams and clamped beams, respectively,

$$f_n = \sin\frac{n\pi x}{l} \quad (n = 1, 2, 3, \ldots, \infty) \tag{12.22}$$

$$f_n = \sin\frac{\pi x}{l}\sin\frac{n\pi x}{l} \quad (n = 1, 2, 3, \ldots, \infty) \tag{12.23}$$

Applying Eqs. (12.20) and (12.21) into Eq. (12.19), the Galerkin equation becomes

$$\delta\bar{w}_n\!: \sum_{\bar{n}=1} \bar{w}_{\bar{n}} \int_0^l \left[f_{\bar{n}}'''' - \frac{1}{r^2}\tilde{w}'\tilde{w}'' f_{\bar{n}}' - \frac{1}{2r^2}(\tilde{w}')^2 f_{\bar{n}}'' \right] f_n \,\mathrm{d}x = \int_0^l \frac{\bar{p}}{EI} f_n \,\mathrm{d}x \quad (n = 1, 2, 3, \ldots) \tag{12.24}$$

Thus, the Galerkin equation reduces to linear, nonhomogeneous, algebraic equations with respect to unknown coefficients $\bar{w}_{\bar{n}}$.

Since Eq. (12.24) contains the known deflection $\tilde{w}$ produced by a dead load $\tilde{p}$, the deflection $\tilde{w}$ must be initially determined. The deflection $\tilde{w}$ subject to a uniformly distributed dead load can be expressed by

$$\tilde{w} = \frac{\tilde{p}l^4}{24EI}\left[a\left(\frac{x}{l}\right) + b\left(\frac{x}{l}\right)^2 - 2\left(\frac{x}{l}\right)^3 + \left(\frac{x}{l}\right)^4\right] \tag{12.25}$$

for simply supported beams and clamped beams, in which the constants a and b take values of $a = 1$ and $b = 0$ for simply supported beams and values of $a = 0$ and $b = 1$ for clamped beams.

On the other hand, the static live loads $\bar{p}$ may be assumed in general form as

$$\bar{p} = \alpha\,\tilde{p}\left(k_0 - k_1 \sin\frac{m\pi x}{l}\right) \tag{12.26}$$

in which the live-loads-to-dead-loads ratio, α, is defined by $\alpha = \bar{p}/\tilde{p}$; and k_0 and k_1 are parameters representing the shape of the static live loads; for example, a uniform load takes $k_0 = 1$ and $k_1 = 0$. Also, $\sin(m\pi x/l)$ indicates a variation in the axial direction x of the live loads, in which m is an integer selected.

The bending moments $\bar{M}$ produced by the live load can be obtained by

$$\bar{M} = -EI\bar{w}'' = -EI\sum_{n=1} \bar{w}_n f_n'' \quad (n = 1, 2, 3, \ldots, \infty) \tag{12.27}$$

Next, we evaluate previously the general displacements $\bar{w}_0$ and general bending moments $\bar{M}_0$ due to the live load $\bar{p}$, excluding the effect of dead load. These quantities are marked with "0" in the subscription to show explicitly the fact excluding the effect of dead loads. They are given from the elementary beam theory satisfying the following governing equation:

$$\bar{w}_0'''' - \frac{\bar{p}}{EI} = 0 \tag{12.28}$$

The deflections $\bar{w}_0$ and bending moments $\bar{M}_0$ for simply supported beams and clamped beams, subjected to a general live load given in Eq. (12.26) can be expressed as

$$\bar{w}_0 = \frac{\bar{p}l^4}{24EI}\left(\frac{x}{l}\right)\left\{k_0\left[1 - 2\left(\frac{x}{l}\right)^2 + \left(\frac{x}{l}\right)^3\right] - \frac{24k_1}{(m\pi)^4}\left(\frac{l}{x}\right)\sin\frac{m\pi x}{l}\right\}$$

(for simply supported beam) (12.29)

$$\bar{w}_0 = \frac{\bar{p}l^4}{24EI}\left(\frac{x}{l}\right)\left[\frac{24k_1}{(m\pi)^3} + \beta_1\left(\frac{x}{l}\right) - \beta_2\left(\frac{x}{l}\right)^2 + k_0\left(\frac{x}{l}\right)^3 - \frac{24k_1}{(m\pi)^4}\left(\frac{l}{x}\right)\sin\frac{m\pi x}{l}\right]$$

(for clamped beam) (12.30)

$$\bar{M}_0 = \bar{p}\left[k_0\left(\frac{lx}{2} - \frac{x^2}{2}\right) - k_1\left(\frac{l}{m\pi}\right)^2 \sin\frac{m\pi x}{l}\right] \quad \text{(for simply supported beam)} \tag{12.31}$$

$$\bar{M}_0 = \bar{p}\left[\frac{\beta_2 lx}{4} - k_0\frac{x^2}{2} - \frac{\beta_1 l^2}{12} - k_1\left(\frac{l}{m\pi}\right)^2 \sin\frac{m\pi x}{l}\right] \quad \text{(for clamped beam)} \tag{12.32}$$

in which β_1 and β_2 take the following values:

$$\beta_1 = k_0 - \frac{24k_1}{(m\pi)^3}[2 + (-1)^{2m-1}]; \quad \beta_2 = 2k_0 - \frac{24k_1}{(m\pi)^3}[1 + (-1)^{2m-1}] \tag{12.33}$$

Next, we examine the effect of dead loads for simply supported beams and clamped beams. The following data are used in numerical calculations; $E = 2.06 \times 10^{11}$ N/m^2, $I = \alpha_I \cdot I_0$, $A = \alpha_A \cdot A_0$, $l = \alpha_l \cdot l_0$, $r = \alpha_r \cdot r_0$, and $\bar{p} = \alpha_{\bar{p}} \cdot \bar{p}_0$, in which α_I, α_A , α_l, α_r, and $\alpha_{\bar{p}}$ are parameters. I_0, A_0, l_0, and $\bar{p}_0$ take the following standard values selected arbitrarily: $I_0 = 2.5 \times 10^{-4}$ m^4, $A_0 = 0.01$ m^2, $l_0 = 8$ m, and $\bar{p}_0 =$ 47 kN/m. The standard live load $\bar{p}_0$ of snow is calculated from that the snow depth is 2 m, the density 2.94×10^4 N/m^3 and that the beam is arranged with the space of 8 m. The sizes of the live loads are 0.5, 1, and 1.5 with ratios $\alpha_{\bar{p}} = \bar{p}/\bar{p}_0$ and only the values of the dead loads vary from 0.1 to 10 with ratios $\bar{p}/\tilde{p}$. The results of four distinct loading patterns for live loads are calculated.

Numerical results are obtained by substituting Eqs. (12.25) and (12.26) into Eq. (12.24). The deflections $\bar{w}_{\max}$ and $\bar{w}_{0\max}$ are the maximum values of the deflections $\bar{w}$ and $\bar{w}_0$, including and excluding the effect of the dead load, respectively.

Figure 12.2 shows the effect of dead loads on the maximum value $\bar{w}_{\max}$. In this figure, the ratio $\bar{w}_{\max}/\bar{w}_{0\max}$ is 1 where $\bar{p}/\tilde{p} \to \infty$. It means that the effect of dead loads is negligible when the dead load is small in comparison with the live load. Conversely, when the ratio $\bar{w}_{\max}/\bar{w}_{0\max}$ is smaller than 1, it means that the deflections $\bar{w}$ due to live loads are minimized by considering the effect of dead loads. Also, the numerical results for the four loading patterns of live loads show the same characteristics for the effect of dead loads.

Now, indicating the abscissa by the standard live load selected arbitrarily to dead load ratio, $\bar{p}_0/\tilde{p}$, Fig. 12.2 reduces to Fig. 12.3. Their relationships for simply supported beams and clamped beams can be expressed by one curve, respectively.

The relationships between ratios $\bar{p}_0/\tilde{p}$ and $\bar{M}_{\max}/\bar{M}_{0\max}$ almost agree with results for $\bar{w}_{\max}/\bar{w}_{0\max}$, in which $\bar{M}_{\max}$ and $\bar{M}_{0\max}$ are the maximum bending moments due to the live loads, including and excluding the effect of dead loads, respectively. Also, the differences among these four loading patterns of the live loads for the effect of dead loads are negligible for the deflections $\bar{w}$ and bending moments $\bar{M}_{\max}$. Accordingly, Fig. 12.3 indicates the characteristic effects of dead loads for the beam having the

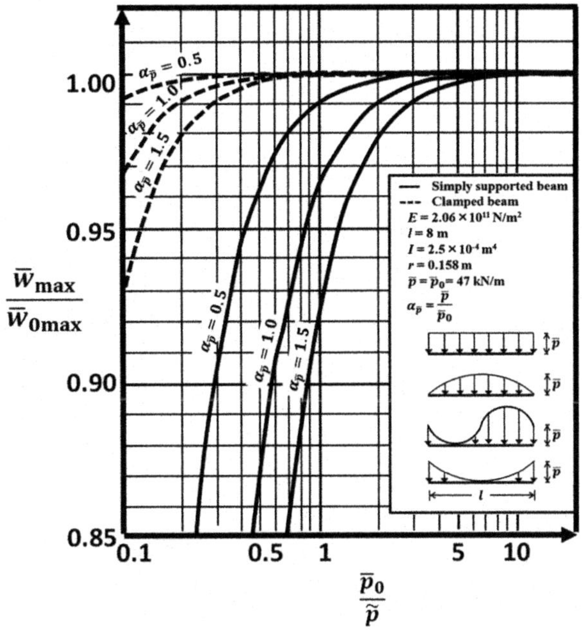

Fig. 12.2 Relationship between $\bar{w}_{\max}/\bar{w}_{0\max}$ and $\bar{p}_0/\tilde{p}$ [2]

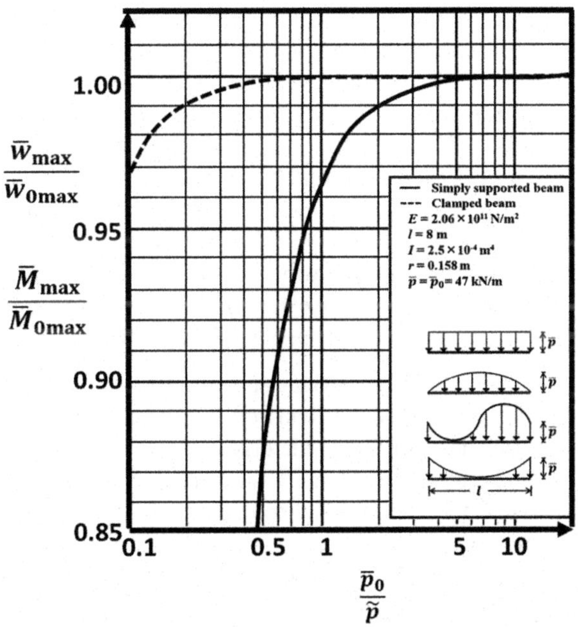

Fig. 12.3 Relationship between $\bar{w}_{\max}/\bar{w}_{0\max}$, $\bar{M}_{\max}/\bar{M}_{0\max}$ and $\bar{p}_0/\tilde{p}$ [2]

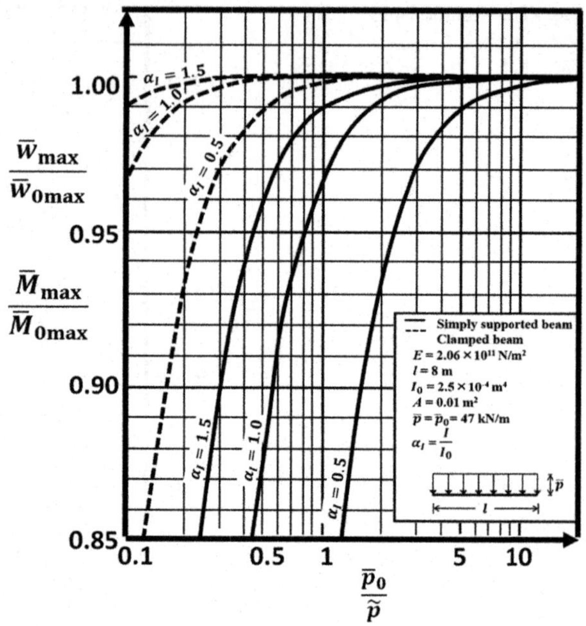

Fig. 12.4 Relationship between $\bar{w}_{max}/\bar{w}_{0max}$, $\bar{M}_{max}/\bar{M}_{0max}$ and $\bar{p}_0/\tilde{p}$ [2]

sectional efficiency used in the numerical calculations. The effect of dead loads for the beam used is determined by the live-load-to-dead-load ratio, $\bar{p}/\tilde{p}$.

In the following numerical results, for simplicity, the live loads are assumed to be a uniformly distributed load $\bar{p} = \bar{p}_0$. Figure 12.4 shows the influence of the moment of inertia on the effect of dead loads by varying only the moments of inertia I while holding the sectional area A constant. Figure 12.5 shows the influence of the span length by varying only the span length l for beams with a constant cross section. Also, Fig. 12.6 shows the relationships between the effect of dead loads and the live-load-to-dead-load ratio $\tilde{p}/\bar{p}$.

From these figures, we can obtain the following results concerning the effect of dead loads on the beams subject to static live loads.

1. The effect of dead loads appears remarkably on beams with heavy dead loads.
2. The effect of dead loads on the deflections and bending moments due to live loads has similar characteristics.
3. The effect of dead loads is determined from both the sectional efficiency of the beam and the live-load-to-dead-load ratio, $\bar{p}/\tilde{p}$, and it is independent of loading pattern of live loads.
4. The effect of dead loads is larger in simply supported beams than in clamped beams.
5. The effect of dead loads increases in any case: when the moment of inertia is small, or the radius of gyration is small, or the span length is large

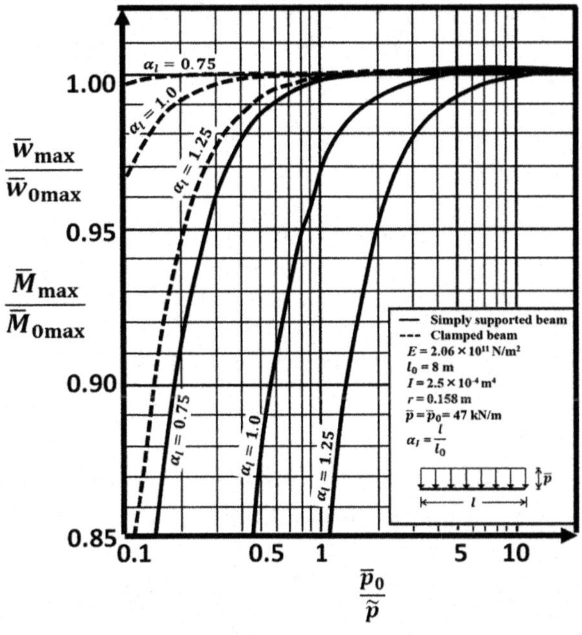

Fig. 12.5 Relationship between $\bar{w}_{max}/\bar{w}_{0max}$, $\bar{M}_{max}/\bar{M}_{0max}$ and $\bar{p}_0/\tilde{p}$ [2]

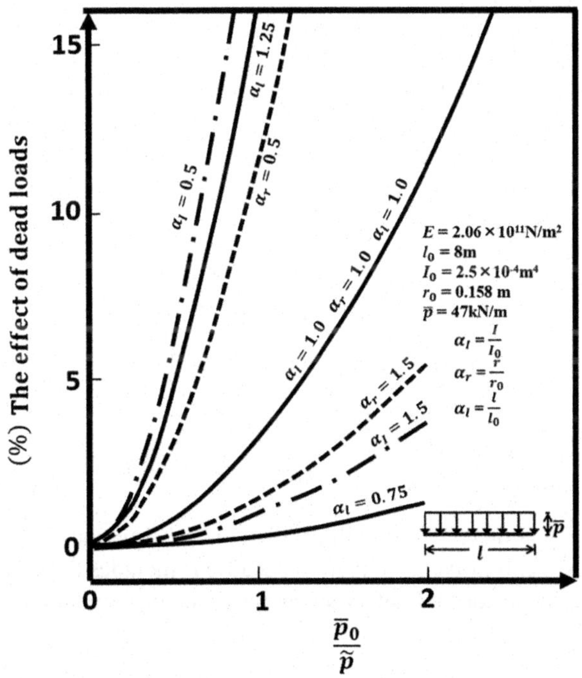

Fig. 12.6 Relationship between the effect of dead loads and $\bar{p}/\tilde{p}$ [2]

The weak beam designed with the smallest cross-sectional property is essentially lower on the load-carrying capacity than the strong beam with the large cross-sectional property, and furthermore, its dead load is made minimum. Although beams with heavy dead loads have the effect of dead loads, beams with light dead loads cannot expect the same effect. If the same degree of safety for all beams is required, lightweight beams are deficient in safety, because the effect of dead loads in lightweight beams is smaller than that in heavyweight beams. It means that the live load in lightweight beams should be increased to reflect the effect of dead loads in heavyweight beams with the same degree of safety. The present chapter does not recommend increasing the dead load of structures; to prevent the collapse of beams due to live loads, it is absolutely imperative that the safety factor for lightweight structures be raised to coincide with the safety factor for heavyweight structures.

12.4 Closed-Form Solutions Including Effect of Dead Loads in Static Problems

In the preceding section, the existence of the effect of dead loads on static live loads has been clarified from numerical results using Eq. (12.24). This section proposes the closed-form approximate solution to solve Eq. (12.24) without numerical calculations and presents key physical factors governing the effect of dead loads in static beams. For that purpose, it is necessary to convert Eq. (12.24) to a form that can be solved analytically using appropriate assumption.

For simplification purposes, we assume that the live loads are uniform. The advanced equilibrium equation including the effect of dead loads for uniform beams subjected to uniformly distributed static live loads can be obtained from Eq. (12.20) as

$$\bar{w}'''' = \frac{\bar{p}}{EI} - R \quad (\bar{p} = \text{constant}) \tag{12.34}$$

where R is a function of unknown deflection $\bar{w}$ and known deflection $\tilde{w}$, as defined by

$$R = -\frac{1}{2r^2}\left[2\,\tilde{w}'\,\tilde{w}''\,\bar{w}' + (\tilde{w}')^2\,\bar{w}''\right] \tag{12.35}$$

Equation (12.34) is a linear differential equation for unknown $\bar{w}$, but it is impossible to easily solve the equation because R includes the unknown displacement $\bar{w}$. So, we replace approximately the unknown displacement $\bar{w}$ in R with the following relation:

$$\bar{w}(x) = \beta\bar{w}_0(x) \tag{12.36}$$

where β represents the ratio of reduction of the deflection due to live loads, occurring from the effect of dead loads. From the above-mentioned numerical results, β is obtained to take a value between from 1 to 0.85 for practical ratios $\bar{p}/\tilde{p}$. For simplicity, assuming that β is a known constant, R given in Eq. (12.35) can be transformed to a known function including all known displacements $\tilde{w}$ and $\bar{w}_0$. Thus, Eq. (12.34) is changed to an analytically solvable form. Substituting Eqs. (12.35) and (12.36) in Eq. (12.34), we have

$$\bar{w}'''' = \frac{\bar{p}}{EI}(1 - \mu) \tag{12.37}$$

in which μ indicates the function of only x, given as

$$\mu = -\frac{EI}{\bar{p}}\frac{\beta}{2r^2}\left[2\,\tilde{w}'\,\tilde{w}''\,\bar{w}_0' + (\tilde{w}')^2\,\bar{w}_0''\right] \tag{12.38}$$

In Eq. (12.37), μ represents the reduction of the live load $\bar{p}$ due to the effect of dead loads. The deflection $\tilde{w}$ is given from Eq. (12.25). The deflection $\bar{w}_0$, excluding the effect of dead loads, due to a uniformly distributed live load $\bar{p}$ is obtained by replacing $\tilde{p}$ in Eq. (12.25) with $\bar{p}$. Hence, μ is given by

$$\mu = k\bar{\mu} \tag{12.39}$$

in which k and $\bar{\mu}$ take the following values:

$$k = \frac{3}{4}\beta\left(\frac{\tilde{p}l^3\lambda}{24EI}\right)^2 \quad \text{(for simply supported beams)} \tag{12.40}$$

$$k = \frac{1}{2}\beta\left(\frac{\tilde{p}l^3\lambda}{24EI}\right)^2 \quad \text{(for clamped beams)} \tag{12.41}$$

and

$$\bar{\mu} = -\left[1 - 6\left(\frac{x}{l}\right)^2 + 4\left(\frac{x}{l}\right)^3\right]^2\left[-\left(\frac{x}{l}\right) + \left(\frac{x}{l}\right)^2\right] \quad \text{(for simply supported beams)} \tag{12.42}$$

$$\bar{\mu} = -\left[\left(\frac{x}{l}\right) - 3\left(\frac{x}{l}\right)^2 + 2\left(\frac{x}{l}\right)^3\right]^2\left[1 - 6\left(\frac{x}{l}\right) + 6\left(\frac{x}{l}\right)^2\right] \quad \text{(for clamped beams)} \tag{12.43}$$

in which the slenderness ratio $\lambda = l/r$.

The distributions of $\bar{\mu}$ for simply supported beams and clamped beams is shown in Fig. 12.7. Since the reduction of live loads due to the effect of dead loads is larger in simply supported beams than in clamped beams, this tendency also gives the same result for the effect of dead loads.

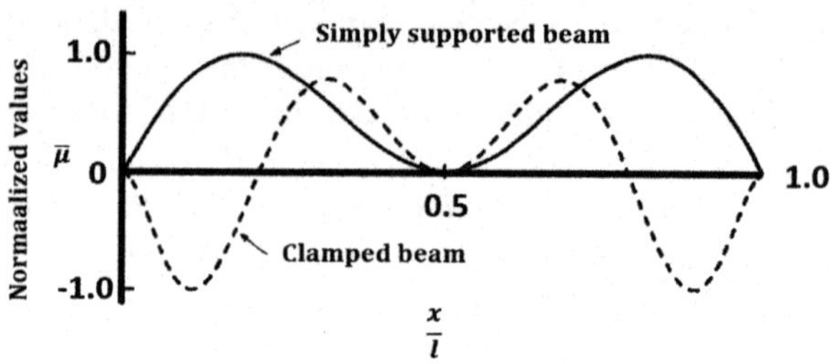

Fig. 12.7 Normalized values of $\bar{\mu}$ [2]

Now, the deflection $\bar{w}$ due to the live load is given by solving Eq. (12.37) as

$$\bar{w} = \frac{C_1x^3}{6} + \frac{C_2x^2}{2} + C_3x + C_4 + \frac{\bar{p}}{EI}\iint\iint(1-\mu)\mathrm{d}x\,\mathrm{d}x\,\mathrm{d}x\,\mathrm{d}x \tag{12.44}$$

in which the integral constants C_1 to C_4 are determined from the boundary conditions. The results for simply supported beams and clamped beams are given as

$$\bar{w} = \frac{C_1x^3}{6} + \frac{C_2x^2}{2} + C_3x + C_4 + \frac{\bar{p}}{EI}\left[\frac{x^4}{24} - kf(x)\right] \tag{12.45}$$

in which the constants C_1 to C_4 take the values

$$C_1 = -\frac{\bar{p}l}{EI}\left\{\frac{1}{2} - k\left[12f(l) + 8lf'(l) + l^2f''(l)\right]\right\} \tag{12.46}$$

$$C_3 = \frac{\bar{p}l^3}{EI}\left\{\frac{1}{24} - k\left[f(l) + \frac{4}{3}lf'(l) + \frac{l^2}{6}f''(l)\right]\right\} \tag{12.47}$$

$$C_2 = C_4 = 0 \tag{12.48}$$

$$f(x) = \frac{\bar{x}}{120} - \frac{\bar{x}^2}{360} - \frac{\bar{x}^3}{70} + \frac{\bar{x}^4}{84} + \frac{\bar{x}^5}{108} - \frac{\bar{x}^6}{60} + \frac{4\bar{x}^7}{495} - \frac{2\bar{x}^8}{1485} \tag{12.49}$$

for simply supported beams and

$$C_1 = -\frac{\bar{p}l}{EI}\left\{\frac{1}{2} - k\left[12f(l) + 6lf'(l)\right]\right\} \tag{12.50}$$

$$C_2 = -\frac{\bar{p}l^2}{EI}\left\{\frac{1}{12} - k\left[2f(l) + 2lf'(l)\right]\right\} \tag{12.51}$$

$$C_3 = C_4 = 0 \tag{12.52}$$

$$f(x) = -\frac{\bar{x}^2}{360} + \frac{\bar{x}^3}{70} - \frac{11\bar{x}^4}{336} + \frac{\bar{x}^5}{24} - \frac{11\bar{x}^6}{360} + \frac{2\bar{x}^7}{165} - \frac{\bar{x}^8}{495} \tag{12.53}$$

for clamped beams where $\bar{x} = x/l$. Hence, the maximum deflections $\bar{w}_{\max}$ including the effect of dead loads at the midspan of beams, where the effect of dead loads is maximum, are obtained as

$$\bar{w}_{max} = \frac{\bar{p}l^4}{EI}\left[\frac{5}{384} - 0.0006425k\right] \quad \text{(for simply supported beams)} \tag{12.54}$$

$$\bar{w}_{max} = \frac{\bar{p}l^4}{EI}\left[\frac{1}{384} - 0.000001797k\right] \quad \text{(for clamped beams)} \tag{12.55}$$

The well-known general deflections $\bar{w}_{0max}$, excluding the effect of dead loads, are easily obtained by neglecting these terms including k in Eqs. (12.54) and (12.55). Now, dividing $\bar{w}_{max}$ by $\bar{w}_{0max}$, Eqs. (12.54) and (12.55) can be expressed as

$$\frac{\bar{w}_{max}}{\bar{w}_{0max}} = 1 - 0.04934k \quad \text{(for simply supported beams)} \tag{12.56}$$

$$\frac{\bar{w}_{max}}{\bar{w}_{0max}} = 1 - 0.0006900k \quad \text{(for clamped beams)} \tag{12.57}$$

Similarly, the maximum values $\bar{M}_{max}$ of stress couples at the midspan can be expressed as

$$\frac{\bar{M}_{max}}{\bar{M}_{0max}} = 1 - 0.04543k \quad \text{(for simply supported beams)} \tag{12.58}$$

$$\frac{\bar{M}_{max}}{\bar{M}_{0max}} = 1 - 0.0006967k \quad \text{(for clamped beams)} \tag{12.59}$$

where $\bar{M}_{0max}$ are the maximum values of stress couples $\bar{M}_0$, excluding the effect of dead loads, due to uniform live loads at the midspan.

Thus, it is clarified that the effect of dead loads on uniformly distributed live loads has been given from Eqs. (12.56) to (12.57) and is physically governed by the value of k. And comparing Eqs. (12.56) and (12.57) with Eqs. (12.58) and (12.59), respectively, the effects of dead loads for deflections and stress couples are shown to be almost equal for beams with the same boundary conditions. This fact agrees with the preceding numerical results.

Next, comparing the closed-form approximate solutions proposed here with the preceding numerical results obtained from Eq. (12.24), good agreement is shown even with $\beta = 1$. Hence, the exactness of the proposed solutions is proven. Thus, the key physical factors governing the effect of dead loads on static live loads is in proportion to the second power of the dead load $\tilde{p}$ and of the slenderness ratio λ, and to the sixth power of span length l and in inverse proportion to the second power of the bending rigidity EI of the beam. By applying similar calculations to other loading patterns, the closed-form solutions can be obtained, although the calculation is complicated.

Next, the effects of dead load on the total maximum static deflections of beams are obtained as

$$\frac{(\bar{w}+\tilde{w})_{\max}}{(\bar{w}_0+\tilde{w})_{\max}} = 1 - \frac{\alpha}{1+\alpha}0.04934k \quad \text{(for simply supported beams)} \tag{12.60}$$

$$\frac{(\bar{w}+\tilde{w})_{\max}}{(\bar{w}_0+\tilde{w})_{\max}} = 1 - \frac{\alpha}{1+\alpha}0.00069k \quad \text{(for clamped beams)} \tag{12.61}$$

Hence, the effects of dead loads on the total maximum deflections are also affected by the factors $\alpha(=\bar{p}/\tilde{p})$ and k.

12.5 Proposal How to Reflect the Effect of Dead Load on Static Beams

We propose here how to apply the effect of dead loads to the modification of static live loads.

We present about a hot-rolled steel beam with the following data: the span length l = 7 m; the moment of inertia I = 6120 cm^4; the radius of gyration r = 8.3 cm; Young's modulus E = 2.06 × 10^{11} N/m^2. The slenderness ratio λ is 84.3. It is accomplished by the following procedure.

Step 1: Evaluate the value of k from Eqs. (12.40) and (12.41) under the assumption of β = 1.

$$k = \frac{3}{4}\left(\frac{700^3 \times 84.3}{24 \times 2.06 \times 10^7 \times 6{,}120}\right)^2 \tilde{p}^2 = 6.85 \times 10^{-5}\tilde{p}^2$$
$$\text{(for simply supported beams)} \tag{12.62}$$

$$k = 4.57 \times 10^{-5}\tilde{p}^2 \quad \text{(for clamped beams)} \tag{12.63}$$

in which $\tilde{p}$ (N/cm) is a uniform dead load in the beam.

Step 2: From Eqs. (12.56) and (12.57), the reduction ratios of the maximum deflections produced by live loads are given as

$$\frac{\bar{w}_{\max}}{\bar{w}_{0\max}} = 1 - 3.32 \times 10^{-6}\tilde{p}^2 \quad \text{(for simply supported beams)} \tag{12.64}$$

$$\frac{\bar{w}_{\max}}{\bar{w}_{0\max}} = 1 - 3.15 \times 10^{-8}\tilde{p}^2 \quad \text{(for clamped beams)} \tag{12.65}$$

Step 3: We have from the previous numerical results that the action of live loads is reduced on heavyweight beams more than on lightweight beams, due to the effect of dead loads. Denoting dead loads of safe beams for the current beam by $\tilde{p}_0$ for the current beam, the reduction ratios of the maximum deflections produced by live loads for safe beams are given from the above equations by

$$\frac{\bar{w}_{\max}}{\bar{w}_{0\max}} = 1 - 3.32 \times 10^{-6} \tilde{p}_0^2 \quad \text{(for simply supported beams)} \tag{12.66}$$

$$\frac{\bar{w}_{\max}}{\bar{w}_{0\max}} = 1 - 3.15 \times 10^{-8} \tilde{p}_0^2 \quad \text{(for clamped beams)} \tag{12.67}$$

The dead load $\tilde{p}_0$ of the safe beam will be obtained from the preventive investigation of structures collapsed by live loads.

Step 4: When the dead load $\tilde{p}$ of the beam is less than the dead load $\tilde{p}_0$ of the safe beam, live loads acting on a beam with dead loads $\tilde{p}$ must rise by

$$\frac{1 - 3.32 \times 10^{-6} \tilde{p}^2}{1 - 3.32 \times 10^{-6} \tilde{p}_0^2} \quad \text{(for simply supported beams)} \tag{12.68}$$

$$\frac{1 - 3.15 \times 10^{-8} \tilde{p}^2}{1 - 3.15 \times 10^{-8} \tilde{p}_0^2} \quad \text{(for clamped beams)} \tag{12.69}$$

To show definitely, assuming that $\tilde{p}_0 = 245$ N/cm and $\tilde{p} = 49$ N/cm, live loads $\bar{p}$ acting on the current beam with $\tilde{p} = 49$ N/cm must be raised 1.239 times for the simply supported beam and 1.002 times for the clamped beam, from Eqs. (12.68) and (12.69).

Step 5: The effect of dead loads on the total deflection of the current beam can be obtained from Eq. (12.60) for arbitrary live loads.

Thus, the increase of live loads depends on span length, bending rigidity, the slenderness ratio of the beams used, and the standard value $\tilde{p}_0$ for dead loads. This standard value of dead loads is determined from experiments or investigation of structures collapsed, as was mentioned. However, for practical use, it is recommended to substitute dead loads of reinforced concrete beams instead of those used commonly for steel beams as the standard value of dead loads. This method will have the same effect on live loads for steel beams and reinforced concrete beams.

12.6 Free Transverse Vibrations of Uniform Beams

In the preceding section, we have clarified the effect of dead loads on static beams. In this section, we consider the effect of dead loads on the natural frequencies of uniform elastic beams. For simplicity, it is assumed that the deflections of supporting ends of the uniform beams do not occur when external loads are applied dynamically. The advanced governing equation for free transverse vibrations including the effect of dead loads can be written from Eq. (12.16) as

$$\rho A \ddot{\bar{w}} + EI \bar{w}'''' - \underline{\frac{1}{2}[EA(\tilde{w}')^2 \bar{w}']'} = 0 \tag{12.70}$$

The well-known equation of motion for free transverse vibrations of beams is obtained by neglecting the underlined term in Eq. (12.70).

Equation (12.70) is linear differential equation with respect to unknown displacement $\bar{w}$ which is function of x and t. To solve Eq. (12.70), employing the method of separation of variable for the dynamic deflection, we have

$$\bar{w}(x,t) = \bar{W}(x)\Phi(t) \tag{12.71}$$

where $\bar{W}(x)$ is a function of x, and $\Phi(t)$ is a function of time t. Applying Eq. (12.71) to Eq. (12.70), we obtain the following two equations satisfied by $\bar{W}(x)$ and $\Phi(t)$:

$$\bar{W}'''' + \bar{R}(x) = k^4 \bar{W} \tag{12.72}$$

$$\ddot{\Phi}(t) + \omega^2 \Phi(t) = 0, \tag{12.73}$$

where $\bar{R}(x)$ is defined by

$$\bar{R}(x) = -\frac{1}{2r^2}\left[2\tilde{w}'\tilde{w}''\bar{W}' + (\tilde{w}')^2\bar{W}''\right] \tag{12.74}$$

$\bar{R}(x)$ represents the effect of dead loads and is a function including the known $\tilde{w}$ and the unknown $\bar{W}$. Also, γ is radius of gyration of the beam; and k^2 is defined as

$$k^2 = \omega\sqrt{\frac{\rho A}{EI}} \tag{12.75}$$

and ω^2 is a constant.

Assuming that current beams are simply supported beams and clamped beams, the deflections $\tilde{w}(x)$ due to uniformly distributed dead loads $\tilde{p}$ are given by Eq. (12.25), as already stated in Sect. 12.3.

Applying Eq. (12.25) into Eq. (12.72), we have

$$\bar{W}'''' + \sum_{i=0}^{6} b_i x^i \bar{W}'' + \sum_{i=1}^{5} a_i x^i \bar{W}' - k^4 \bar{W} = 0 \tag{12.76}$$

where the notations a_i and b_i are as defined in Appendix.

12.7 Analytical Method for Free Transverse Vibrations of Beams Using Galerkin Method

This section presents an analytical method to solve Eq. (12.76) by means of the Galerkin method. So, $\bar{W}(x)$ can be expressed in a power series expansion as follows

$$\bar{W}(x) = \sum_{n=1}^{\infty} \bar{W}_n f_n(x) \quad (n = 1, 2, 3, \ldots, \infty) \tag{12.77}$$

where $\bar{W}_n$ are unknown displacement coefficients; and $f_n(x)$ are shape functions satisfying the boundary conditions of the beams. We use the following functions represent f_n for simply supported beams and clamped beams:

$$f_n = \sin\frac{n\pi x}{l} \quad \text{(for simply supported beams)} \tag{12.78}$$

$$f_n = \sin\frac{\pi x}{l}\sin\frac{n\pi x}{l} \quad \text{(for clamped beams)} \tag{12.79}$$

The Galerkin equation of Eq. (12.76) can be obtained by

$$\int_0^l Q \cdot \delta \bar{W} \mathrm{d}x = 0 \tag{12.80}$$

in which, for simplicity, the notation Q represents Eq. (12.76). Using Eqs. (12.76) and (12.77), the Galerkin equation can be expressed by the following linear simultaneous algebraic equations with respect to the unknown $\bar{W}_m$

$$\delta \bar{W}_n : \quad \sum_{m=1} \bar{W}_m (A_{mn} - k^4 B_{mn}) = 0 \tag{12.81}$$

where the coefficient matrixes A_{mn} and B_{mn} are given as

$$A_{mn} = \int_0^l \left(f_m'''' + \sum_{i=0}^{6} b_i x^i f_m'' + \sum_{i=1}^{5} a_i x^i f_m' \right) f_n \mathrm{d}x \tag{12.82}$$

$$B_{mn} = \int_0^l f_m f_n \mathrm{d}x \tag{12.83}$$

For simply supported beams, the shape functions f_m (m =1, 2, …), given by Eq. (12.78), have orthogonal relation and the values of B_{mn} are

$$B_{mn} = \frac{1}{2}\delta_{mn} \tag{12.84}$$

in which δ_{mn} is the Kronecker delta which takes the value of 1 for $m = n$ and of zero for $m \neq n$. Thus, the nonzero values of B_{mn} appear in only the diagonal terms of the coefficient On the other hand, since the shape functions f_m for clamped beams, given by Eq. (12.79) have not orthogonal relation, the nonzero values of B_{mn} appear in all terms.

From the condition that Eq. (12.81) has nontrivial solutions for $\bar{W}_m$, the determinant of the coefficients must equal zero. The result reduces to a high-order algebraic equation with respect to the unknown k^4. By solving this equation for k^4 and substituting the i-th value k_i^4 into Eq. (12.75), the i-th natural frequency ω_i can be determined from

$$\omega_i = k_i^2\sqrt{\frac{EI}{\rho A}} \quad (i = 1, 2, \ldots) \tag{12.85}$$

Thus, the natural frequencies including the effect of dead loads for simply supported and clamped beams are evaluated for desired order of m.

12.8 Numerical Results for Free Transverse Vibrations of Beams

In this section, we conduct numerical computations corresponding free transverse vibrations for the following simply supported beams and clamped beams: $E = 2.06 \times 10^{11}$ N/m^2; the reference moment of inertia $I_0 = 2.5 \times 10^{-4}$m^4; reference radius of gyration $r_0 = 0.158$ m; and reference span length $l_0 = 8$ m. A reference dead load per unit length of the beams $\tilde{p}_0 = 6.7$ kN/m for steel beams without a concrete slab. The variations of I, r, l, and $\tilde{p}$ are given by parameters α_I, α_r, α_l, and $\alpha_{\tilde{p}}$ as follows

$$I = \alpha_I I_0;\ r = \alpha_r r_0;\ l = \alpha_l l_0;\ \tilde{p} = \alpha_{\tilde{p}}\tilde{p}_0 \tag{12.86}$$

Numerical computations result from varying these parameters.

Figure 12.8 shows the effect of dead loads on the natural frequencies of simply supported and clamped beams found by varying the ratio $\tilde{p}_0/\tilde{p}$ from 0.1 to 10. In this figure, solid lines and broken lines represent the results for simply supported beams and clamped beams, respectively, and the ordinate represents the nondimensioned value, in which ω_n is the n-th natural frequency including the effect of dead loads; and ω_{0n} denotes the n-th well-known natural frequency excluding the effect of dead loads. The well-known natural frequencies ω_{0n} are given by

$$\omega_{0n} = \left(\frac{n\pi}{l}\right)^2\sqrt{\frac{El}{\rho A}}(n = 1, 2, 3, \ldots, \infty)\ \text{(for simply supported beam)} \tag{12.87}$$

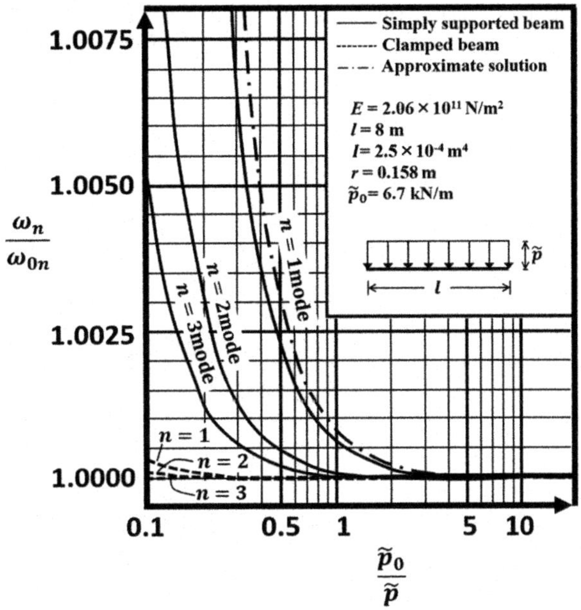

Fig. 12.8 Relationship between ω_n/ω_{0n} and $\tilde{p}_0/\tilde{p}$

$$\omega_{0n} = k_{0n}^2 \sqrt{\frac{El}{\rho A}} \quad (n = 1, 2, 3, \ldots, \infty) \quad \text{(for clamped beam)} \tag{12.88}$$

from Volterra and Zachmanoglou [5]. Here k_{0n} are solutions of the following frequency equations

$$\cos\,(k_{0n}l)\cosh\,(k_{0n}l) = 1 \tag{12.89}$$

Figure 12.8 shows that the natural frequencies of beams are slightly increased by the effect of dead loads. This effect is remarkable at higher $\tilde{p}$ (i.e., low values of the ratio $\tilde{p}_0/\tilde{p}$). It must be noticed that this phenomenon quite separates from the normal fluctuation of natural frequencies due to the change of mass per unit length of the beams. When $\tilde{p}$ is larger than $\tilde{p}_0$ selected arbitrarily, the ratios ω_n/ω_{0n} are greater than 1. Similarly, Figs. 12.9, 12.10, and 12.11 demonstrate the effect of dead loads on the first natural frequency depending on the moments of inertia, only the radii of gyration under a constant moment of inertia, and the span length, respectively.

We have the following conclusions about the effect of dead loads on free vibration of a uniform beam from numerical results.

(1) The natural frequencies of beams increase when the effect of dead loads is considered; this increase is substantial at lower natural frequencies of heavyweight beams. The effect is more apparent in simply supported beams than in clamped beams.

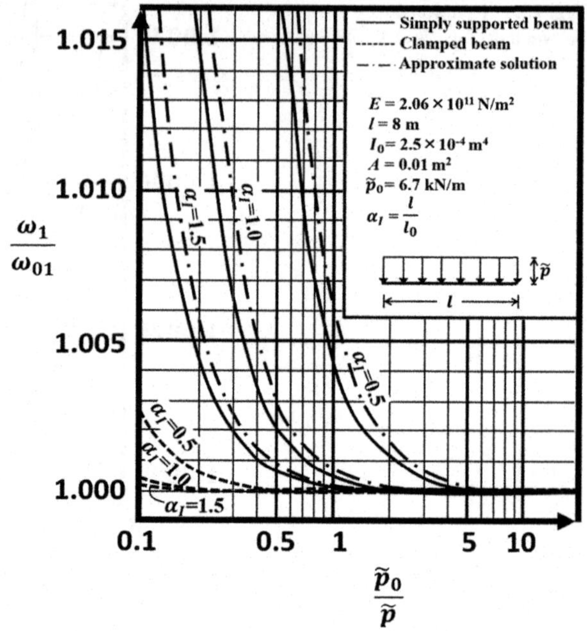

Fig. 12.9 Relationship between ω_1/ω_{01} and $\tilde{p}_0/\tilde{p}$

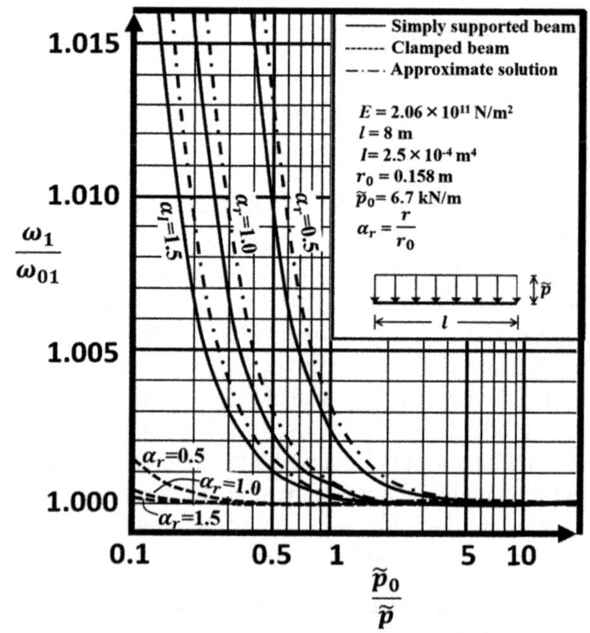

Fig. 12.10 Relationship between ω_1/ω_{01} and $\tilde{p}_0/\tilde{p}$

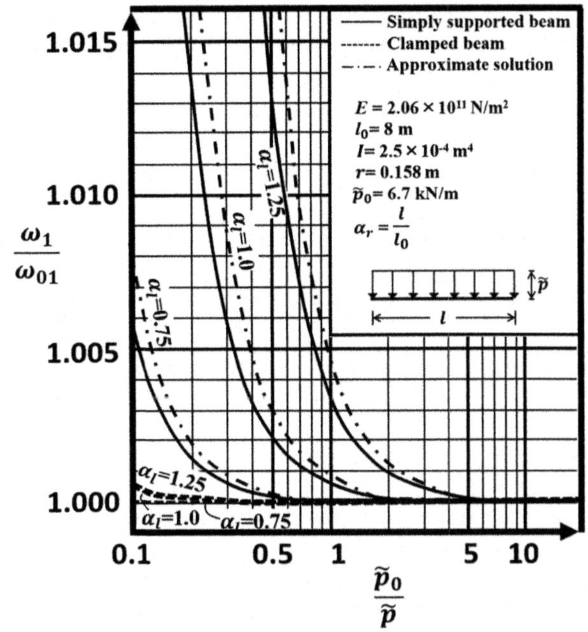

Fig. 12.11 Relationship between ω_1/ω_{01} and $\tilde{p}_0/\tilde{p}$

(2) The effect of dead loads on the natural frequencies of beams is remarkable in cases where the moment of inertia is small, or the radius of gyration is small for a constant moment of inertia, or the span is long.

12.9 Closed-Form Approximate Solutions for Natural Frequencies

In the preceding section, the existence of the effect of dead loads on natural frequencies has been clarified from numerical computations using Eq. (12.81) by means of the Galerkin method. This section presents a closed-form approximate solution for free vibration problem on beams, including the effect of dead loads. Also, the key physical factors governing the effect of dead loads are explicitly presented without the use of numerical computations. This can be accomplished by analytically solving, with suitable assumptions, the advanced equation of free vibrations with the effect of dead loads namely, Eq. (12.72). The method is to replace the unknown $\bar{W}$ in $\bar{R}(x)$ with an appropriate known function. A closed-form solution for only simply supported uniform beams is considered because it has been shown from the preceding results of the numerical computations that the effect of dead loads on the natural frequencies is substantial in these beams, but negligible in clamped beams, as shown in Figs. 12.8, 12.9, 12.10, and 12.11.

Firstly, we present the well-known equation of free vibration excluding the effect of dead loads. Indicating a quantity excluding the effect of dead loads with the subscript 0, the well-known equation for $\bar{W}_0$, excluding the effect of dead loads, is given from the following equation

$$\bar{W}_0'''' = k_0^4 \bar{W}_0 \tag{12.90}$$

For simply supported beams, the solution of Eq. (12.90) becomes

$$\bar{W}_0(x) = \sin \frac{m\pi x}{l} \tag{12.91}$$

Substituting Eq. (12.91) into Eq. (12.90), k_{0m}^4, for integers m, is

$$k_{0m}^4 = \left(\frac{m\pi}{l}\right)^4 \tag{12.92}$$

and the corresponding natural frequencies excluding the effect of dead loads, ω_{0m}, are given by Eq. (12.87).

Next, we replace the unknown $\bar{W}(x)$ in $\bar{R}(x)$ with an appropriate known function to solve analytically Eq. (12.72). It has shown from the preceding numerical computations that the rise in natural frequencies due to the effect of dead loads is a small value between 1 and 1.015 in terms of the ratios ω_i/ω_{0i}, for practical values of dead loads. So, the unknown $\bar{W}$ in $\bar{R}(x)$ given by Eq. (12.74) can be assumed to be

$$\bar{W}(x) = \beta\, \bar{W}_0(x) \tag{12.93}$$

in which β is a known constant indicating the effect of dead loads. Thus, $\bar{R}(x)$ can be changed from a function $\bar{R}(\bar{W}, \tilde{w})$, including both the unknown $\bar{W}$ and the known $\tilde{w}$, to a function $\bar{R}(\bar{W}_0, \tilde{w})$, including only the known $\bar{W}_0$ and $\tilde{w}$. Consequently, Eq. (12.72) is converted to an analytically solvable form. Then, $\bar{R}(x)$ can be expanded in a Fourier series as

$$\bar{R}(x) = \sum_{m=1} \bar{R}_m \sin \frac{m\pi x}{l} \tag{12.94}$$

in which $\bar{R}_m$ are the Fourier coefficients given by

$$\bar{R}_m = \frac{2}{l} \int_0^l \bar{R}(x) \sin \frac{m\pi x}{l} \mathrm{d}x \tag{12.95}$$

Assuming the following expression for $\bar{W}(x)$ in Eq. (12.72)

$$\bar{W}(x) = \sum_{m=1} \bar{W}_m \sin \frac{m\pi x}{l} \tag{12.96}$$

Substituting Eqs. (12.94) and (12.96) into Eq. (12.72), and using the relation of Eq. (12.92), k_m^4 for a selected integer m can be obtained as

$$k_m^4 = k_{0m}^4 + \bar{R}_m \tag{12.97}$$

Applying Eq. (12.97) to Eq. (12.85), the natural frequencies including the effect of dead loads, ω_m, can be given by

$$\omega_m = \omega_{0m}\sqrt{1 + \frac{\bar{R}_m}{k_{0m}^4}} \tag{12.98}$$

The natural frequencies including the effect of dead loads, ω_m, have the same unity $\sqrt{EI/\rho A}$ as the well-known natural frequencies ω_{0m}, excluding the effect of dead loads. The effect of dead loads on the natural frequencies is clarified to be reflected by $\bar{R}_m$ in Eq. (12.98). Therefore, by specifying $\bar{R}_m$ specifically, the physical factor that dominates the effect becomes clear. We present concrete $\bar{R}_m$ below.

For simply supported uniform beams, $\tilde{w}$ in $\bar{R}(x)$ can be expanded in a Fourier series as

$$\tilde{w} = \sum_{m=1} \tilde{w}_m \sin \frac{m\pi x}{l} \tag{12.99}$$

with the Fourier coefficients $\tilde{w}_m$ given by

$$\tilde{w}_m = \tilde{w}_m^* \tilde{p} \frac{l^4}{EI} \tag{12.100}$$

where $\tilde{w}_m^*$ denote the nondimensioned values of $\tilde{w}_m$ as given by

$$\tilde{w}_m^* = \frac{\tilde{p}_m^*}{(m\pi)^4} \tag{12.101}$$

Here, the nondimensioned coefficients $\tilde{p}_m^*$ have the following value for a uniform dead load $\tilde{p}$.

$$\tilde{p}_m^* = \frac{4}{m\pi} \quad (\text{for } m = 1, 3, 5, \ldots) \tag{12.102}$$

Then, substituting Eqs. (12.74), (12.91), (12.93), and (12.99) into Eq. (12.95), $\bar{R}_m$ can be written

$$\bar{R}_m = \frac{\beta l^4 \tilde{p}^2}{(rEI)^2} g_m \tag{12.103}$$

in which g_m are nondimensioned quantities for the m-th frequency mode of the beam, given by

$$g_m = \sum_{\tilde{n}=1} \sum_{\hat{n}=1} \sum_{\bar{m}=1} \frac{16\bar{m}}{\pi^6(\tilde{n}\hat{n})^4} \left(2\hat{n} \int_0^1 \sin(m\pi\xi)\,\sin(\hat{n}\pi\xi)\,\cos(\tilde{n}\pi\xi)\,\cos(\bar{m}\pi\xi)\,d\xi \right.$$
$$\left. +\bar{m} \int_0^1 \sin(m\pi\xi)\,\sin(\bar{m}\pi\xi)\,\cos(\tilde{n}\pi\xi)\,\cos(\hat{n}\pi\xi)\,d\xi \right) \quad (12.104)$$

where ξ is a nondimensioned variable of x ($\xi = x/l$). Substituting Eq. (12.103) into Eq. (12.98) gives

$$\omega_m = \omega_{0m}\sqrt{1 + k_\omega G_m} \quad (12.105)$$

in which k_ω and G_m are defined as follows

$$k_\omega = \beta\left[\frac{l^4\tilde{p}}{rEI}\right]^2 \quad (12.106)$$

$$G_m = \frac{g_m}{(m\pi)^4} \quad (12.107)$$

The values of G_m are found to be

$$\left.\begin{array}{ll} G_1 = 1.52503 \times 10^{-4} & \text{(for the first mode } m = 1) \\ G_2 = 0.22901 \times 10^{-4} & \text{(for the second mode } m = 2) \\ G_3 = 0.10016 \times 10^{-4} & \text{(for the third mode } m = 3) \\ G_4 = 0.0034186 \times 10^{-4} & \text{(for the fourth mode } m = 4) \\ G_5 = 0.0022899 \times 10^{-4} & \text{(for the fifth mode } m = 5) \end{array}\right\} \quad (12.108)$$

Subsequently, the exactness of the closed-form approximate solution proposed for the simply supported beams is examined by comparing the results obtained from Eq. (12.105) with the preceding results obtained from Eq. (12.81) by the use of the Galerkin method, in which β is assumed to be 1. The excellent agreement is obtained except in the first natural frequency, as shown in Figs. 12.8, 12.9, 12.10 and 12.11. The differences in the first natural frequency may be negligible in practical applications. Since the discrepancy results produced from taking $\beta = 1$, the solution for the first natural frequency will be improved by using a suitable function of x for β. The excellent agreement shown in the higher natural frequencies (i.e., above the first) is a consequence of the influence of β being countervailed by the sinusoidal shape of the vibrations.

Thus, it is clarified that the key physical factors governing the effect of dead loads on the natural frequencies of simply supported beams are represented in k_ω given in Eq. (12.106). It means that the effect of dead loads is influenced by the eighth power of the span length and the second power of the dead load, and inversely the second power of the radius of gyration and of the bending rigidity of the beam.

12.10 Conclusions

The effect of dead loads on static and transverse free vibration of elastic beams has been presented: First, the advanced governing equation of beams inclusive of the effect of dead loads has been formulated in general form; second, the existence of the effect of dead loads has been clarified from numerical calculations of beams; third, the closed-form approximate solutions for beams subject to uniformly distributed static live loads have been proposed; key factors governing the effect of dead loads on static elastic beams have been explicitly shown; And the simple method applying the effect of dead loads to live loads has been presented.

Next, the natural frequencies of simply supported and clamped beams have been demonstrated to increase by considering the effect of dead loads. The effect is more apparent in simply supported beams than clamped beams. So, the natural frequencies for simply supported beams have been presented in closed-form approximate solution. Also, the key factors governing the effect of dead loads on the natural frequencies have been clearly presented.

The effect of dead loads will be a significant factor in the safety of structures. The sectional efficiency of beams and the dead loads are important in estimating the safety of structures subjected to snow loads. This chapter proposes that in order to prevent the collapse of beams due to live loads, the safety factor for lightweight beams must be increased to equal that of the safety factor for heavyweight beams.

Appendix: Evaluation of a_i, and b_i

The sums defined in Eq. (12.76) are expressed explicitly as follows:

$$\sum_{i=1}^{5} a_i x^i = a_1 x + a_2 x^2 + a_3 x^3 + a_4 x^4 + a_5 x^5 \tag{12.109}$$

$$\sum_{i=0}^{6} b_i x^i = b_0 + b_1 x + b_2 x^2 + b_3 x^3 + b_4 x^4 + b_5 x^5 + b_6 x^6 \tag{12.110}$$

where the coefficients a_i and b_i are defined as follows:

$$\begin{gathered}a_1 = -(-12a + 4b^2)\bar{\beta};\ a_2 = -(12a - 36b)\tfrac{\bar{\beta}}{l};\ a_3 = -(72 + 32b)\tfrac{\bar{\beta}}{l^2}\\ a_4 = 120\tfrac{\bar{\beta}}{l^3};\ a_5 = -48\tfrac{\bar{\beta}}{l^4}\\ b_0 = -\tfrac{l^2}{2}a^2\bar{\beta};\ b_1 = 0;\ b_2 = -(-6a + 2b^2)\bar{\beta};\ b_3 = -(4a - 12b)\tfrac{\bar{\beta}}{l};\\ b_4 = -(18 + 8b)\tfrac{\bar{\beta}}{l^2};\ b_5 = 24\tfrac{\bar{\beta}}{l^3};\ b_6 = -8\tfrac{\bar{\beta}}{l^4}\end{gathered} \tag{12.111}$$

in which $\bar{\beta}$ takes the value

$$\bar{\beta} = \left(\frac{\bar{p}l^2}{24EIr}\right)^2 \tag{12.112}$$

and the constants a and b take the following values

$$\begin{gathered}a = 1 \text{ and } b = 0 \text{ (for simply supported beams)}\\ a = 0 \text{ and } b = 1 \text{ (for clamped beams)}\end{gathered} \tag{12.113}$$

References

1. An investigative report to damage by a very heavy snowfall in 1981. Architectural Institute of Japan, Tokyo, Japan (1981)
2. Takabatake, H.: Effects of dead loads in static beams. J. Struct. Eng. ASCE **116**(4), 1102–1120 (1990)
3. Takabatake, H.: Effect of dead loads on natural frequencies of beams. J. Struct. Eng. ASCE **117**(4), 1039–1052 (1991)
4. Washizu, K.: Variational Methods in Elasticity and Plasticity, 3rd edn. Pergamon Press, New York (1982)
5. Volterra, E., Zachmanoglou, E.C.: Dynamics of Vibrations. Charles, E., Merrill Books, Inc., Columbus, Ohio (1965)

Chapter 13
Effects of Dead Loads on Dynamic Problems of Beams

13.1 Introduction

In Chap. 12, the effect of dead loads on the static and free vibration problems of elastic beams has been demonstrated. This chapter is to present the effect of dead loads on dynamic problems of elastic beams subject to unmoving and moving dynamic live loads. First, the effect of dead loads on elastic beams subject to unmoving dynamic live loads is clarified from numerical computations using Galerkin method to the advanced governing equation of beams including the effect of dead loads. Second, a closed-form approximate solution for dynamic deflections of beams is presented from the advanced governing equation by introducing suitable assumptions. Third, a simple method that reflects the effect of dead loads on unmoving dynamic live loads is presented by an example. Forth, the effect of dead loads on elastic beams subjected to moving loads is clarified from the results of numerical computations applying the Galerkin method and Wilson-θ method to the advanced governing equation of beams. Sixth, a closed-form approximate solution for dynamic deflections of elastic beams subject to moving live loads is presented from the advanced governing equation including the effect of dead loads. Finally, it will be clarified that the effect of dead loads on elastic uniform beams subjected to unmoving and moving live loads restricts the transverse vibration for both the cases without and with the additional mass due to moving loads. The methodology proposed in this chapter is mainly based on Refs. [1, 2] proposed by the author.

13.2 Dynamic Analyses of Beams Subject to Unmoving Dynamic Live Loads

The advanced equation of motion of beams which includes the effect of dead loads is given from Eq. (12.16) in Chap. 12 by

H. Takabatake, *Simplified Analytical Methods of Elastic Plates*,
https://doi.org/10.1007/978-981-13-0086-8_13

$$\rho A\ddot{\bar{w}} + c\dot{\bar{w}} + (\mathrm{EI}\,\bar{w}'')'' - \frac{1}{2}[EA(\tilde{w}')^2\bar{w}']' = \bar{p} \tag{13.1}$$

in which $\bar{w}(x,t)$ represents the dynamic deflections produced by dynamic live loads and is measured from the reference state subjected to only dead loads.

We also use the Galerkin method to solve Eq. (13.1). The method of separation of variables is also employed assuming that

$$\bar{w}(x,t) = \sum_{n=1} \bar{w}_n(t) f_n(x) \tag{13.2}$$

where $\bar{w}_n(t)$ are the unknown displacement coefficients with respect to time t, and $f_n(x)$, shape functions satisfying the specified boundary conditions of beams. The following functions represent $f_n(x)$ for simply supported beams and clamped beams

$$f_n(x) = \sin\frac{n\pi x}{l} \quad \text{(for simply supported beams)} \tag{13.3}$$

$$f_n(x) = \sin\frac{\pi x}{l}\sin\frac{n\pi x}{l} \quad \text{(for clamped beams)} \tag{13.4}$$

Employing Eq. (13.2) into Eq. (13.1), the Galerkin equation can be written as

$$\delta\bar{w}_n : \sum_{\bar{n}=1}^{\infty} \bar{w}_{\bar{n}}[A_{n\bar{n}}] + \sum_{\bar{n}=1}^{\infty}\left[\ddot{\bar{w}}_{\bar{n}}\frac{\rho A}{\mathrm{EI}} + \dot{\bar{w}}_{\bar{n}}\frac{c}{\mathrm{EI}}\right]\int_0^l f_n f_{\bar{n}}\,\mathrm{d}x = \int_0^l \frac{\bar{p}}{\mathrm{EI}} f_n\,\mathrm{d}x \tag{13.5}$$

where $A_{n\bar{n}}$ are defined as

$$A_{n\bar{n}} = \int_0^l f_n f_{\bar{n}}''''\,\mathrm{d}x - \frac{1}{r^2}\int_0^l \bar{w}'\tilde{w}'' f_n f_{\bar{n}}''\,\mathrm{d}x - \frac{1}{2r^2}\int_0^l (\tilde{w}')^2 f_n f_{\bar{n}}''\,\mathrm{d}x \tag{13.6}$$

Equation (13.5) can be solved by means of the step-by-step integration methods. Once the dynamic displacements, $\bar{w}$, are determined from Eq. (13.2), the bending moments $\bar{M}$ due to the live load are determined from $\bar{M} = -\mathrm{EI}\,\bar{w}''$.

13.3 Numerical Results for Beams Subject to Unmoving Dynamic Live Loads

We examine the effect of dead loads on dynamic beams subject to unmoving dynamic live loads from numerical computations using Eq. (13.5). Numerical computations are conducted for the following data: Young's modulus $E = 2.06 \times 10^{11}$ N/m^2; span length $l = 8$ m; $I = 2.04 \times 10^{-4}$ m^4; and the radius of gyration $r = 0.131$ m for simply supported beams, while $I = 0.612 \times 10^{-4}$ m^4 and $r = 0.104$ m for clamped

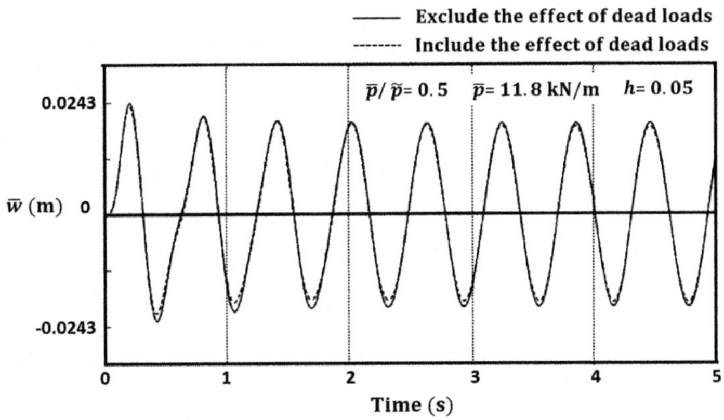

Fig. 13.1 Dynamic deflections $\bar{w}$ of the simply supported beam ($\bar{p}/\tilde{p} = 0.5$) [1]

beams. An external live load without mass is assumed to be the following harmonic load

$$\bar{p}\,(x, t) = \bar{p}_x \sin \omega_p t \tag{13.7}$$

and the maximum live load $\bar{p} = 11.8\,\text{kN/m}$, in which $\bar{p}_x(x)$ is a function of x and ω_p is a frequency of external load. The frequency ω_p is assumed to be half the first natural frequency of each beam. For uniformly distributed dynamic live loads whose intensity is $\bar{p}_0$, the relation $\bar{p}_x = \bar{p}_0$ exists.

Figures 13.1 and 13.2 show the time histories of dynamic deflections $\bar{w}$ at the midpoint of simply supported beams with the live-load-to-dead-load ratios $\bar{p}/\tilde{p} = 0.5$ and $\bar{p}/\tilde{p} = 0.1$, respectively, in which the damping constant $h = 0.05$ and the mass density of the beams is only the dead-load mass. The results demonstrate that the dynamic deflections produced by the dynamic live loads are minimized by considering the effect of dead loads. The effect of dead loads is larger in heavyweight beams than in lightweight beams.

Figure 13.3 represents the time histories of dynamic deflections $\bar{w}$ at the midpoint of clamped beams with $\bar{p}/\tilde{p} = 0.1$ and $h = 0.05$, in which the mass density of the beams is also only the dead-load mass. The results represent that the dynamic deflections are also decreased by considering the effect of dead loads, but the effect is not significant than that in simply supported beams.

Although the above results have been represented for beams with only the dead-load mass for the mass density, similar results may be obtained for beams with the mass density consisting of dead-load mass and live-load mass. Next, varying continuously the magnitude of dead loads in simply supported beams subjected to the above harmonic live load, we examine the effects of dead loads on dynamic deflections in beams. The results with a damping constant $h = 0.05$ are shown in Figs. 13.4, 13.5 and 13.6, in which $\bar{w}_{max}$ is the maximum value of dynamic deflections

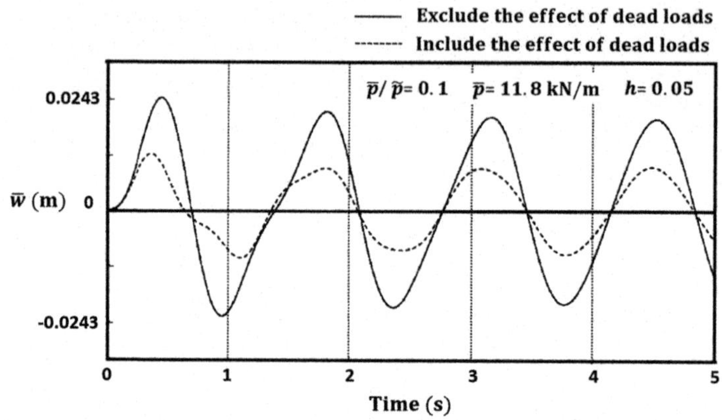

Fig. 13.2 Dynamic deflections $\bar{w}$ of the simply supported beam ($\bar{p}/\tilde{p} = 0.1$) [1]

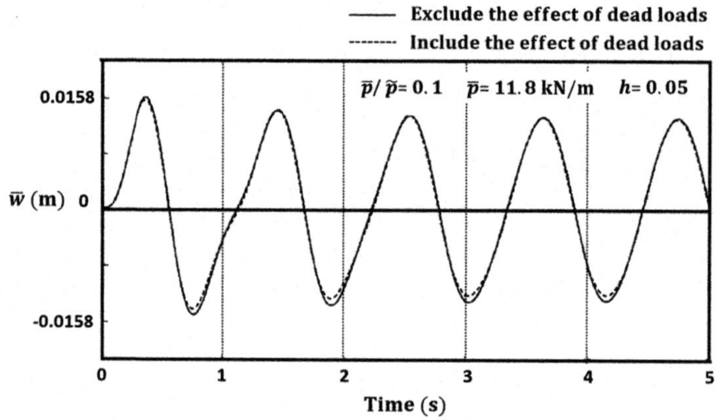

Fig. 13.3 Dynamic deflections $\bar{w}$ of the clamped beam ($\bar{p}/\tilde{p} = 0.1$) [1]

including the effect of dead loads, and $\bar{w}_{0\max}$ is the maximum value of dynamic deflections excluding the effect of dead loads.

Figure 13.4 represents the effect of dead loads in beams with three different moments of inertia, in which the mass density of beams excludes the live-load mass. The results demonstrate that the effect of dead loads is significant on the beam with the smallest moment of inertia. For example, this figure shows that for the beam with $\alpha_I = 1$ and the live-load-to-dead-load ratio $\bar{p}/\tilde{p} = 0.5$, the maximum values of the dynamic deflections and bending moments reduce to be 0.967 times of that without considering the dead-load effect. Similarly, Fig. 13.5 shows the effect of dead loads produced by the variation of the span length. The effect of dead loads is more remarkable on long span beam. Figure 13.6 shows the effect of dead loads for beams subjected to asymmetric, sinusoidal live loads. Thus, the effect of dead loads on the dynamic problem is clarified to be more significant in the weak beam.

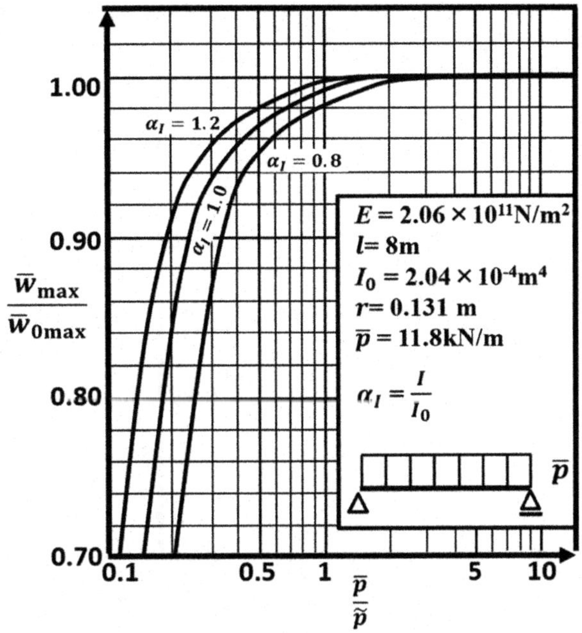

Fig. 13.4 Relationship between $\bar{w}_{max}/\bar{w}_{0\,max}$ and $\bar{p}/\tilde{p}$ [1]

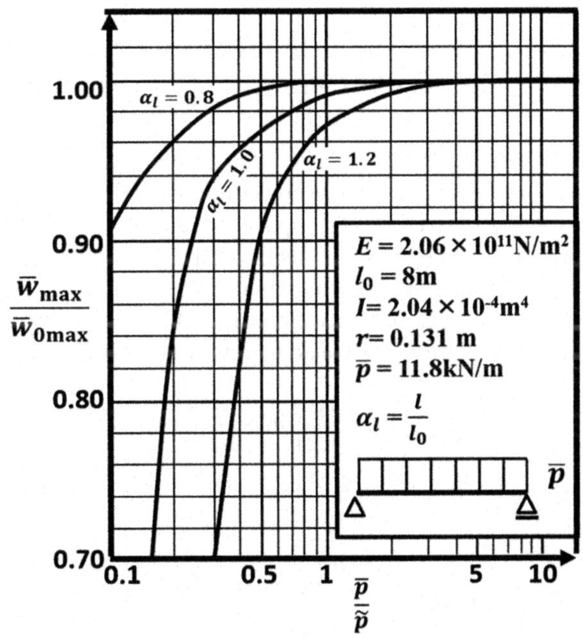

Fig. 13.5 Relationship between $\bar{w}_{max}/\bar{w}_{0max}$ and $\bar{p}/\tilde{p}$ [1]

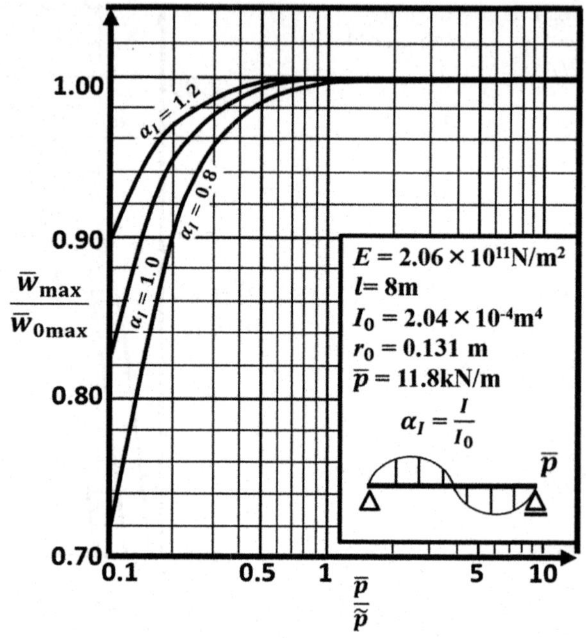

Fig. 13.6 Relationship between $\bar{w}_{\max}/\bar{w}_{0\,\max}$ and $\bar{p}/\tilde{p}$ [1]

13.4 Approximate Solutions for Simply Supported Beams Subject to Unmoving Dynamic Live Loads

The effect of dead loads on dynamic beams subjected to unmoving dynamic live loads has been shown from the preceding numerical results. For practical applications in structural analyses, we propose the closed-form approximate solution by employing appropriate assumptions into Eq. (13.1). The preceding results have shown that the effect of dead loads is remarkable in simply supported beams but negligible in clamped beams so that we consider the closed-form solution for simply supported beams. Equation (13.1) can be rewritten in simplified form as

$$\frac{\rho A}{\mathrm{EI}}\ddot{\bar{w}} + \frac{c}{\mathrm{EI}}\dot{\bar{w}} + \bar{w}'''' + R(x,t) = \frac{\bar{p}}{\mathrm{EI}} \tag{13.8}$$

in which $R(x, t)$ is defined as

$$R(x,t) = -\frac{1}{2r^2}[2\tilde{w}'\tilde{w}''\bar{w}' + (\tilde{w}')^2\bar{w}''] \tag{13.9}$$

We consider an approximate dynamic solution including the effect of dead loads for uniform elastic simply supported beams subject to unmoving dynamic live loads. Using the method of separation of variables for $R(x, t)$ and $\bar{w}\,(x, t)$, we have

$$R(x,t) = \sum_{n=1} \bar{R}_n(x)\ \Phi_n(t) \tag{13.10}$$

$$\bar{w}\,(x,t) = \sum_{n=1} \bar{W}_n(x)\,\Phi_n(t) \tag{13.11}$$

in which $\bar{W}_n(x)$ are natural functions satisfying the following differential equation for natural frequencies

$$\frac{\mathrm{EI}}{\rho A}[\bar{W}_n(x)'''' + \bar{R}_n(x)] = \omega_n^2 \bar{W}_n(x) \tag{13.12}$$

and $\Phi_n(t)$ is function with respect to time. Substituting Eqs. (13.10) and (13.11) into Eq. (13.8) and employing the relation of Eq. (13.12) into the result, we have

$$\sum_{n=1}^{\infty}\left[\bar{W}_n(x)\,\ddot{\Phi}_n(t) + \frac{c}{\rho A}\bar{W}_n(x)\ \dot{\Phi}_n(t) + \omega_n^2\ \bar{W}_n(x)\,\Phi_n(t)\right] = \frac{\bar{p}(x,t)}{\rho A} \tag{13.13}$$

The natural frequency ω_n including the effect of dead loads is given from Eq. (12.105) in the preceding chapter. Since the n-th natural function of simply supported beam is $\sin(n\pi x/l)$, multiplying Eq. (13.13) by $\bar{W}_m(x)$ and integrating it from 0 to l, we have

$$\ddot{\Phi}_n(t) + 2h_n\omega_n\dot{\Phi}_n(t) + \omega_n^2\Phi_n(t) = \frac{2}{\rho Al}Q_n(t) \tag{13.14}$$

in which the n-th damping constant h_n corresponding to the n-th mode and the notation $Q_n(t)$ are defined as

$$\frac{c}{\rho A} = 2h_n\omega_n \tag{13.15}$$

$$Q_n(t) = \int_0^l \bar{p}(x,t)\,\bar{W}_n\,\mathrm{d}x \tag{13.16}$$

The general solution of Eq. (13.14) is given as

$$\begin{aligned}\Phi_n(t) &= \mathrm{e}^{(-h_n\omega_n t)}[A_n \sin\omega_{Dn}t + B_n\cos\omega_{Dn}t] \\ &+ \frac{2}{\rho Al\omega_{Dn}}\int_0^t Q_n(\tau)\,\mathrm{e}^{[-h_n\omega_n(t-\tau)]}\sin[\omega_{Dn}(t-\tau)]\,\mathrm{d}\tau\end{aligned} \tag{13.17}$$

in which A_n and B_n are constants, and ω_{Dn} are given by the well-known relation

$$\omega_{Dn} = \omega_n\sqrt{1-h_n^2} \tag{13.18}$$

For unmoving harmonic live loads as given in Eq. (13.7), Eq. (13.16) becomes

$$Q_n(t) = \sin \omega_p t \; Q_n^* \tag{13.19}$$

in which Q_n^* are defined as

$$Q_n^* = \int_0^l \bar{p}_x(x)\, \bar{W}_n(x)\, \mathrm{d}x \tag{13.20}$$

Now, for simplicity, assuming the following initial conditions

$$\bar{w}\,(x, 0) = 0; \quad \dot{\bar{w}}(x, 0) = 0 \tag{13.21}$$

the dynamic deflections $\bar{w}(x, t)$ can be written as

$$\begin{aligned} \bar{w}(x,t) = & \frac{1}{\rho A l} \sum_{n=1} \frac{Q_n^*}{\omega_{Dn}} \left\{ \frac{h_n \omega_n \cos \omega_p t + (\omega_p + \omega_{Dn}) \sin \omega_p t}{(h_n \omega_n)^2 + (\omega_p + \omega_{Dn})^2} \right. \\ & - \frac{h_n \omega_n \cos \omega_p t + (\omega_p - \omega_{Dn}) \sin \omega_p t}{(h_n \omega_n)^2 + (\omega_p - \omega_{Dn})^2} \\ & + \mathrm{e}^{(-h\omega_n t)} \left[\frac{(\omega_p + \omega_{Dn}) \sin \omega_{Dn} t - h_n \omega_n \cos \omega_{Dn} t}{(h_n \omega_n)^2 + (\omega_p + \omega_{Dn})^2} \right. \\ & \left.\left. + \frac{(\omega_p - \omega_{Dn}) \sin \omega_{Dn} t + h_n \omega_n \cos \omega_{Dn} t}{(h_n \omega_n)^2 + (\omega_p - \omega_{Dn})^2} \right] \right\} \bar{W}_n \end{aligned} \tag{13.22}$$

The above solution is formulated in general form for damped beams. The deflections for undamped beams are easily obtained by putting $h_n = 0$ in Eq. (13.22). The well-known dynamic deflections excluding the effect of dead loads on dynamic beams subject to unmoving harmonic live loads are obtained by replacing ω_n with ω_{0n}. The damping constant h_n for n-th natural mode is $h_n = h_1 \cdot \omega_n / \omega_1$, in which ω_1 is the first natural frequency (rad/s).

From the relationships between the results obtained using the proposed approximate solution Eq. (13.22) and the preceding results obtained using Eq. (13.5), the proposed solution can show well the dynamic behavior of beams with the effect of dead loads. Figure 13.7 shows the relationships between numerical results obtained using Eqs. (13.5) and (13.22) about the time history of dynamic deflections of beams with $h = 0.05$ and $\bar{p}/\tilde{p} = 0.1$.

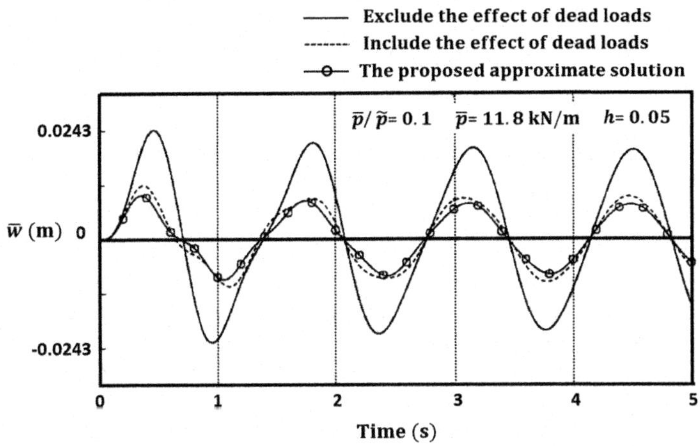

Fig. 13.7 Relationships between the approximate solution and numerical results [1]

13.5 How to Import the Effect of Dead Loads for Dynamic Beams Subject to Unmoving Dynamic Live Loads

The effect of dead loads on dynamic behavior of elastic beams is reflected on the modification of dynamic live loads by the following simple procedure:

Step 1: We consider to design a steel beam subjected to dead loads $\tilde{p}$ and dynamic live loads $\bar{p}$. Since live loads acting on current beam are prescribed from design codes, the designer cannot change freely the value. However, dead loads of the beam can be determined by the designer. When the structural cost is lowered, a beam is designed with the smallest cross-sectional property, and its dead loads are made minimum. The maximum value $\bar{w}_{\max}(\tilde{p}, \bar{p})$ of dynamic deflections $\bar{w}(x, t)$ for the beam, subjected to dead loads $\tilde{p}$ and dynamic live loads $\bar{p}$, is determined from Eq. (13.22). The maximum value includes the effect of dead loads produced by the selected dead loads $\tilde{p}$. If for the current beam figures like Figs. 13.4, 13.5 and 13.6 are previously prepared, the maximum value is obtained easily from the figures without numerical computation.

Step 2: Due to the effect of dead loads, the action of dynamic live loads is reduced on heavyweight beams more than on lightweight beams. It is assumed that a reference ratio of dynamic live loads to dead loads for current beam is already presented, with regard to the live-load problem to design safe beams. This ratio will be determined from wide investigation of structures collapsed by live loads or experiments. However, in the present state it is unclear. Therefore, for practical use, it is recommended that the reference magnitude of dead loads which gives the reference ratio of live loads to dead loads substitutes for the magnitude of dead loads of reinforced concrete beams subjected to the same dynamic live load. The reference magnitude of dead loads of safe beams is denoted by $\tilde{p}_0$. Thus, the maximum value $\bar{w}_{\max}(\tilde{p}_0, \bar{p})$

of dynamic deflections for the beam, subjected to the reference dead loads $\tilde{p}_0$ and dynamic live loads $\bar{p}$, is also determined from Eq. (13.22). Its value also includes the effect of dead loads produced by the reference dead loads $\tilde{p}_0$.

Step 3: When the dead loads $\tilde{p}$ which are selected by the designer are smaller than the reference dead loads $\tilde{p}_0$, the maximum dynamic deflection $\bar{w}_{\max}(\tilde{p}, \bar{p})$ of the beam with $\tilde{p}$ is larger than $\bar{w}_{\max}(\tilde{p}_0, \bar{p})$ of the safe beam with $\tilde{p}_0$, because the dynamic deflections produced by dynamic live loads are subjected to the effect of dead loads. This implies that live loads acting on the lightweight beam designed with light dead loads $\tilde{p}$ must rise by the ratio, $\bar{w}_{\max}(\tilde{p}, \bar{p})/\bar{w}_{\max}(\tilde{p}_0, \bar{p})$ to require the same degree of safety for heavyweight and lightweight beams. Thus, the magnitude of live loads may be estimated with considering the effect of dead loads. The beam with the dead loads $\tilde{p}$ is designed by live loads with this modified magnitude. On the other hand, when $\tilde{p}$ is larger than $\tilde{p}_0$, the magnitude of live loads is unchanged.

Thus, it will be possible to change the present value of live loads to an accurate value that has a positive effect on beams. Consequently, this method will have the same effect to live loads for steel beams and reinforced concrete beams. The present chapter does not recommend increasing the dead loads of structures to prevent the collapse of beams due to dynamic live loads. It is absolutely imperative that the safety factor for lightweight beams is raised to coincide with the safety factor for heavyweight beams.

13.6 Dynamic Analyses Using the Galerkin Method on Dynamic Beams Subject to Moving Live Loads

In the preceding sections, we have discussed the effect of dead loads on dynamic elastic beams subject to dynamic live loads without both the moving mass and moving loads. We consider the same problem on dynamic beams subject to the moving mass and moving loads in hereafter sections.

The dynamic displacement $\bar{w}_n(t)$ is obtained from Eq. (13.5) by means of the step-by-step integration method. Once the dynamic displacements $\bar{w}$ are determined from Eq. (13.2), the bending moments $\bar{M}$ due to live loads are determined from $\bar{M} = -\text{EI}\,\bar{w}''$. It must be noticed that these dynamic values are influenced by the effect of the dead loads.

13.7 Various Moving Loads

We consider moving loads for the dynamic live loads $\bar{p}(x, t)$. The moving loads $\bar{p}(x, t)$ are explained in a Fourier series as

$$\bar{p}(x, t) = \sum_{n=1} \bar{p}_n(t) \sin \frac{n\pi x}{l} \tag{13.23}$$

Volterra and Zachmanoglou [3] presented the various moving loads. Employing the Dirac delta function into the treatment of Volterra and Zachmanoglou, the Fourier coefficients $\bar{p}_n$ for various moving loads may be obtained as follows:

Case 1: A Concentrated Traveling Load
Consider the case of a concentrated traveling load $\bar{p}$ advancing along a beam with constant velocity v_0 as shown in Fig. 13.8a. At the instant t, the moving load $\bar{p}$ is at a distance $a = v_0 t$ from the left support. Hence, the moving load $\bar{p}(x, t)$ is given in

$$\bar{p}(x, t) = \bar{p}_0 \, \delta(x - v_0 t) \tag{13.24}$$

in which $\delta(x - v_0 t)$ is the Dirac delta function and $\bar{p}_0$ indicates the magnitude of the concentrated live load which is unmoving. The Fourier coefficients $\bar{p}_n$ of the moving load become

$$\bar{p}_n = \frac{1}{2} \int_0^l \bar{p}_0 \sin \frac{n\pi(v_0 t)}{l} \, \mathrm{d}x \tag{13.25}$$

Case 2: A Distributed Traveling Load
Consider the case of a uniformly distributed live load $\bar{p}_0$ per unit length advancing along the beam with constant velocity v_0. When the head of the moving load has reached a distance, $a = v_0 t$, from the left support as shown in Fig. 13.8b, the moving load distribution at this instant is given by

$$\begin{aligned} \bar{p}(x, t) &= \bar{p}_0 \ \text{for}\ 0 \le x \le a \\ \bar{p}(x, t) &= 0 \quad \text{for}\ a < x \end{aligned} \tag{13.26}$$

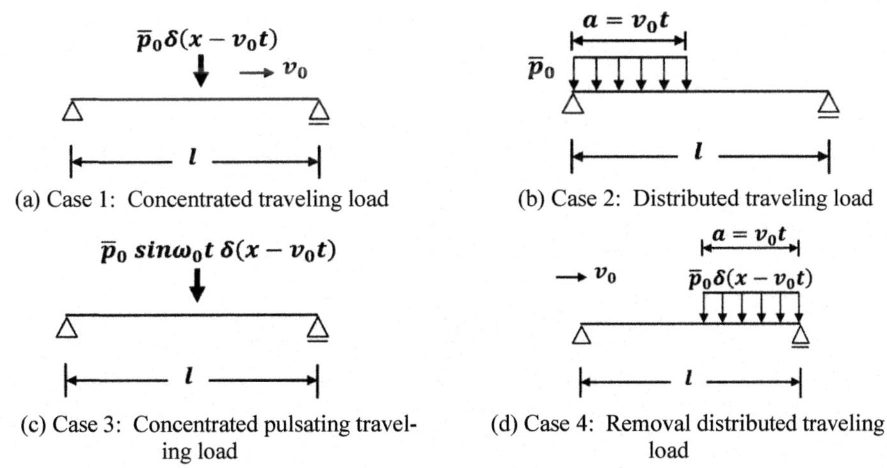

(a) Case 1: Concentrated traveling load

(b) Case 2: Distributed traveling load

(c) Case 3: Concentrated pulsating traveling load

(d) Case 4: Removal distributed traveling load

Fig. 13.8 Four kinds of moving loads: **a** Case 1, **b** Case 2, **c** Case 3, and **d** Case 4 [2] where $a = v_0 t$. Then, the Fourier coefficients $\bar{p}_n$ become

$$\bar{p}_n = \frac{2\bar{p}_0}{n\pi}\left[1 - \cos\frac{n\pi(v_0 t)}{l}\right] \tag{13.27}$$

Case 3: A Concentrated Pulsating Traveling Load

Consider the case of a concentrated alternating force $\bar{p}_0 \sin\omega_0 t$ advancing along the beam with constant velocity v_0, as shown in Fig. 13.8c. The moving load $\bar{p}(x, t)$ is given in

$$\bar{p}(x, t) = \bar{p}_0 \sin\omega_0 t\, \delta(x - v_0 t) \tag{13.28}$$

The Fourier coefficients $\bar{p}_n$ become

$$\bar{p}_n = \frac{1}{2}\bar{p}_0 \sin\omega_0 t \sin\frac{(n\pi\, v_0 t)}{l} \tag{13.29}$$

Case 4: A Removal Distributed Traveling Load

Consider the case that a uniformly distributed live load $\bar{p}_0$ per unit length advancing along the beam with constant velocity v_0 leaves out, as shown in Fig. 13.8d. This case is inverse with Case 2. This case is found out the sliding of snow on the roof. The moving load $\bar{p}(x, t)$ is

$$\begin{aligned} \bar{p}(x, t) &= 0 \quad \text{for } 0 \le x < a \\ \bar{p}(x, t) &= \bar{p}_0 \quad \text{for } a \le x \end{aligned} \tag{13.30}$$

in which $a = v_0 t$. The Fourier coefficients $\bar{p}_n$ become

$$\bar{p}_n = -\frac{2\bar{p}_0}{n\pi}\left[-(-1)^n \cos\frac{n\pi(v_0 t)}{l}\right] \tag{13.31}$$

Thus, the above-mentioned cases 1– 4 indicate the representative moving loads which are really interesting for structural engineering. The dynamic response including the effect of dead loads due to moving loads is calculated by substituting Eq. (13.23) into the external load term in Eq. (13.5).

13.8 Additional Mass Due to Moving Loads

The mass in the beam theory including the effect of dead loads is composed of the mass $\tilde{m}$ of only dead loads $\tilde{p}$ and the mass $\bar{m}(x, t)$ of only live loads $\bar{p}(x, t)$, as given

$$m = \tilde{m} + \bar{m}(x, t) \tag{13.32}$$

The mass $\tilde{m}$ is invariant and independent of the time. On the other hand, the mass $\bar{m}(x, t)$ is influenced on the location of moving loads. In beam theory, these masses depend on the dead loads and live loads as given in $\tilde{m} = \tilde{p}/g$ and $\bar{m} = \bar{p}/g$, respectively, in which g is gravity acceleration.

The additional mass $\bar{m}(x, t)$ may be expressed in the same treatment as the moving loads $\bar{p}(x, t)$. The moving additional mass $\bar{m}(x, t)$ is also explained in a Fourier series as

$$\bar{m}(x, t) = \sum_{n=1} \bar{m}_n \sin \frac{n\pi x}{l} \tag{13.33}$$

in which the Fourier coefficients $\bar{m}_n$ are given from the moving loads of four cases 1–4 stated in the Sect. 13.7.

$$\text{Case 1} \quad \bar{m}_n = \frac{2}{l} \bar{m}_0 \sin \frac{n\pi(v_0 t)}{l} \tag{13.34}$$

$$\text{Case 2} \quad \bar{m}_n = \frac{2\bar{m}_0}{n\pi} \left[1 - \cos \frac{n\pi(v_0 t)}{l} \right] \tag{13.35}$$

$$\text{Case 3} \quad \bar{m}_n = \frac{2}{l} \bar{m}_0 \sin \omega_0 t \, \sin \frac{n\pi(v_0 t)}{l} \tag{13.36}$$

$$\text{Case 4} \quad \bar{m}_n = \frac{2\bar{m}_0}{n\pi} \left[(-1)^n + \cos \frac{n\pi(v_0 t)}{l} \right] \tag{13.37}$$

Here $\bar{m}_0 = \bar{p}_0/g$.

Hence, in the calculation of step-by-step integration methods of Eq. (13.5) used Wilson-θ method, the variation with respect to the time t must be considered both the mass $m(x, t)$ and the moving loads $\bar{p}(x, t)$. However, the variation of moving additional mass $\bar{m}(x, t)$ during infinitesimal time may be assumed to be negligible.

13.9 Approximate Solutions of Beams Subject to Moving Live Loads

The effect of dead loads on dynamic beams subjected to moving loads is given from the numerical results using Eq. (13.5). For the practical use, we consider the closed-form approximate solution. Equation (13.5) is coupled form with respect to n and $\bar{n}$. Now assuming the uncoupled form of Eq. (13.5) upon the approximate orthogonal relation of the used shape functions, the Galerkin equation becomes

$$\ddot{\bar{w}}_n \frac{m}{\mathrm{EI}} \int_0^l f_n f_n \,\mathrm{dx} + \dot{\bar{w}}_n \frac{c}{\mathrm{EI}} \int_0^l f_n f_n \,\mathrm{dx} + \bar{w}_n A_{nn} = \int_0^l \frac{\bar{p}}{\mathrm{EI}} f_n \,\mathrm{dx} \tag{13.38}$$

The following notation is defined as

$$\int_0^l f_n f_n \, \mathrm{d}x = \beta_n \tag{13.39}$$

in which β_n is a constant. The substitution of Eq. (13.39) into Eq. (13.38) becomes

$$\ddot{\bar{w}}_n m + \dot{\bar{w}}_n c + \bar{w}_n A_{nn} \frac{\mathrm{EI}}{\beta_n} = \frac{1}{\beta_n} \int_0^l \bar{p} \, f_n \, \mathrm{d}x \tag{13.40}$$

The above equation is exactly the differential equation with variable coefficient because the additional mass $\bar{m}(x, t)$ due to moving loads depends on the time. When the additional mass $\bar{m}(x, t)$ is far smaller than unmoving mass $\tilde{m}$ due to the dead loads, the mass m may be considered as a constant which is independent of time. Using this assumption, Eq. (13.40) may reduce to the differential equation with constant coefficients.

$$\ddot{\bar{w}}_n + 2a_n \dot{\bar{w}}_n + b_n \bar{w} = q_n(t) \tag{13.41}$$

in which

$$a_n = \frac{c}{2m}; \quad b_n = \frac{\mathrm{EI}}{m \, \beta_n} A_{nn}; \quad q_n(t) = \frac{1}{m \, \beta_n} \int_0^t \bar{p}(x, t) \, f_n \, \mathrm{d}x \tag{13.42}$$

The general solution of Eq. (13.41) becomes

$$\bar{w}_n = \mathrm{e}^{-a_n t} [C_1 \sin \alpha_{0n} t + C_2 \cos \alpha_{0n} t] + \frac{1}{\alpha_{0n} t} \int_0^t \mathrm{e}^{-a_n (t-\tau)} \sin \alpha_{0n} (t - \tau) \, q_n(\tau) \, \mathrm{d}\tau \tag{13.43}$$

with $\alpha_{0n} = \sqrt{b_n - a_n^2}$.

13.10 Numerical Results for Beams Subject to Moving Live Loads

We examine the effect of dead loads on dynamic behavior of a simply supported elastic beams subject to the above-mentioned four kinds of moving loads. The beam has the following properties: $E = 2.06 \times 10^{11}$ N/m^2, $I = 2.04 \times 10^{-4}$ m^4, $l = 8$ m, $A = 0.11887$ m^2, $v_0 = 15$ m/s, $\omega_0 = 4\pi$, $\tilde{p} = 1960$ N/m, and $r = 0.131$ m. The

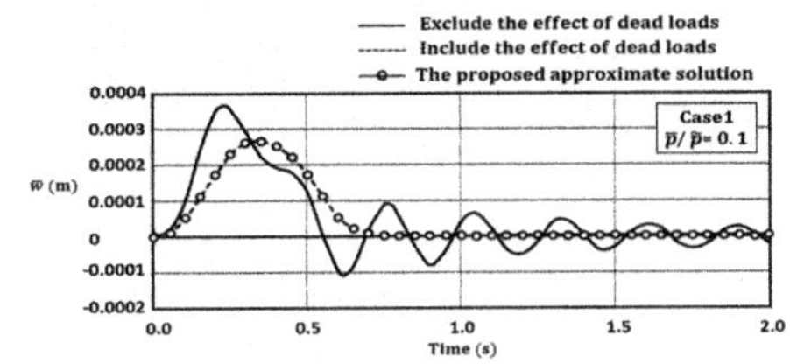

(a) $\bar{p}/\tilde{p}$ =0.1

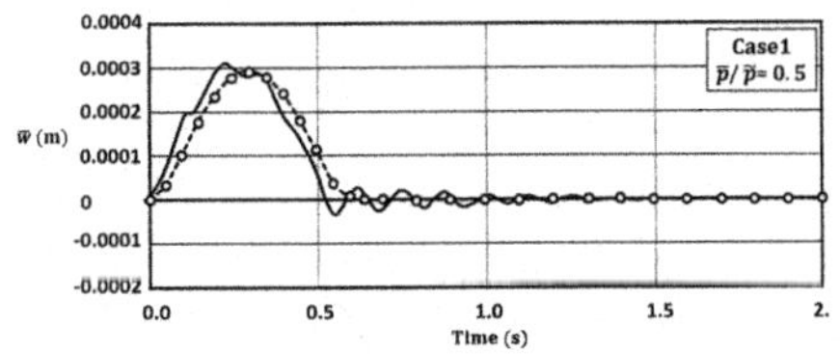

(b) $\bar{p}/\tilde{p}$ =0.5

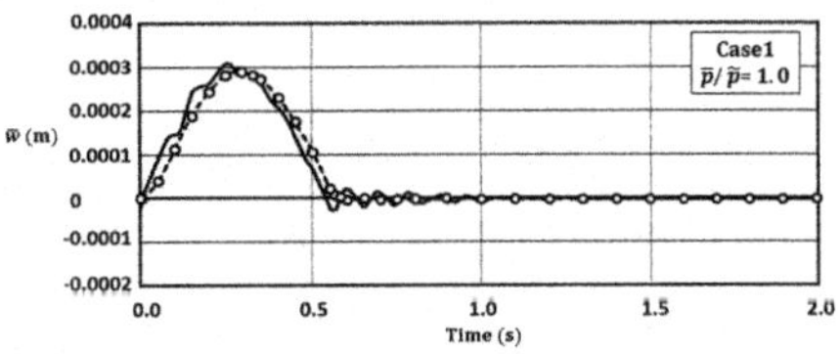

(c) $\bar{p}/\tilde{p}$ =1

Fig. 13.9 Time histories of dynamic deflection $\bar{w}$ at the midspan for Case 1; **a** $\bar{p}/\tilde{p} = 0.1$, **b** $\bar{p}/\tilde{p} = 0.5$, and **c** $\bar{p}/\tilde{p} = 1$ [2]

damping constant h_1 for the first natural mode of the beam is 0.02. The damping constant h_n for the higher n-th natural mode is $h_n = h_1 \cdot \omega_n/\omega_1$, in which ω_n is the n-th natural frequency (rad/s).

The effect of dead loads on the simply supported beam subject to moving loads is examined by varying the magnitude of the moving loads under the constant dead load. This is reflected by the load ratio ($\bar{p}/\tilde{p}$). The magnitude of the moving load in the current beam of load ratio 1 indicates 10 times larger than one in the load ratio 0.1. Figure 13.9a–c shows the time histories of dynamic deflection $\bar{w}$ at the midspan for Case 1, in which the load ratios take the values of 0.1, 0.5, and 1. The solid line displays the dynamic deflection excluded the effect of dead loads and agrees with the deflection obtained from the well-known fundamental beam theory subject to the moving load. The broken line indicates the deflection including the effect of dead loads. On the other hand, the line with circle represents the proposed approximate solution. It is clarified from these figures that the effect of dead loads is remarkable in the case that the load ratio is small. This implies that the dynamic behavior due to moving loads in the heavyweight beam is remarkably reduced by the effect of dead loads. Also, the approximated solution proposed here agrees with the numerical result including the effect of dead loads which is indicated by the broken line.

Figure 13.10a–c show the time histories of dynamic deflection $\bar{w}$ at the midspan for Case 2. Since the uniformly distributed moving load moves from the left side to the right side with a constant velocity v_0, the dynamic deflection does not cause the vibration. The effect of dead load is more remarkable in heavyweight beams than in lightweight ones.

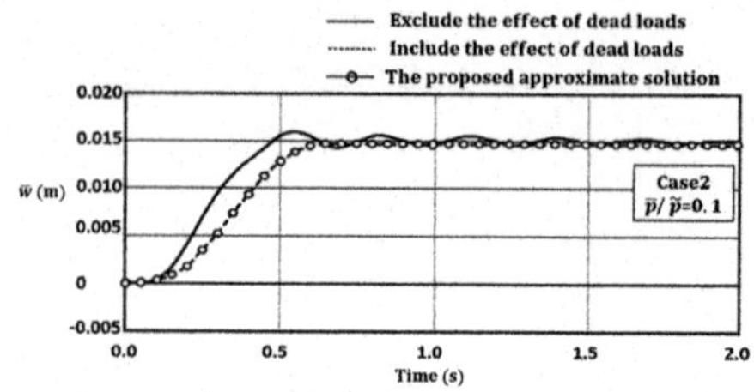

(a) $\overline{p}/\widetilde{p}$ =0.1

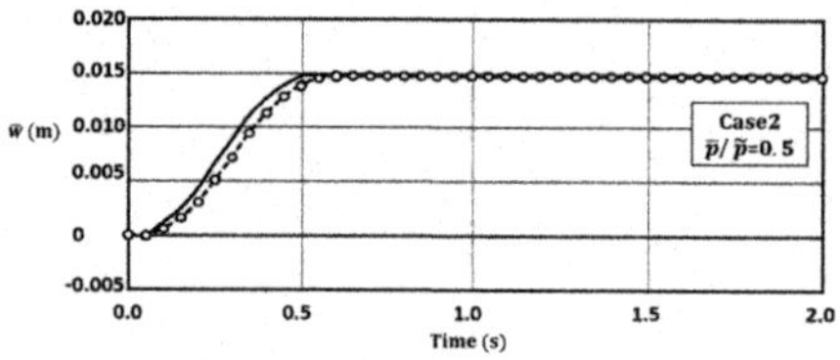

(b) $\overline{p}/\widetilde{p}$ =0.5

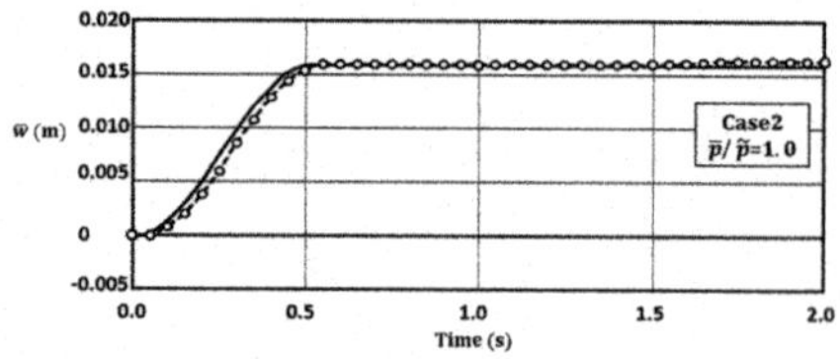

(c) $\overline{p}/\widetilde{p}$ =1

Fig. 13.10 Time histories of dynamic deflection $\bar{w}$ at the midspan for Case 2; **a** $\bar{p}/\tilde{p} = 0.1$, **b** $\bar{p}/\tilde{p} = 0.5$, and **c** $\bar{p}/\tilde{p} = 1$ [2]

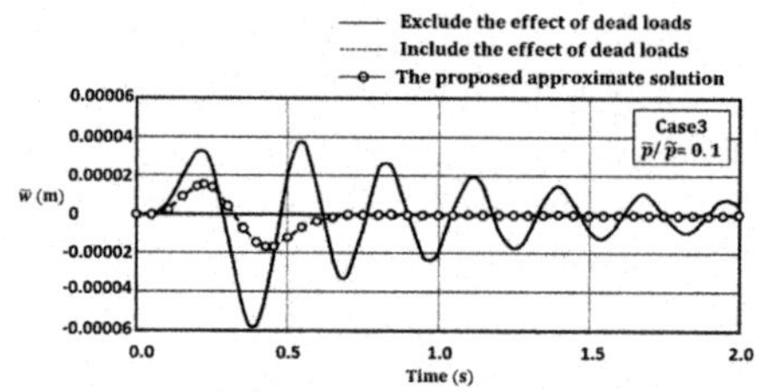

(a) $\overline{p}/\widetilde{p}$ =0.1

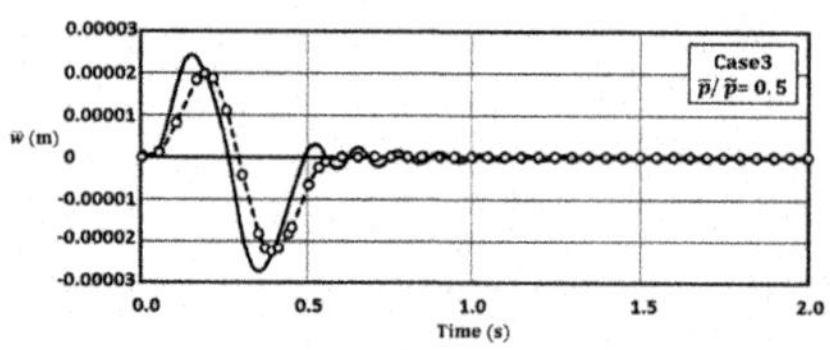

(b) $\overline{p}/\widetilde{p}$ =0.5

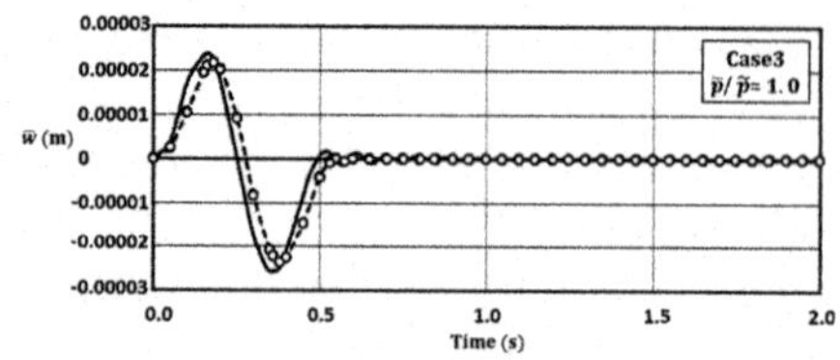

(c) $\overline{p}/\widetilde{p}$ =1

Fig. 13.11 Time histories of dynamic deflection $\bar{w}$ at the midspan for Case 3; **a** $\bar{p}/\tilde{p} = 0.1$, **b** $\bar{p}/\tilde{p} = 0.5$, and **c** $\bar{p}/\tilde{p} = 1$ [2]

Figure 13.11a–c shows the time histories of dynamic deflection $\bar{w}$ at the midspan for Case 3. The effect of dead loads is more remarkable on heavyweight beams than on lightweight ones. Its effect acts on the direction which constrains the vibration due to the moving loads in the heavyweight beam.

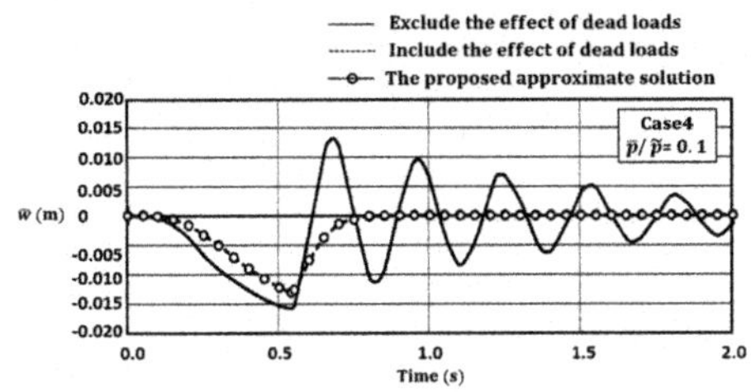

(a) $\bar{p}/\tilde{p}$ =0.1

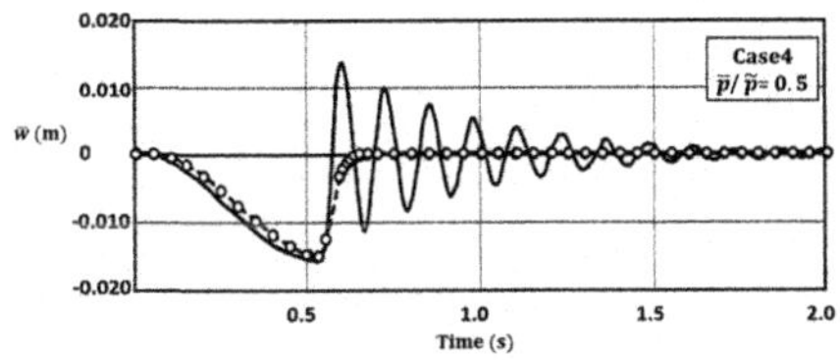

(b) $\bar{p}/\tilde{p}$ =0.5

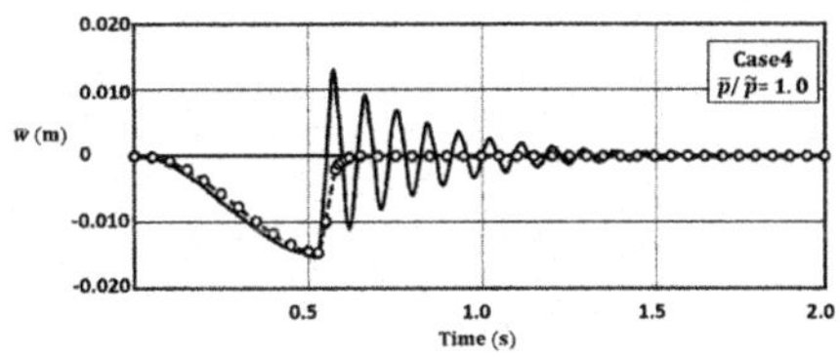

(c) $\bar{p}/\tilde{p}$ =1

Fig. 13.12 Time histories of dynamic deflection $\bar{w}$ at the midspan for Case 4; **a** $\bar{p}/\tilde{p} = 0.1$, **b** $\bar{p}/\tilde{p} = 0.5$, and **c** $\bar{p}/\tilde{p} = 1$ [2]

Also, Fig. 13.12a–c shows the time histories of dynamic deflection $\bar{w}$ at the midspan of Case 4. The numerical results in the case are the same behavior as the above-mentioned results.

Thus, it has been clarified that the effect of dead loads has the behavior which restrains the vibration of the dynamic deflection due to moving loads. The effect is more remarkable on the heavyweight beams than on lightweight ones. The above-mentioned results have been considered in the case that the additional mass due to the moving loads is negligible.

Next, we present the effect of dead loads considered the additional mass due to the moving loads. The numerical results including the effect of the additional mass are also shown in the figure used for the time history of the dynamic response described above.

Figure 13.13a, b shows the time histories of the dynamic deflection $\bar{w}$ at the midspan for Case 1, in which the load ratios take the value of 0.1 and 1, respectively. The solid line displays the dynamic deflection $\bar{w}$ excluding the effect of the dead loads. The broken line indicates the dynamic deflection including the effect of dead loads. These lines include the influence of the additional mass due to moving loads. It is clarified that the effect of dead loads on the dynamic deflection which includes the additional mass due to the moving loads restrains the vibration of dynamic defection due to the moving loads and that the effect is more remarkable on heavyweight beams than on lightweight ones.

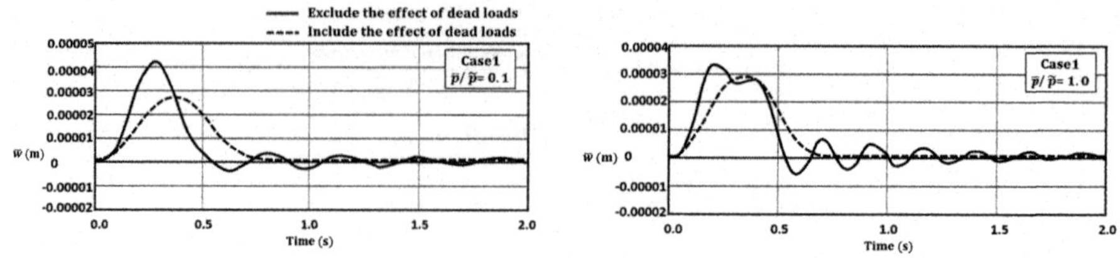

(a) $\overline{p}/\widetilde{p}$ =0.1 (b) $\overline{p}/\widetilde{p}$ =1

Fig. 13.13 Time histories of dynamic deflection $\bar{w}$ at the midspan for Case 1; **a** $\bar{p}/\tilde{p} = 0.1$ and **b** $\bar{p}/\tilde{p} = 1$ [2]

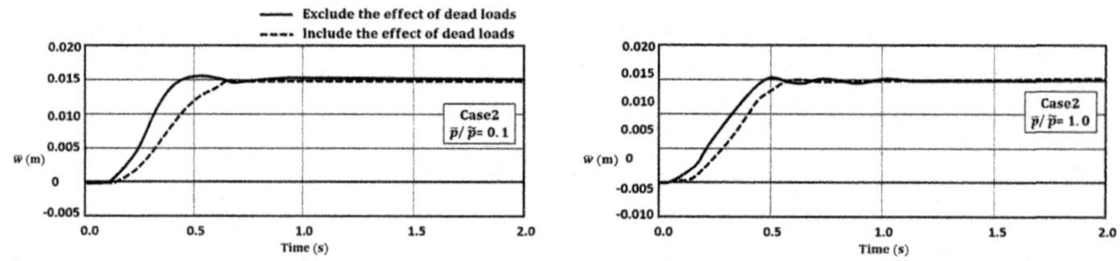

(a) $\overline{p}/\widetilde{p}$ =0.1 (b) $\overline{p}/\widetilde{p}$ =1

Fig. 13.14 Time histories of dynamic deflection $\bar{w}$ at the midspan for Case 2; **a** $\bar{p}/\tilde{p} = 0.1$ and **b** $\bar{p}/\tilde{p} = 1$ [2]

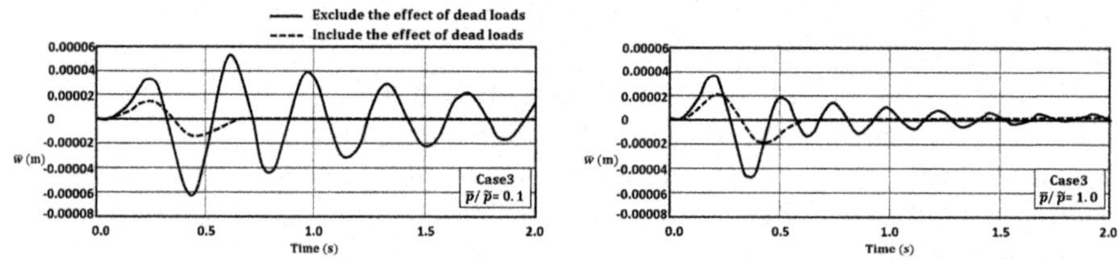

(a) $\overline{p}/\widetilde{p}$ =0.1 (b) $\overline{p}/\widetilde{p}$ =1

Fig. 13.15 Time histories of dynamic deflection $\bar{w}$ at the midspan for Case 3; **a** $\bar{p}/\tilde{p} = 0.1$ and **b** $\bar{p}/\tilde{p} = 1$. [2]

Figure 13.14a, b shows the time histories of dynamic deflection $\bar{w}$ at the midspan for Case 2. Figures 13.15a, b and 13.16a, b indicate the time histories of dynamic deflection $\bar{w}$ at the midspan for Cases 3 and 4, respectively. It is clarified from the above-mentioned numerical results that the effect of dead loads which includes the additional mass due to moving loads is more remarkable on heavyweight beams than on lightweight ones.

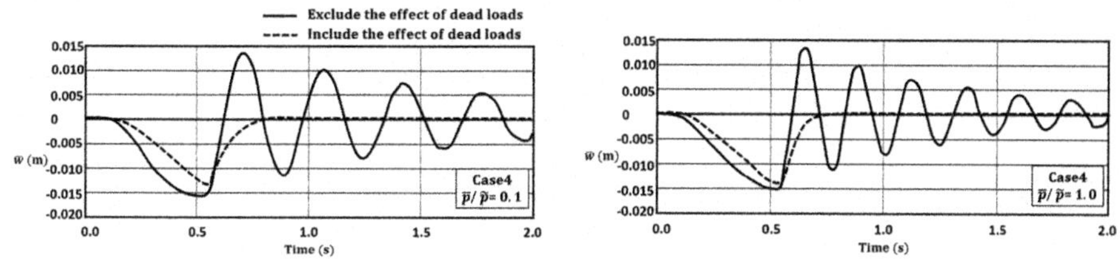

(a) $\bar{p}/\tilde{p}$ =0.1 (b) $\bar{p}/\tilde{p}$ =1

Fig. 13.16 Time histories of dynamic deflection $\bar{w}$ at the midspan for Case 4; **a** $\bar{p}/\tilde{p} = 0.1$ and **b** $\bar{p}/\tilde{p} = 1$ [2]

13.11 Conclusions

The effect of dead loads on elastic beams subjected to unmoving and moving live loads has been presented by the advanced moving equation including the influence of dead loads. From the results of numerical computation using the Galerkin method and from the closed-form approximate solution including the effect of dead loads, it has been clarified that the effect of dead loads reduces the action of dynamic live loads acting on elastic beams and is larger on heavyweight beams than on lightweight beams. It is also explained that this effect of dead loads restrains the transverse dynamic deflection $\bar{w}$ in the both cases with and without the additional mass due to moving loads. Last, the approximate solutions proposed here for the case without the additional mass due to unmoving and moving live loads show excellent agreement with the corresponding numerical results. This study will raise an important point to dynamic responses of bridges subject to moving loads. Since the action due to moving loads is reduced more effectively on heavyweight beams than on lightweight beams, the author proposes that the safety factor for lightweight beams should be raised to coincide with the safety factor for heavyweight beams in order to estimate the same action of beams due to moving loads.

References

1. Takabatake, H.: Effects of dead loads on dynamic analyses of beams. Earthq. Struct. **1**(4), 411–425 (2010)
2. Takabatake, H.: Effects of dead loads on dynamic analysis of beams subject to moving loads. Earthq. Struct. **5**(5), 589–605 (2013)
3. Volterra, E., Zachmanoglou, E.C.: Dynamics of Vibrations. Charles, E. Merrill Books, Inc., Columbus, Ohio (1965)

Part VI
Recent Topics of Plate Analysis

Chapter 14
Refined Plate Theory in Bending Problem of Rectangular Normal Plates

14.1 Introduction

The object of this book is to present the practical analytical methodology to simply solve the static and dynamic problems of various plates widely used in civil, such as plates with voids, cellular plates, stiffened with beams, and plates with stepped thickness. These plates have many distinguishing traits which are heavy-weight dead loads, the variable stiffness and variable mass; relatively long span, as compared with plates used in mechanical, aerospace, nuclear, and electronics. Also plates used in civil and buildings are mainly composed of reinforced concrete to bring down the influence of noise and oscillation produced by live loads. For plates with long span, structural designers select the most rational structural form in which the ratio of structural stiffness to the dead loads is extremely high rate, after considering the cost and execution of construction. Although these plates are considered to be composed of isotropic material in broad perspective, the analysis of plates with variable stiffness and variable mass is complicated.

Therefore, in the practical design the static and dynamic analyses to these plates almost depend on analytical method used FEM because of tremendous progress of personal computer (PC) power and the extension of FEM software. However, it is very difficult for structural engineers to solve exactly an originally three-dimensional and complicated plate structure by means of FEM-used solid element and shell element.

So, many structural engineers demand a simple and exact analytical method and the closed-form solutions, which are usable in the preliminary stage of design to such a new plate structure with unexperienced matters. These simplified analytical methods based on two-dimensional approach and simple expression are also necessary to examine a new structural form of plates and the exactness of computational results obtained from FEM. Thus, the many analytical methodologies proposed in the preceding all chapters are effective to simply solve the various plate in civil. This chapter presents recent topics of plates analysis based on two-dimensional approach.

H. Takabatake, *Simplified Analytical Methods of Elastic Plates*,
https://doi.org/10.1007/978-981-13-0086-8_14

14.2 Various Plate Theories

In order to simply analyze the static and dynamic lateral behaviors of plates being originally three-dimensional structures, many researchers proposed many effective methodologies based on the two-dimensional approach by making suitable assumptions concerning the kinematic of deformation and/or the state of stress through the thickness of the plate.

Table 14.1 comprehends the various problems of plates in 2-dimensional approach.

The present remarkable development depends on many comprehensive reviews of plate theory. The representative reviews of various plate theories are given as follows:

Table 14.1 Various problems of plates in the 2-D approach

Cause	Division
Deformation	Linear theory (small displacement) Nonlinear theory
Thickness	Thin plate Thick plate
Material	Isotropic plate Orthotropic plate Anisotropic plate
Plate	Isotropic plate Homogeneous plate Sandwich plate Laminated plate Functionally graded plate (FGP)
Shear deformation	Classical thin plate theory (CPT) Kirchhoff-Love hypothesis First-order shear deformation theory (FSDT) Mindlin-Reissner hypothesis Higher-order shear deformation theory (HSDT) Trigonometric shear deformation plate theory (TSDT) Two-variable refined plate theory (RPT) Advanced two-variable refined plate theory (ARPT)
Boundary condition	Edge support Inner support Elastic foundation
Load	Static Free vibration Forced vibration Moving load
Stiffness and mass	Uniform stiffness and mass Variable stiffness and mass

Vasilév [1]	Reviewed topic analysis of the classical theory of thin isotropic plates
Altenbach and Altenbach [2]	Engineering plate theory
Ghugal and Shimpi [3]	Refined shear deformation theories of isotropic and anisotropic laminated plates
Vijayakumar [4]	Plate theory in bending
Carrera and Brischetto [5]	Analysis of sandwich plates
Reddy and Lin [6]	Analysis of laminated plates
Liew et al. [7]	Thick plate vibration

The structural elements of constructions on mechanical structural analysis in civil engineering are in general single-layered plates composed of metal or reinforced concrete materials. Since their mechanical properties may be regarded as a macroscopic level homogeneously through the thickness, the structural response may be considered as isotropic or orthotropic.

Nowadays, various plate elements consisting of three (sandwich plate) or a much greater number of layers (multilayered plate and laminated plate) are widely used in various fields of engineering, in aircraft and rocket constructions, machine building, automotive industry, oil production, civil engineering, energetics, sports industry, etc.

Due to recent astonishing advances on new structural materials, functionally graded materials are classified as novel composite materials. The material properties exhibit a continuous change from one surface to another by gradually varying the volume fraction of constituent materials.

The most significant point in bending of plate theory is how to express rationally and simply the transverse shear deformation and transverse shear stress distributed along the thickness from an engineering viewpoint.

Although many plate theories, which take the transverse deformation, stress, and strain into, are reported in the literature, as shown in References [1–42], the basic concept may be classified as shown in Table 14.1.

(1) Classical plate theory (CPT)

The classical plate theory is based on Kirchhoff-Love hypothesis, which the lines being normal to the neutral surface (middle surface) before deformation remain straight and normal to the neutral surface after deformation, and neglects the transverse shear deformation. This assumption causes underestimation for deflection and over-estimation for natural frequencies. However, this classical plate theory is the simplest theory of plate and is effective for thin plate.

The displacement field based on this theory may be expressed as follows:

$$\left.\begin{aligned} U(x, y, z, t) &= -zw_{,x} \\ V(x, y, z, t) &= -zw_{,y} \\ W(x, y, z, t) &= w(x, y, t) \end{aligned}\right\} \tag{14.1}$$

in which U, V, and W indicate displacement components on the general points; and w is the only unknown parameter in this theory which represents the deflection of the middle surface.

The reduced transverse shear strains γ_{xz} and γ_{yz} are zero. The transverse stresses τ_{xz} and τ_{yz} distribute parabolic over the thickness h_0.

(2) First-order shear deformation theory (FSDT)

This theory is called generically as Mindlin-Reissner plate theory and considers the transverse shear deformation, in which the transverse shear strain is assumed to be constant. Since the distribution of the transverse shear stress is constant through the thickness of the plate, this assumption cannot satisfy zero stress conditions on the free surfaces. Therefore, the shear correction factor is needed to modify the value of maximum shear stress with that of exact value. This theory has three variables concerning displacements, in which the deflection of the middle surface, w, and two rotation angles, ϕ_x and ϕ_y, of the middle surface. This theory corresponds to Timoshenko beam theory in beams.

The displacement field based on this theory may be expressed as follows:

$$\left.\begin{aligned} U(x, y, z, t) &= z\phi_x(x, y, t) \\ V(x, y, z, t) &= z\phi_y(x, y, t) \\ W(x, y, z, t) &= w(x, y, t) \end{aligned}\right\} \tag{14.2}$$

in which ϕ_x, ϕ_y, and w are the unknown parameters in this theory.

The reduced transverse shear strains γ_{xz} and γ_{yz} are constant. The transverse stresses τ_{xz} and τ_{yz} distribute constant over the thickness h_0. Since zero stress conditions on the surfaces for the distribution of the transverse shear stress do not satisfy, the shear correction factor is needed to modify the value of the maximum shear stress.

(3) High-order shear deformation theory (HSDT)

This theory assumes high-order terms in Taylor's expansions of the transverse shear deformations in the thickness coordinate, such as second-order shear deformation theory, third-order shear deformation theory, more higher-order shear deformation, etc., in order to avoid the use of the shear correction factor and to resolve the problem of free surface concerning the transverse stress. The variable increases corresponding to high-order shear deformation.

The displacement field based on this theory may be expressed as follows:

$$\left.\begin{aligned} U(x, y, z, t) &= z \overset{(1)}{\phi}{}_x(x, y, t) + z^2 \overset{(2)}{\phi}{}_x(x, y, t) + z^3 \overset{(3)}{\phi}{}_x(x, y, t) + \cdots \\ V(x, y, z, t) &= z \overset{(1)}{\phi}{}_y(x, y, t) + z^2 \overset{(2)}{\phi}{}_y(x, y, t) + z^3 \overset{(3)}{\phi}{}_y(x, y, t) + \cdots \\ W(x, y, z, t) &= w(x, y, t) \end{aligned}\right\} \tag{14.3}$$

in which $\overset{(n)}{\phi}_x (n = 1, 2, \ldots), \overset{(n)}{\phi}_y (n = 1, 2, \ldots)$, and w are the unknown parameters in this theory.

For multilayered laminated plates involving layers with low transverse shear moduli, the shear deformation effects become more pronounced, so that the higher-order terms in power series will be necessary.

Meanwhile, a refined equivalent single-layer theory which replaces the multilayered laminated plate with single-layer plates expresses displacement field as follows, instead of power series with respect to thickness coordinate:

$$\left.\begin{aligned} U(x, y, z, t) &= -zw_{,x} + f(z)\phi_x(x, y, t) \\ V(x, y, z, t) &= -zw_{,y} + f(z)\phi_y(x, y, t) \\ W(x, y, z, t) &= w(x, y, t) \end{aligned}\right\} \tag{14.4}$$

in which ϕ_x, ϕ_y, and w are the unknown parameters associated with the shear slopes in this theory. $f(z)$ is known function depending on z and is assumed to be sinusoidal or exponential shear deformation with respect to thickness coordinates.

$$f(z) = \frac{h_0}{\pi} \sin \frac{\pi z}{h_0} \quad \text{or} \quad z \cdot \exp\left[-2\left(\frac{z}{h_0}\right)^2\right] \tag{14.5}$$

in which h_0 is the thickness of the plate.

(4) Trigonometric shear deformation plate theory (TSDT)

The in-plane displacements consist of two components: the component produced by bending and the component produced by transverse shear deformation being sinusoidal distribution with respect to thickness coordinate.

The displacement field based on this theory may be expressed as follows:

$$\left.\begin{aligned} U(x, y, z, t) &= -zw_{,x} + \tfrac{h_0}{\pi} \sin \tfrac{\pi z}{h_0} \phi_x(x, y, t) \\ V(x, y, z, t) &= -zw_{,y} + \tfrac{h_0}{\pi} \sin \tfrac{\pi z}{h_0} \phi_y(x, y, t) \\ W(x, y, z, t) &= w(x, y, t) + \tfrac{h_0}{\pi} \cos \tfrac{\pi z}{h_0} \phi_z(x, y, t) \end{aligned}\right\} \tag{14.6}$$

in which ϕ_x,ϕ_y,ϕ_z, and w are the unknown parameters in this theory. For multilayered laminated plates, this theory is a displacement-based refined equivalent single-layer theory.

(5) Two-variable refined plate theory (RPT)

This theory contains only two unknown variables to satisfy the zero stress conditions on the surfaces for the transverse stress and not to need the shear correction factor. This theory is proposed originally by Shimpi [8–11] for isotropic plates. Thereafter this theory is extended for various problems of orthotropic plates, laminated composite plates, functionally graded plates, etc. In the two-variable refined plate theory proposed by Shimpi, the transverse displacement of plates is represented by two components, viz. bending component and shear component. The governing equations to flexure of shear-deformable isotropic plates have two corresponding to two variables.

The displacement field based on this theory may be expressed as follows:

$$\left.\begin{aligned} U(x, y, z, t) &= -zw_{\mathrm{b},x} + f(z)w_{\mathrm{s},x} \\ V(x, y, z, t) &= -zw_{\mathrm{b},y} + f(z)w_{\mathrm{s},y} \\ W(x, y, z, t) &= w_{\mathrm{b}}(x, y, z, t) + w_{\mathrm{s}}(x, y, z, t) \end{aligned}\right\} \quad (14.7)$$

in which w_{b} and w_{s} are the unknown parameters in this theory. The subscripts "b" and "s" indicate to be quantites produced by bending and transverse shear, respectively. $f(z)$ is known function depending on z.

(6) Advanced two-variable refined plate theory (ARPT)

This theory translates from two coupled governing equations obtained from two-variable refined plate theory in terms of the bending deflection component w_{b} and the shear deflection component w_{s} to one uncoupled governing equation including only w_{b} by expressing w_{s} with w_{b}.

The displacement field based on this theory may be expressed as follows:

$$\left.\begin{aligned} U(x, y, z, t) &= -zw_{\mathrm{b},x} + f(z)w_{\mathrm{s},x} \\ V(x, y, z, t) &= -zw_{\mathrm{b},y} + f(z)w_{\mathrm{s},y} \\ W(x, y, z, t) &= w_{\mathrm{b}}(x, y, z, t) + w_{\mathrm{s}}(x, y, z, t) \end{aligned}\right\} \quad (14.8)$$

in which w_{b} and w_{s} are the unknown parameters in this theory. $f(z)$ is known function depending on z. The displacement component w_{s} is replaced with w_{b} by the use of relation based on the equilibrium equations of three-dimensional elasticity theory.

$$w_{\mathrm{s}} = w_{\mathrm{s}}(w_{\mathrm{b}}) \quad (14.9)$$

14.3 Analysis of Isotropic Plates Using Refined Plate Theory

This section introduces expression generalized refined plate theory proposed by Shimpi and coworkers [8–11] for rectangular isotropic plates. The notations and coordinate system are the same as representation used in the preceding chapters concerning rectangular plates.

Assumptions used in current refined plate theory may be summarized as follows:

(1) The displacements are small and strains involved are infinitesimal.

(2) The lateral displacement W is composed of two components: bending component w_b and shear component w_s, which both the components are functions of coordinate x and y and of time t. Hence, the deflection W is expressed by only the deflection $W = w_b + w_s$ on the middle surface.

$$W(x, y, z, t) = w_b(x, y, t) + w_s(x, y, t) \tag{14.10}$$

(3) The transverse normal stress σ_z is negligible in comparison with σ_x and σ_y. For a linearly elastic, homogenous, and isotropic material, in-plane stresses σ_x and σ_y are related to ε_x and ε_y.

$$\sigma_x = \frac{E}{1-\nu^2}(\varepsilon_x + \nu\varepsilon_y); \; \sigma_y = \frac{E}{1-\nu^2}(\varepsilon_y + \nu\varepsilon_x) \tag{14.11}$$

(4) The displacements U and V in x- and y-directions, respectively, each consist of bending components, U_b and V_b, and shear components, U_s and V_s, respectively.

$$U(x, y, z, t) = U_b(x, y, z, t) + U_s(x, y, z, t) \tag{14.12}$$

$$V(x, y, z, t) = V_b(x, y, z, t) + V_s(x, y, z, t) \tag{14.13}$$

The bending components U_b and V_b of the displacements U and V, respectively, are assumed to be given by, in which the effect of the transverse shear deformation is neglected.

$$U_b = -z w_{b,x}; \quad V_b = -z w_{b,y} \tag{14.14}$$

On the other hand, the shear components U_s and V_s of the displacements U and V, respectively, are determined from the following conditions in conjunction with w_s. The first condition is the parabolic variations of shear stresses τ_{xz} and τ_{yz} across the cross section of the plates. The second condition is zero shear stresses τ_{xz} and τ_{yz} at $z = -h_0/2$ and at $z = h_0/2$. From these two conditions, the shear components U_s and V_s are given as

$$U_s = -f(z) w_{s,x}; \quad V_s = -f(z) w_{s,y} \tag{14.15}$$

in which $f(z)$ is defined as

$$f(z) = -\frac{1}{4}z + \frac{5}{3}z\left(\frac{z}{h_0}\right)^2 \tag{14.16}$$

Hence, the displacements U and V are expressed as

$$U(x, y, z, t) = -z w_{b,x} - f(z) w_{s,x} \tag{14.17}$$

$$V(x, y, z, t) = -z w_{b,y} - f(z) w_{s,y} \tag{14.18}$$

In bending problem of plates, the strain–displacement relations can be expressed as

$$\left.\begin{aligned} &\varepsilon_x = U_{,x};\quad \varepsilon_y = V_{,y};\quad \varepsilon_z = W_{,z} \\ &\gamma_{xy} = V_{,x} + U_{,y};\quad \gamma_{yz} = W_{,y} + V_{,z};\quad \gamma_{xz} = W_{,x} + U_{,z} \end{aligned}\right\} \tag{14.19}$$

The stress–strain relations are

$$\left.\begin{aligned} &\sigma_x = \frac{E}{1-\nu^2}(\varepsilon_x + \nu\varepsilon_y);\quad \sigma_y = \frac{E}{1-\nu^2}(\varepsilon_y + \nu\varepsilon_x);\quad \sigma_z = 0 \\ &\tau_{xy} = G\gamma_{xy};\quad \tau_{yz} = G\gamma_{yz};\quad \tau_{xz} = G\gamma_{xz} \end{aligned}\right\} \tag{14.20}$$

Substituting Eqs. (14.10), (14.17), and (14.18) into Eq. (14.19) becomes

$$\varepsilon_x = -zw_{b,xx} - f(z)w_{s,xx} \tag{14.21}$$

$$\varepsilon_y = -zw_{b,yy} - f(z)w_{s,yy} \tag{14.22}$$

$$\varepsilon_z = 0 \tag{14.23}$$

$$\gamma_{xy} = -2zw_{b,xy} - 2f(z)w_{s,xy} \tag{14.24}$$

$$\gamma_{yz} = g(z)w_{s,y} \tag{14.25}$$

$$\gamma_{xz} = g(z)w_{s,x} \tag{14.26}$$

in which $g(z)$ is defined as

$$g(z) = 1 - f(z)_{,z} = \frac{5}{4} - 5\left(\frac{z}{h_0}\right)^2 \tag{14.27}$$

$g(z)$ takes the values 5/4 at $z = 0$, corresponding to the location of the middle surface, and 0 at $z = \pm h_0/2$, corresponding to the upper and lower surfaces. Hence, the condition for the distribution of the transverse shear stress is satisfied.

The substitution of Eqs. (14.21)–(14.26) into Eq. (14.20) becomes

$$\sigma_x = -\frac{Ez}{1-\nu^2}(w_{b,xx} + \nu w_{b,yy}) - \frac{E}{1-\nu^2}f(z)(w_{s,xx} + \nu w_{s,yy}) \tag{14.28}$$

$$\sigma_y = -\frac{Ez}{1-\nu^2}(w_{b,yy} + \nu w_{b,xx}) - \frac{E}{1-\nu^2}f(z)(w_{s,yy} + \nu w_{s,xx}) \tag{14.29}$$

$$\tau_{xy} = -\frac{Ez}{1+\nu}w_{b,xy} - \frac{E}{1+\nu}f(z)w_{s,xy} \tag{14.30}$$

$$\tau_{yz} = \frac{E}{2(1+\nu)}g(z)w_{s,y} \tag{14.31}$$

$$\tau_{xz} = \frac{E}{2(1+\nu)}g(z)w_{s,x} \tag{14.32}$$

The moment and transverse shear forces are defined as

$$M_x^{\mathrm{b}} = \int_{-h_0/2}^{h_0/2} \sigma_x z\, \mathrm{d}z; \quad M_y^{\mathrm{b}} = \int_{-h_0/2}^{h_0/2} \sigma_y z\, \mathrm{d}z\,; \quad M_{xy}^{\mathrm{b}} = \int_{-h_0/2}^{h_0/2} \tau_{xy} z\, \mathrm{d}z \tag{14.33}$$

$$M_x^{\mathrm{s}} = \int_{-h_0/2}^{h_0/2} \sigma_x f(z)\mathrm{d}z; \quad M_y^{\mathrm{s}} = \int_{-h_0/2}^{h_0/2} \sigma_y f(z)\mathrm{d}z; \quad M_{xy}^{\mathrm{s}} = \int_{-h_0/2}^{h_0/2} \tau_{xy} f(z)\mathrm{d}z \tag{14.34}$$

$$Q_x^{\mathrm{s}} = \int_{-h_0/2}^{h_0/2} \tau_{xz}\mathrm{d}z; \quad Q_y^{\mathrm{s}} = \int_{-h_0/2}^{h_0/2} \tau_{yz}\mathrm{d}z \tag{14.35}$$

The substitution of Eqs. (14.28)–(14.32) into Eqs. (14.33)–(14.35) becomes

$$\begin{aligned} M_x^{\mathrm{b}} &= -D_0(w_{\mathrm{b},xx} + \nu w_{\mathrm{b},yy}); \; M_y^{\mathrm{b}} = -D_0(w_{\mathrm{b},yy} + \nu w_{\mathrm{b},xx}); \\ M_{xy}^{\mathrm{b}} &= -D_0(1-\nu)w_{\mathrm{b},xy} \end{aligned} \tag{14.36}$$

$$\begin{aligned} M_x^{\mathrm{s}} &= -D_{\mathrm{s}}(w_{\mathrm{s},xx} + \nu w_{\mathrm{s},yy}); \; M_y^{\mathrm{s}} = -D_{\mathrm{s}}(w_{\mathrm{s},yy} + \nu w_{\mathrm{s},xx}); \\ M_{xy}^{\mathrm{s}} &= -D_{\mathrm{s}}(1-\nu)w_{\mathrm{s},xy} \end{aligned} \tag{14.37}$$

$$Q_x^{\mathrm{s}} = \frac{5Eh_0}{12(1+\nu)} w_{\mathrm{s},x} = \frac{5}{6} G h_0 w_{\mathrm{s},x}; \; Q_y^{\mathrm{s}} = \frac{5Eh_0}{12(1+\nu)} w_{\mathrm{s},y} = \frac{5}{6} G h_0 w_{\mathrm{s},y} \tag{14.38}$$

in which D_{s} is defined as

$$D_{\mathrm{s}} = \frac{E}{1-\nu^2} \int_{-h_0/2}^{h_0/2} (f(z))^2 \mathrm{d}z \tag{14.39}$$

Hence, the moments M_x, M_y, and M_{xy} are the sum of the components produced by bending deflection component and shear deflection component as follows:

$$M_x = M_x^{\mathrm{b}} + M_x^{\mathrm{s}} \tag{14.40}$$

$$M_y = M_y^{\mathrm{b}} + M_y^{\mathrm{s}} \tag{14.41}$$

$$M_{xy} = M_{xy}^{\mathrm{b}} + M_{xy}^{\mathrm{s}} \tag{14.42}$$

The shear forces are expressed by the following relations:

$$Q_x = Q_x^{\mathrm{s}}; \; Q_y = Q_y^{\mathrm{s}} \tag{14.43}$$

From Eqs. (14.38) and (14.43) for the transverse shear forces of RPT, the shear correction factor κ ($\kappa = 5/6$) agrees with the value 5/6 proposed by Reissner [41] and is very close the value $\pi^2/12$ proposed by Mindlin [42].

Shimpi [8] assumes in the moments M_x, M_y, and M_{xy}, the shear components U_s and V_s are not contribution. Hence, M_x^s, M_y^s, and M_{xy}^s are neglected.

$$M_x \approx M_x^b;\ M_y \approx M_y^b;\ M_{xy} \approx M_{xy}^b \tag{14.44}$$

14.4 The Governing Equation in RPT

The governing equations of rectangular plates subject to dynamic loads may be obtained using Hamilton's principle, as presented in the preceding chapters. The kinematic energy T, strain energy $\bar{U}$, and potential energy $\bar{V}$ are given by

$$T = \int_0^{l_x}\int_0^{l_y}\int_{-h_0/2}^{h_0/2} \frac{1}{2}\rho\left[\underline{(\dot{U})^2 + (\dot{V})^2} + (\dot{W})^2\right]\mathrm{d}x\mathrm{d}y\mathrm{d}z \tag{14.45}$$

$$\bar{U} = \int_0^{l_x}\int_0^{l_y}\int_{-h_0/2}^{h_0/2} \frac{1}{2}(\sigma_x\varepsilon_x + \sigma_y\varepsilon_y + \tau_{xy}\gamma_{xy} + \tau_{yz}\gamma_{yz} + \tau_{zx}\gamma_{zx})\mathrm{d}x\mathrm{d}y\mathrm{d}z \tag{14.46}$$

$$\begin{aligned}\bar{V} = &-\iint p(x,y,t)W\mathrm{d}x\mathrm{d}y + \iint p_D W\mathrm{d}x\mathrm{d}y - \int_c \left[m_x^* w_{b,xx} + m_{xy}^* w_{b,yy}\right]_0^{l_x}\mathrm{d}y\\ &- \int_c \left[m_y^* w_{b,yyx} + m_{yx}^* w_{b,xx}\right]_0^{l_y}\mathrm{d}x - \int_c \left[v_x^*(w_b + w_s)\right]_0^{l_x}\mathrm{d}y\\ &- \int_c \left[v_y^*(w_b + w_s)\right]_0^{l_y}\mathrm{d}x\end{aligned} \tag{14.47}$$

in which p_D is damping force given approximately by the relation $p_D = c(\dot{w}_b + \dot{w}_s) \approx c\dot{w}_b$. Equations (14.45) and (14.46) may be expressed by employing Eq. (14.10) and Eqs. (14.17)–(14.32) as follows:

$$\begin{aligned}T = &\frac{\rho h^3}{24}\int_0^{l_x}\int_0^{l_y}[(\dot{w}_{b,x})^2 + (\dot{w}_{b,y})^2]\mathrm{d}x\mathrm{d}y + \underline{\frac{\rho h^3}{2016}\int_0^{l_x}\int_0^{l_y}[(\dot{w}_{s,x})^2 + (\dot{w}_{s,y})^2]\mathrm{d}x\mathrm{d}y}\\ &+ \frac{\rho h}{2}\int_0^{l_x}\int_0^{l_y}(\dot{w}_b + \dot{w}_s)^2\mathrm{d}x\mathrm{d}y\end{aligned} \tag{14.48}$$

$$\bar{U} = \frac{Eh^3}{24(1-\nu^2)} \int_0^{l_x}\int_0^{l_y} [(w_{b,xx})^2 + (w_{b,yy})^2 + 2\nu\, w_{b,xx} w_{b,yy} + 2(1-\nu)(w_{b,xy})^2] dx dy$$
$$+ \frac{5Eh}{24(1+\nu)} \int_0^{l_x}\int_0^{l_y} [(w_{s,x})^2 + (w_{s,y})^2] dx dy$$
$$+ \frac{Eh^3}{2016(1-\nu^2)} \int_0^{l_x}\int_0^{l_y} [(w_{s,xx})^2 + (w_{s,yy})^2 + 2\nu\, w_{s,xx} w_{s,yy} + 2(1-\nu)(w_{s,xy})^2] dx dy \tag{14.49}$$

Employing the following Hamilton's principle

$$\delta I = \delta \int_{t_1}^{t_2} (T - \bar{U} - \bar{V}) dt = 0, \tag{14.50}$$

we have the equations of motion in RPT may be obtained as follows:

$$D_0 \nabla^2\nabla^2 w_b - \frac{\rho h_0^3}{12} \nabla^2 \ddot{w}_b + m_0(\ddot{w}_b + \ddot{w}_s) + c\dot{w}_b = p \tag{14.51}$$

$$\frac{1}{84} D_0 \nabla^2\nabla^2 w_s - \frac{5Eh_0}{12(1+\nu)} \nabla^2 w_s - \frac{m_0 h_0^2}{1008} (\nabla^2 \ddot{w}_s) + m_0(\ddot{w}_b + \ddot{w}_s) = p \tag{14.52}$$

The above two equations of motion for w_b and w_s are only inertial coupling. The boundary conditions of the plate are given as follows:

(1) At corners $(x = 0, y = 0)$, $(x = 0, y = l_y)$, $(x = l_x, y = 0)$, and $(x = l_x, y = l_y)$, the following conditions hold:

(a) For the conditions involving bending component w_b of lateral displacement

$$M_{xy}^b = -D_0(1-\nu) w_{b,xy} = 0 \quad \text{or} \quad w_b = w_b^* \tag{14.53}$$

(b) For the conditions involving for shear component w_s of lateral displacement

$$-D(1-\nu) w_{s,xy} = 0 \quad \text{or} \quad w_s = w_s^* \tag{14.54}$$

(2) On edges at $x = 0$ and l_x, the following conditions hold:

(a) For the conditions involving bending component w_b of lateral displacement

$$-D_0[w_{b,xxx} + (2-\nu) w_{b,xyy}] + \frac{\rho h_0^3}{12} \ddot{w}_{b,x} = v_x^* \quad \text{or} \quad w_b = w_b^* \tag{14.55}$$

$$M_x^{\mathrm{b}} = -D_0(w_{\mathrm{b},xx} + \nu w_{\mathrm{b},xy}) = m_x^* \quad \text{or} \quad w_{\mathrm{b},x} = w_{\mathrm{b},x}^* \tag{14.56}$$

(b) For the conditions involving for shear component w_{s} of lateral displacement

$$\underline{\frac{5Eh}{12(1+\nu)} w_{\mathrm{s},x} - \frac{1}{84} D_0[w_{\mathrm{s},xxx} + (2-\nu) w_{\mathrm{s},xyy}] + \frac{\rho h_0^3}{1008} \ddot{w}_{\mathrm{s},x} = 0} \quad \text{or} \quad w_{\mathrm{s}} = w_{\mathrm{s}}^* \tag{14.57}$$

$$\underline{M_x^{\mathrm{s}} = -D_0(w_{\mathrm{s},xx} + \nu w_{\mathrm{s},yy}) = 0} \quad \text{or} \quad w_{\mathrm{s},x} = w_{\mathrm{s},x}^* \tag{14.58}$$

(3) On edges at $y = 0$ and l_y, the following conditions hold:

(a) For the conditions involving bending component w_{b} of lateral displacement

$$-D_0[w_{\mathrm{b},yyy} + (2-\nu) w_{\mathrm{b},xxy}] + \frac{\rho h_0^3}{12} \ddot{w}_{\mathrm{b},y} = v_y^* \quad \text{or} \quad w_{\mathrm{b}} = w_{\mathrm{b}}^* \tag{14.59}$$

$$M_y^{\mathrm{b}} = -D_0(w_{\mathrm{b},yy} + \nu w_{\mathrm{b},xx}) = m_y^* \quad \text{or} \quad w_{\mathrm{b},y} = w_{\mathrm{b},y}^* \tag{14.60}$$

(b) For the conditions involving for shear component w_{s} of lateral displacement

$$\underline{\frac{5Eh}{12(1+\nu)} w_{\mathrm{s},y} - \frac{1}{84} D_0[w_{\mathrm{s},yyy} + (2-\nu) w_{\mathrm{s},xxy}] + \frac{\rho h_0^3}{1008} \ddot{w}_{\mathrm{s},y} = 0} \quad \text{or} \quad w_{\mathrm{s}} = w_{\mathrm{s}}^* \tag{14.61}$$

$$\underline{M_y^{\mathrm{s}} = -D_0(w_{\mathrm{s},yy} + \nu\, w_{\mathrm{s},xx}) = 0} \quad \text{or} \quad w_{\mathrm{s},y} = w_{\mathrm{s},y}^* \tag{14.62}$$

In Eqs. (14.53)–(14.62), the quantity with subscript * indicates to be the quantity specified by the boundary.

14.5 Simplified RPT

Although the equations of motion and boundary conditions for unknown w_{b} and w_{s}, proposed in the preceding section, are less than three variables of FSDT in number of variable, the reduced equations are complicated. So, we consider to translate these equations to more simplified equations by the use of the following assumptions:

(1) In the kinetic energy, the rotatory inertia components produced by the displacement components U and V are negligible because of small as compared with w.
(2) Since the displacement components U_{s} and V_{s} do not contribute toward moment M_x, M_y, and M_{xy}, as given in Eqs. (14.40) to (14.42), respectively, the contribution from terms involving w_{s} is insignificant in the bending energy component of the strain energy.

Using the above-mentioned assumptions, the underlined terms indicated in Eqs. (14.45), (14.48), (14.49), (14.52), (14.54), (14.57), (14.58), (14.61), and (14.62) vanish. Hence, the simplified equations of RPT reduce to as follows:

(1) **Simplified equations of motion**

$$D_0\nabla^2\nabla^2 w_b - \frac{\rho h_0^3}{12}\nabla^2\ddot{w}_b + m_0(\ddot{w}_b + \ddot{w}_s) + c\dot{w}_b = p \tag{14.63}$$

$$-\frac{5Eh_0}{12(1+\nu)}\nabla^2 w_s + m_0(\ddot{w}_b + \ddot{w}_s) = p \tag{14.64}$$

(2) **Boundary conditions**

① At corners $x = 0$, $y = 0$, $(x = 0,\ y = l_y)$, $(x = l_x,\ y = 0)$, and $(x = l_x,\ y = l_y)$, the following equation holds:

$$-D_0(1-\nu)w_{b,xy} = 0 \quad \text{or } w_b = w_b^* \tag{14.65}$$

② On edges at $x = 0$ and l_x, the following conditions hold:

(a) For the conditions involving bending component w_b of lateral displacement

$$-D_0[w_{b,xxx} + (2-\nu)w_{b,xyy}] + \frac{\rho h_0^3}{12}\ddot{w}_{b,x} = v_x^* \quad \text{or } w_b = w_b^* \tag{14.66}$$

$$M_x^b = -D_0(w_{b,xx} + \nu\, w_{b,yy}) = m_x^* \quad \text{or } w_{b,x} = w_{b,x}^* \tag{14.67}$$

(b) For the conditions involving for shear component w_s of lateral displacement

$$w_{s,x} = w_{s,x}^* \quad \text{or } w_s = w_s^* \tag{14.68}$$

③ On edges at $y = 0$ and l_y, the following conditions hold:

(a) For the conditions involving bending component w_b of lateral displacement

$$-D_0[w_{b,yyy} + (2-\nu)w_{b,xxy}] + \frac{\rho h_0^3}{12}\ddot{w}_{b,y} = v_y^* \quad \text{or } w_b = w_b^* \tag{14.69}$$

$$M_y^b = -D_0(w_{b,yy} + \nu w_{b,xx}) = m_y^* \quad \text{or } w_{b,y} = w_{b,y}^* \tag{14.70}$$

(b) For the conditions involving for shear component w_s of lateral displacement

$$w_{s,y} = w_{s,y}^* \quad \text{or } w_s = w_s^* \tag{14.71}$$

14.6 Static Analysis Used Simplified RPT

We consider static problem of rectangular plates subject to lateral load p by the use of simplified RPT. The equilibrium equations are obtained from Eqs. (14.63) and (14.64).

$$D_0 \nabla^2 \nabla^2 w_b - p = 0 \tag{14.72}$$

$$-\frac{5Eh_0}{12(1+\nu)} \nabla^2 w_s = p \rightarrow \nabla^2 w_s + \frac{h_0^2}{5(1-\nu)} \frac{p}{D_0} = 0 \tag{14.73}$$

The equilibrium equations become uncoupled form for static problem.

First we consider a simple problem of simply supported plates which may be easily solved in closed-form solution by the use of Naive method.

(1) Analytical example for simply supported plates subject to sinusoidal lateral load

We consider static problem of a simply supported plate to understand the simplified RPT.

The lateral load is assumed to be the following sinusoidal distributed load with the maximum magnitude p_0.

$$p(x, y) = p_0 \sin \frac{\pi x}{l_x} \sin \frac{\pi y}{l_y} \tag{14.74}$$

The current equilibrium equations are from Eqs. (14.72) and (14.73) as follows:

$$\nabla^2 \nabla^2 w_b = \frac{p_0}{D_0} \sin \frac{\pi x}{l_x} \sin \frac{\pi y}{l_y} \tag{14.75}$$

$$\nabla^2 w_s = -\frac{h_0^2}{5(1-\nu)} \frac{p_0}{D_0} \sin \frac{\pi x}{l_x} \sin \frac{\pi y}{l_y} \tag{14.76}$$

The boundary conditions can be stated as

$$w_b = 0 \quad \text{on } x = 0,\ l_x \tag{14.77}$$

$$-D_0(w_{b,xx} + \nu\, w_{b,yy}) = 0 \quad \text{on } x = 0,\ l_x \tag{14.78}$$

$$w_s = 0 \quad \text{on } x = 0,\ l_x \tag{14.79}$$

$$w_b = 0 \quad \text{on } y = 0,\ l_y \tag{14.80}$$

$$-D_0(w_{b,yy} + \nu\, w_{b,xx}) = 0 \quad \text{on } y = 0,\ l_y \tag{14.81}$$

$$w_s = 0 \quad \text{on } y = 0,\ l_y \tag{14.82}$$

If w_b and w_s are assumed to be Navier-type solution, Eqs. (14.75)–(14.82) are satisfied. Hence, from Eqs. (14.75) and (14.76), w_b and w_s are obtained easily as follows:

$$w_{\mathrm{b}} = \frac{5}{\frac{5\pi^4}{12(1-\nu^2)}\left[\left(\frac{h_0}{l_x}\right)^2 + \left(\frac{h_0}{l_y}\right)^2\right]} \frac{p_0 h_0}{E} \sin\frac{\pi x}{l_x} \sin\frac{\pi y}{l_y} \tag{14.83}$$

$$w_{\mathrm{s}} = \frac{\pi^2}{(1-\nu)} \cdot \frac{1}{(1-\nu)\frac{5\pi^4}{12(1-\nu^2)}\left[\left(\frac{h_0}{l_x}\right)^2 + \left(\frac{h_0}{l_y}\right)^2\right]} \frac{p_0 h_0}{E} \sin\frac{\pi x}{l_x} \sin\frac{\pi y}{l_y} \tag{14.84}$$

Hence, the deflection W is given by

$$W = w_{\mathrm{b}} + w_{\mathrm{s}} = \left(5 + \frac{\pi^2}{1-\nu}\right) \frac{1}{\frac{5\pi^4}{12(1-\nu^2)}\left[\left(\frac{h_0}{l_x}\right)^2 + \left(\frac{h_0}{l_y}\right)^2\right]} \frac{p_0 h_0}{E} \sin\frac{\pi x}{l_x} \sin\frac{\pi y}{l_y} \tag{14.85}$$

(2) General static analysis of simplified RPT using the Galerkin method

In the preceding many chapters, we have proposed a general analytical methodology by means of the Galerkin method. So, we present a general analytical approach based on the simplified RPT for static problems of uniform rectangular plates by means of the Galerkin method.

The deflections w_{b} and w_{s} are expressed by a power series expansion as follows:

$$w_{\mathrm{b}}(x, y) = \sum_{m=1}\sum_{n=1} w_{\mathrm{b}\,mn} f_{\mathrm{b}\,mn}(x, y) \tag{14.86}$$

$$w_{\mathrm{s}}(x, y) = \sum_{m=1}\sum_{n=1} w_{\mathrm{s}\,mn} f_{\mathrm{s}\,mn}(x, y) \tag{14.87}$$

in which $f_{\mathrm{b}\,mn}$ and $f_{\mathrm{s}\,mn}$ are shape functions satisfying specified boundary conditions. The Galerkin equations for static problems can be written from Eqs. (14.63) and (14.73) as

$$\int_0^{l_x}\int_0^{l_y} Q_{\mathrm{b}}^* \cdot \delta w_{\mathrm{b}} \mathrm{d}x\mathrm{d}y = 0 \tag{14.88}$$

$$\int_0^{l_x}\int_0^{l_y} Q_{\mathrm{s}}^* \cdot \delta w_{\mathrm{s}} \mathrm{d}x\mathrm{d}y = 0 \tag{14.89}$$

in which Q_{b}^* and Q_{s}^* represent Eqs. (14.72) and (14.73), respectively. Substituting Eqs. (14.86) and (14.87) into Eqs. (14.88) and (14.89), the Galerkin equations become

$$\delta w_{\mathrm{b}\,mn}: \quad \sum_{m=1}\sum_{n=1} w_{\mathrm{b}\,mn} A_{\mathrm{b}\,\bar{m}\bar{n}mn} = P_{\mathrm{b}\,\bar{m}\bar{n}} \tag{14.90}$$

$$\delta w_{\mathrm{s}\,mn}: \quad \sum_{m=1}\sum_{n=1} w_{\mathrm{s}\,mn} A_{\mathrm{s}\,\bar{m}\bar{n}mn} = P_{\mathrm{s}\,\bar{m}\bar{n}} \tag{14.91}$$

in which $A_{\mathrm{b}\,\bar{m}\bar{n}mn}$, $A_{\mathrm{s}\,\bar{m}\bar{n}mn}$, $P_{\mathrm{b}\,\bar{m}\bar{n}}$, and $P_{\mathrm{s}\,\bar{m}\bar{n}}$ are defined as follows:

$$A_{\mathrm{b}\,\bar{m}\bar{n}mn} = \int_0^{l_x}\int_0^{l_y} (f_{\mathrm{b}\,mn,xxxx} + 2 f_{\mathrm{b}\,mn,xxyy} + f_{\mathrm{b}\,mn,yyyy}) f_{\mathrm{b}\,\bar{m}\bar{n}} \mathrm{d}x\mathrm{d}y \tag{14.92}$$

$$A_{\mathrm{s}\,\bar{m}\bar{n}mn} = \int_0^{l_x}\int_0^{l_y} (f_{\mathrm{s}\,mn,xx} + f_{\mathrm{s}\,mn,yy}) f_{\mathrm{s}\,\bar{m}\bar{n}} \mathrm{d}x\mathrm{d}y \tag{14.93}$$

$$P_{\mathrm{b}\,\bar{m}\bar{n}} = \frac{1}{D_0}\int_0^{l_x}\int_0^{l_y} P f_{\mathrm{b}\,\bar{m}\bar{n}} \mathrm{d}x\mathrm{d}y \tag{14.94}$$

$$P_{\mathrm{s}\,\bar{m}\bar{n}} = -\frac{h^2}{5(1-\nu)}\frac{1}{D_0}\int_0^{l_x}\int_0^{l_y} P f_{\mathrm{s}\,\bar{m}\bar{n}} \mathrm{d}x\mathrm{d}y \tag{14.95}$$

The unknown displacement coefficients $w_{\mathrm{b}\,mn}$ and $w_{\mathrm{s}\,mn}$ are obtained from solving each set of algebraic linear equations given by Eqs. (14.90) and (14.91), respectively, in which $\bar{m}$ and $\bar{n}$ indicate the row of the set of algebraic linear equations and on the other hand m and n give the column in the square matrixes $A_{\mathrm{b}\,\bar{m}\bar{n}mn}$ and $A_{\mathrm{s}\,\bar{m}\bar{n}mn}$, and the column vectors $w_{\mathrm{b}\,mn}$ and $w_{\mathrm{s}\,mn}$ and $P_{\mathrm{b}\,\bar{m}\bar{n}}$ and $P_{\mathrm{s}\,\bar{m}\bar{n}}$, respectively. The unknown displacement coefficients $w_{\mathrm{b}\,mn}$ and $w_{\mathrm{s}\,mn}$ are easily obtained from solving each set of simultaneous equations with multiple unknowns. This calculation can be carried out by the use of computer if suitable shape functions which satisfy the prescribed boundary conditions of the plate are selected previously.

In general, the square matrixes $A_{\mathrm{b}\,\bar{m}\bar{n}mn}$ and $A_{\mathrm{s}\,\bar{m}\bar{n}mn}$ have nonzero value in the diagonal and nondiagonal terms. However, when the suitable shape functions which satisfy the following orthogonal relations are selected, the square matrixes $A_{\mathrm{b}\,\bar{m}\bar{n}mn}$ and $A_{\mathrm{s}\,\bar{m}\bar{n}mn}$ have the value in only the diagonal terms,

$$\int_0^{l_x} f_{\mathrm{b}m}(x) f_{\mathrm{b}\bar{m}}(x)\,\mathrm{d}x = \delta_{m\bar{m}} a_{\mathrm{b}m} \qquad \int_0^{l_x} f_{\mathrm{s}m}(x) f_{\mathrm{s}\bar{m}}(x)\,\mathrm{d}x = \delta_{m\bar{m}} a_{\mathrm{s}m} \tag{14.96}$$

$$\int_0^{l_y} f_{\mathrm{b}n}(y) f_{\mathrm{b}\bar{n}}(y)\,\mathrm{d}y = \delta_{n\bar{n}} a_{\mathrm{b}n} \qquad \int_0^{l_y} f_{\mathrm{s}n}(y) f_{\mathrm{s}\bar{n}}(y)\,\mathrm{d}y = \delta_{n\bar{n}} a_{\mathrm{s}n} \tag{14.97}$$

in which $\delta_{m\bar{m}}$ and $\delta_{n\bar{n}}$ are the Kronecker deltas; and $a_{\mathrm{b}m}$, $a_{\mathrm{s}m}$, $a_{\mathrm{b}n}$, and $a_{\mathrm{s}n}$ are constant integral values. The Kronecker delta $\delta_{m\bar{m}}$ takes the value of 1 for $m = \bar{m}$ and the value of 0 for $m \neq \bar{m}$. Thus, Eqs. (14.90) and (14.91), respectively, reduce to uncoupled form and the displacement coefficients $w_{\mathrm{b}mn}$ and $w_{\mathrm{s}mn}$ are given in closed form as

$$w_{\text{b}mn} = \frac{1}{A_{\text{b}mnmn}} P_{\text{b}mn} \tag{14.98}$$

$$w_{\text{s}mn} = \frac{1}{A_{\text{s}mnmn}} P_{\text{s}mn} \tag{14.99}$$

The above solutions are also considered to be the approximate solutions in the case of the selection of shape functions which do not satisfy the orthogonal relations given in Eqs. (14.96) and (14.97).

14.7 Selection of Shape Functions for Static Problems

The shape functions $f_{mn}(x, y)$ consist of shape functions $f_{\text{b}m}(x)$ and $f_{\text{s}m}(x)$ in the x-direction and $f_{\text{b}n}(y)$ and $f_{\text{s}n}(y)$ in the y-direction. They have the relation $f_{\text{b}mn}(x, y) = f_{\text{b}m}(x)\ f_{\text{b}n}(y)$ and $f_{\text{s}mn}(x, y) = f_{\text{s}m}(x)\ f_{\text{s}n}(y)$. The shape functions $f_{\text{b}mn}(x, y)$ and $f_{\text{s}mn}(x, y)$ represented in terms of the series expansion must satisfy the geometrical boundary at the least. We consider the shape functions which satisfy the boundary conditions at $x = 0$ and $x = l_x$. For the boundary conditions at $y = 0$ and $y = l_y$, the similar results are obtained.

(1) Shape function for simply supported boundary condition at $x = 0$ and l_x:

$$f_{\text{b}m}(x) = f_{\text{s}m}(x) = \sin \frac{m\pi x}{l_x} \quad (m = 1, 2, 3, \ldots) \tag{14.100}$$

(2) Shape function for clamped boundary condition at $x = 0$ and l_x:

$$f_{\text{b}m}(x) = f_{\text{s}m}(x) = \frac{1}{2}\left(1 - \cos \frac{2m\pi x}{l_x}\right) \quad (m = 1, 3, 5, \ldots) \tag{14.101}$$

$$f_{\text{b}m}(x) = f_{\text{s}m}(x) = \sin \frac{\pi x}{l_x} \sin \frac{m\pi x}{l_x} \tag{14.102}$$

Once the displacement coefficients $w_{\text{b}mn}$ and $w_{\text{s}mn}$ are determined from Eqs. (14.90) and (14.91) or Eqs. (14.98) and (14.99), respectively, the deflection components w_b, w_s, and W are obtained from Eqs. (14.86), (14.87), and (14.10), respectively. Also, the stress couples M_x, M_y, and $M_{xy} = M_{yx}$, par unit length of the middle surface are derived from Eqs. (14.36), (14.37), and (14.44) respectively. The transverse shearing forces Q_x and Q_y are given from Eqs. (14.38) and (14.43).

14.8 Free Transverse Vibrations of Plates Without Damping

The equation of free transverse vibrations of a rectangular plate without damping is expressed by modifying Eqs. (14.63) and (14.64) as

$$D_0\nabla^2\nabla^2 w_b - \frac{\rho h_0^3}{12}\nabla^2\ddot{w}_b + m_0(\ddot{w}_b + \ddot{w}_s) = 0 \tag{14.103}$$

$$-\frac{5Eh_0}{12(1+\nu)}\nabla^2 w_s + m_0(\ddot{w}_b + \ddot{w}_s) = 0 \tag{14.104}$$

in which m_0 is mass per unit area. The deflections w_b and w_s are function of x, y, and time t. The method of separation of variables is employed, assuming that

$$w_b(x, y, t) = \bar{W}_b(x, y)\,e^{i\omega t} \tag{14.105}$$

$$w_s(x, y, t) = \bar{W}_s(x, y)\,e^{i\omega t} \tag{14.106}$$

in which $\bar{W}_b(x, y)$ and $\bar{W}_s(x, y)$ are a function of x and y. Substituting Eqs. (14.105) and (14.106) into Eqs. (14.103) and (14.104), the two equations are obtained.

$$\bar{W}_{b,xxxx} + 2\bar{W}_{b,xxyy} + \bar{W}_{b,yyyy} + \frac{\omega^2\rho h_0^3}{12D_0}(\bar{W}_{b,xx} + \bar{W}_{b,yy}) - \frac{m_0\omega^2}{D_0}(\bar{W}_b + \bar{W}_s) = 0 \tag{14.107}$$

$$(\bar{W}_{s,xx} + \bar{W}_{s,yy}) + \frac{12(1+\nu)}{5Eh_0}m_0\omega^2(\bar{W}_b + \bar{W}_s) = 0 \tag{14.108}$$

in which ω is a constant.

The natural frequencies of a normal plate are presented by means of the Galerkin method. W_{bmn} and W_{smn} are expressed as follows:

$$\bar{W}_b(x, y) = W_{bmn}\,f_{bmn}(x, y) \tag{14.109}$$

$$\bar{W}_s(x, y) = W_{smn}\,f_{smn}(x, y) \tag{14.110}$$

in which W_{bmn} and W_{smn} are the unknown displacement coefficients, and $f_{bmn}(x, y)$ and $f_{smn}(x, y)$ are functions satisfying the specified boundary conditions of the plates.

Applying the Galerkin method to Eqs. (14.107) and (14.108) are

$$\int_0^{l_x}\int_0^{l_y} \delta\bar{W}_b Q_b^*\,dxdy = 0 \tag{14.111}$$

$$\int_0^{l_x}\int_0^{l_y} \delta\bar{W}_s Q_s^*\,dxdy = 0 \tag{14.112}$$

in which the notations Q_b^* and Q_s^* indicate Eqs. (14.107) and (14.108), respectively, for simple expression.

The Galerkin equations of Eqs. (14.111) and (14.112) become

$$\delta W_{b\bar{m}\bar{n}}:\; W_{bmn}(A_{b\bar{m}\bar{n}mn} - \omega^2 F_{bb\bar{m}\bar{n}mn}) - \omega^2 W_{smn}F_{bs\bar{m}\bar{n}mn} = 0 \tag{14.113}$$

$$\delta W_{\mathrm{s}\bar{m}\bar{n}} : \; -\omega^2 W_{\mathrm{b}mn} F_{\mathrm{sb}\bar{m}\bar{n}mn} + W_{\mathrm{s}mn}(A_{\mathrm{s}\bar{m}\bar{n}mn} - \omega^2 F_{\mathrm{ss}\bar{m}\bar{n}mn}) = 0 \tag{14.114}$$

in which the notations $A_{\mathrm{b}\bar{m}\bar{n}mn}$, $A_{\mathrm{s}\bar{m}\bar{n}mn}$, $F_{\mathrm{bb}\bar{m}\bar{n}mn}$, $F_{\mathrm{bs}\bar{m}\bar{n}mn}$, $F_{\mathrm{sb}\bar{m}\bar{n}mn}$, and $F_{\mathrm{ss}\bar{m}\bar{n}mn}$ are defined as follows:

$$A_{\mathrm{b}\bar{m}\bar{n}mn} = \int_0^{l_x}\int_0^{l_y} (f_{\mathrm{b}mn,xxxx} + 2 f_{\mathrm{b}mn,xxyy} + f_{\mathrm{b}mn,yyyy}) f_{\mathrm{b}\bar{m}\bar{n}} \mathrm{d}x\mathrm{d}y \tag{14.115}$$

$$A_{\mathrm{s}\bar{m}\bar{n}mn} = \int_0^{l_x}\int_0^{l_y} (f_{\mathrm{s}mn,xx} + f_{\mathrm{s}mn,yy}) f_{\mathrm{s}mn} \mathrm{d}x\mathrm{d}y \tag{14.116}$$

$$F_{\mathrm{bb}\bar{m}\bar{n}mn} = -\frac{\rho h_0^3}{12 D_0} \int_0^{l_x}\int_0^{l_y} (f_{\mathrm{b}mn,xx} + f_{\mathrm{b}mn,yy}) f_{\mathrm{b}\bar{m}\bar{n}} \mathrm{d}x\mathrm{d}y + \frac{m_0}{D_0} \int_0^{l_x}\int_0^{l_y} f_{\mathrm{b}mn} f_{\mathrm{b}\bar{m}\bar{n}} \mathrm{d}x\mathrm{d}y \tag{14.117}$$

$$F_{\mathrm{bs}\bar{m}\bar{n}mn} = \frac{m_0}{D_0} \int_0^{l_x}\int_0^{l_y} f_{\mathrm{s}mn} f_{\mathrm{b}\bar{m}\bar{n}} \mathrm{d}x\mathrm{d}y \tag{14.118}$$

$$F_{\mathrm{sb}\bar{m}\bar{n}mn} = -\frac{12(1+\nu)m_0}{5Eh_0} \int_0^{l_x}\int_0^{l_y} f_{\mathrm{b}mn} f_{\mathrm{s}\bar{m}\bar{n}} \mathrm{d}x\mathrm{d}y \tag{14.119}$$

$$F_{\mathrm{ss}\bar{m}\bar{n}mn} = -\frac{12(1+\nu)m_0}{5Eh_0} \int_0^{l_x}\int_0^{l_y} f_{\mathrm{s}mn} f_{\mathrm{s}\bar{m}\bar{n}} \mathrm{d}x\mathrm{d}y \tag{14.120}$$

A necessary and sufficient condition for the existence of a nontrivial solution for $W_{\mathrm{b}mn}$ and $W_{\mathrm{s}mn}$ is that the determinant of the coefficient matrix of Eqs. (14.113) and (14.114) is zero, namely

$$\begin{vmatrix} A_{\mathrm{b}\bar{m}\bar{n}mn} - \omega^2 F_{\mathrm{bb}\bar{m}\bar{n}mn} & -\omega^2 F_{\mathrm{bs}\bar{m}\bar{n}mn} \\ -\omega^2 F_{\mathrm{sb}\bar{m}\bar{n}mn} & A_{\mathrm{s}\bar{m}\bar{n}mn} - \omega^2 F_{\mathrm{ss}\bar{m}\bar{n}mn} \end{vmatrix} = 0 \tag{14.121}$$

The above condition gives following quadratic characteristic equation.

$$\begin{aligned}&(F_{\mathrm{bb}\bar{m}\bar{n}mn} F_{\mathrm{ss}\bar{m}\bar{n}mn} - F_{\mathrm{bs}\bar{m}\bar{n}mn} F_{\mathrm{sb}\bar{m}\bar{n}mn})\omega^4 \\ &- (A_{\mathrm{b}\bar{m}\bar{n}mn} F_{\mathrm{ss}\bar{m}\bar{n}mn} + A_{\mathrm{s}\bar{m}\bar{n}mn} F_{\mathrm{bb}\bar{m}\bar{n}mn})\omega^2 + A_{\mathrm{b}\bar{m}\bar{n}mn} A_{\mathrm{s}\bar{m}\bar{n}mn} = 0\end{aligned} \tag{14.122}$$

The above characteristic equation has two frequencies ω_{mn} for m and n modes. The lower solution of these two solutions obtained for ω_{mn} corresponds to bending mode, whereas the higher one is shear mode.

14.9 Forced Vibration of Plates in Simplified RPT

We consider the forced vibration of a uniform rectangular plate based on simplified RPT. The equations of motion for uniform rectangular plates with damping in simplified RPT are given from Eqs. (14.63) and (14.64).

The general solutions of Eqs. (14.63) and (14.64) are assumed to be of the form

$$w_b(x, y, t) = \sum_{m=1}\sum_{n=1} f_{bmn}(x, y)\Phi_{bmn}(t) \tag{14.123}$$

$$w_s(x, y, t) = \sum_{m=1}\sum_{n=1} f_{smn}(x, y)\Phi_{smn}(t) \tag{14.124}$$

in which $\Phi_{bmn}(t)$ and $\Phi_{smn}(t)$ are unknown functions of time t; and $f_{bmn}(x, y)$ and $f_{smn}(x, y)$ are the natural functions satisfying both Eqs. (14.107) and (14.108) and the specified boundary conditions of current plate. The damping coefficient c is assumed to be also series expansion c_{mn} corresponding to the shape function f_{bmn}. Substituting Eqs. (14.123) and (14.124) into Eqs. (14.63) and (14.64), we have

$$\begin{aligned}\sum_{m=1}\sum_{n=1}&[m_0(f_{bmn}\ddot{\Phi}_{bmn} + f_{smn}\ddot{\Phi}_{smn}) - \frac{\rho h_0^3}{12}(f_{bmn,xx} + f_{bmn,yy})\ddot{\Phi}_{bmn}\\ &+ c_{mn}f_{bmn}\dot{\Phi}_{bmn} + D_0(f_{bmn,xxxx} + 2f_{bmn,xxyy} + f_{bmn,yyyy})\Phi_{bmn}] - p(x, y, t) = 0\end{aligned} \tag{14.125}$$

$$\begin{aligned}\sum_{m=1}\sum_{n=1}&[m_0(f_{bmn}\ddot{\Phi}_{bmn} + f_{smn}\ddot{\Phi}_{smn}) - \frac{5Eh_0}{12(1+\nu)}(f_{smn,xx} + f_{smn,yy})\Phi_{smn}]\\ &- p(x, y, t) = 0\end{aligned} \tag{14.126}$$

The shape functions f_{bmn} and f_{smn} must be satisfied Eqs. (14.107) and (14.108), respectively, as

$$\begin{aligned}&f_{bmn,xxxx} + 2f_{bmn,xxyy} + f_{bmn,yyyy} + \frac{\omega_{mn}^2\rho h_0^3}{12D_0}(f_{bmn,xx} + f_{bmn,yy})\\ &- \frac{m_0\omega_{mn}^2}{D_0}(f_{bmn} + f_{smn}) = 0\end{aligned} \tag{14.127}$$

$$(f_{smn,xx} + f_{smn,yy}) + \frac{12(1+\nu)}{5Eh_0}m_0\omega_{mn}^2(f_{bmn} + f_{smn}) = 0 \tag{14.128}$$

The substitutions of Eqs. (14.127) and (14.128) into Eqs. (14.125) and (14.126) become

$$\sum_{m=1}\sum_{n=1}\Big[m_0(f_{bmn}\ddot{\Phi}_{bmn}+f_{smn}\ddot{\Phi}_{smn})-\frac{\rho h_0^3}{12}(f_{bmn,xx}+f_{bmn,yy})\ddot{\Phi}_{bmn}$$
$$+c_{mn}f_{bmn}\dot{\Phi}_{bmn}-\frac{\omega_{mn}^2\rho h_0^3}{12D_0}(f_{bmn,xx}+f_{bmn,yy})\Phi_{bmn}$$
$$+m_0\omega_{mn}^2(f_{bmn}+f_{smn})\Phi_{bmn}\Big]-p(x,y,t)=0 \quad (14.129)$$

$$\sum_{m=1}\sum_{n=1}[m_0(f_{bmn}\ddot{\Phi}_{bmn}+f_{smn}\ddot{\Phi}_{smn})+m_0\omega_{mn}^2(f_{bmn}+f_{smn})\Phi_{smn}]$$
$$-p(x,y,t)=0 \quad (14.130)$$

The relationships between the damping coefficient c_{mn} and the damping constant h_{mn} for the m and nth modes' shape of the plate, corresponding the shape function f_{bmn}, may be written as

$$\frac{c_{mn}}{m_0}=2h_{mn}\omega_{mn} \quad (14.131)$$

Applying Eq. (14.131) into Eq. (14.129), we obtain

$$\sum_{m=1}\sum_{n=1}\Big[m_0(f_{bmn}\ddot{\Phi}_{bmn}+f_{smn}\ddot{\Phi}_{smn})-\frac{m_0h_0^2}{12}(f_{bmn,xx}+f_{bmn,yy})\ddot{\Phi}_{bmn}$$
$$+2h_{mn}\omega_{mn}m_0f_{bmn}\dot{\Phi}_{bmn}-\frac{\omega_{mn}^2m_0h_0^2}{12}(f_{bmn,xx}+f_{bmn,yy})\Phi_{bmn}$$
$$+m_0\omega_{mn}^2(f_{bmn}+f_{smn})\Phi_{bmn}\Big]-p(x,y,t)=0 \quad (14.132)$$

Representing Eqs. (14.132) and (14.130) with simple notations Q_b^* and Q_s^*, respectively, for simplicity, the Galerkin equations may be expressed in

$$\int_0^{l_x}\int_0^{l_y}\delta w_b\cdot Q_b^*\mathrm{d}x\mathrm{d}y=0 \quad (14.133)$$

$$\int_0^{l_x}\int_0^{l_y}\delta w_s\cdot Q_s^*\mathrm{d}x\mathrm{d}y=0 \quad (14.134)$$

The variations of Eqs. (14.123) and (14.124) for w_b and w_s are written as

$$\delta w_b=\sum_{\bar{m}=1}\sum_{\bar{n}=1}f_{b\bar{m}\bar{n}}(x,y)\delta\Phi_{b\bar{m}\bar{n}}(t) \quad (14.135)$$

$$\delta w_s=\sum_{\bar{m}=1}\sum_{\bar{n}=1}f_{s\bar{m}\bar{n}}(x,y)\delta\Phi_{s\bar{m}\bar{n}}(t) \quad (14.136)$$

Hence, the Galerkin equations of Eqs. (14.133) and (14.134) become

$$\sum_{m=1}\sum_{m=1} m_0\big[\ddot{\Phi}_{\mathrm{b}mn}\{F_{\mathrm{b}x}(\bar{m},0;m,0)F_{\mathrm{b}y}(\bar{n},0;n,0)-\frac{h_0^2}{12}[F_{\mathrm{b}x}(\bar{m},0;m,2)F_{\mathrm{b}y}(\bar{n},0;n,0)$$

$$+F_{\mathrm{b}x}(\bar{m},0;m,0)F_{\mathrm{b}y}(\bar{n},0\,;\,n,2)]\}+\ddot{\Phi}_{smn}[F_{\mathrm{s}x}(\bar{m},0;m,0)F_{\mathrm{s}y}(\bar{n},0;n,0)]$$

$$+\dot{\Phi}_{\mathrm{b}mn}[2h_{mn}\omega_{mn}F_{\mathrm{b}x}(\bar{m},0;m,0)F_{\mathrm{b}y}(\bar{n},0;n,0)]$$

$$+\Phi_{\mathrm{b}mn}\Bigg\{-\frac{\omega_{mn}^2 m_0 h_0^2}{12}[F_{\mathrm{b}x}(\bar{m},0;m,2)F_{\mathrm{b}y}(\bar{n},0;n,0)]+[F_{\mathrm{b}x}(\bar{m},0;m,0)F_{\mathrm{b}y}(\bar{n},0;n,2)]$$

$$+\omega_{mn}^2[F_{\mathrm{b}x}(\bar{m},0;m,0)F_{\mathrm{b}y}(\bar{n},0;n,0)]+F_{\mathrm{s}x}(\bar{m},0;m,0)F_{\mathrm{s}y}(\bar{n},0;n,0)]\Bigg\}\Bigg]$$

$$-\int_0^{l_x}\int_0^{l_y}p(x,y,t)f_{\mathrm{b}\bar{m}\bar{n}}\mathrm{d}x\mathrm{d}y=0 \tag{14.137}$$

$$\sum_{m=1}\sum_{n=1} m_0\big[\ddot{\Phi}_{\mathrm{b}mn}[F_{\mathrm{b}x}(\bar{m},0;m,0)F_{\mathrm{b}y}(\bar{n},0;n,0)]+\ddot{\Phi}_{smn}[F_{\mathrm{s}x}(\bar{m},0;m,0)F_{\mathrm{s}y}(\bar{n},0;n,0)]$$

$$+\omega_{mn}^2\Phi_{smn}[F_{\mathrm{b}x}(\bar{m},0;m,0)F_{\mathrm{b}y}(\bar{n},0;n,0)]+F_{\mathrm{s}x}(\bar{m},0;m,0)F_{\mathrm{s}y}(\bar{n},0;n,0)]\big]$$

$$-\int_0^{l_x}\int_0^{l_y}p(x,y,t)f_{\mathrm{s}\bar{m}\bar{n}}\mathrm{d}x\mathrm{d}y=0 \tag{14.138}$$

where the notations $F_{\mathrm{b}x}(\bar{m},0;m,0)$, $F_{\mathrm{b}y}(\bar{n},0;n,0)$, $F_{\mathrm{s}x}(\bar{m},0;m,0)$, and $F_{\mathrm{s}y}(\bar{n},0;n,0)$ are defined in general form by

$$\left.\begin{aligned}F_{\mathrm{b}x}(\bar{m},k_1;m,k_2)&=\int_0^{l_x}f_{\mathrm{b}\bar{m}}^{(k_1)}f_{\mathrm{b}m}^{(k_2)}\mathrm{d}x\\ F_{\mathrm{b}y}(\bar{n},k_1;n,k_2)&=\int_0^{l_y}f_{\mathrm{b}\bar{n}}^{(k_1)}f_{\mathrm{b}n}^{(k_2)}\mathrm{d}y\end{aligned}\right\} \tag{14.139}$$

in which k_1 and k_2 enclosed within parentheses indicate the order of differential with respect x or y.

The above equations are a set of differential equation with coupled form for $\Phi_{\mathrm{b}mn}$ and Φ_{smn}. Its treatment is very troublesome. So, we consider a simplified equation of motion by approximately replacing from the coupled form of Eqs. (14.137) and (14.138) to uncoupled form concerning coupling among the different modes, except for the coupled relation between $\Phi_{\mathrm{b}mn}$ and Φ_{smn}. Using the normalized shape functions for $f_{\mathrm{b}mn}(x,y)$ and $f_{smn}(x,y)$ and employing the following orthogonal relations for the normalized shape functions,

$$\int_0^{l_x}f_{\mathrm{b}m}(x)f_{\mathrm{b}\bar{m}}(x)\mathrm{d}x=\delta_{m\bar{m}};\quad \int_0^{l_y}f_{\mathrm{b}n}(y)f_{\mathrm{b}\bar{n}}(y)\mathrm{d}y=\delta_{n\bar{n}} \tag{14.140}$$

$$\int_0^{l_x} f_{sm}(x) f_{s\bar{m}}(x)\mathrm{d}x = \delta_{m\bar{m}}; \quad \int_0^{l_y} f_{sn}(y) f_{s\bar{n}}(y)\mathrm{d}y = \delta_{n\bar{n}} \tag{14.141}$$

the simplified equations of motion may be obtained in uncoupled form concerning indexes m and n as

$$\begin{aligned}
&\ddot{\Phi}_{bmn}\left\{F_{bx}(m,0;m,0)F_{by}(n,0;n,0) - \frac{h_0^2}{12}[F_{bx}(m,0;m,2)F_{by}(n,0;n,0)\right.\\
&\left.+F_{bx}(m,0;m,0)F_{by}(n,0;n,2)]\right\} + \ddot{\Phi}_{smn}[F_{sx}(m,0;m,0)F_{sy}(n,0;n,0)]\\
&+\dot{\Phi}_{bmn}[2h_{mn}\omega_{mn}F_{bx}(m,0;m,0)F_{by}(n,0;n,0)]\\
&+\Phi_{bmn}\left\{-\frac{\omega_{mn}^2 m_0 h_0^2}{12}[F_{bx}(m,0;m,2)F_{by}(n,0;n,0)\right.\\
&+F_{bx}(m,0;m,0)F_{by}(n,0;n,2)] + \omega_{mn}^2[F_{bx}(m,0;m,0)F_{by}(n,0;n,0)\\
&\left.+F_{sx}(m,0;m,0)F_{sy}(n,0;n,0)]\right\} - Q_{bmn}^*(t) = 0
\end{aligned} \tag{14.142}$$

$$\begin{aligned}
&\ddot{\Phi}_{bmn}[F_{bx}(m,0;m,0)F_{by}(n,0;n,0)] + \ddot{\Phi}_{smn}[F_{sx}(m,0;m,0)F_{sy}(n,0;n,0)]\\
&+\omega_{mn}^2\Phi_{smn}[F_{bx}(m,0;m,0)F_{by}(n,0;n,0)\\
&+F_{sx}(m,0;m,0)F_{sy}(n,0;n,0)] - Q_{smn}^*(t) = 0
\end{aligned} \tag{14.143}$$

in which the terms of external loads indicating with the simple notations Q_{bmn}^* and Q_{smn}^* are defined as

$$Q_{bmn}^*(t) = \frac{1}{m_0}\int_0^{l_x}\int_0^{l_y} p(x,y,t) f_{bmn}(x,y)\mathrm{d}x\mathrm{d}y \tag{14.144}$$

$$Q_{smn}^*(t) = \frac{1}{m_0}\int_0^{l_x}\int_0^{l_y} p(x,y,t) f_{smn}(x,y)\mathrm{d}x\mathrm{d}y \tag{14.145}$$

and are a function of time t. The unknown variables $\Phi_{bmn}(t)$ and $\Phi_{smn}(t)$ may be obtained by solving two coupled Eqs. (14.142) and (14.143) by the use of step-by-step integration methods.

14.10 Advanced Transformation of Uncoupled Form in Simplified RPT

The above numerical calculation is accomplished by the appropriate computer programing. Now comparing both Eqs. (14.142) and (14.143), Φ_{smn} can be expressed by Φ_{bmn}. So, we propose advanced RPT with uncoupled form expressed by only Φ_{bmn}. For simple expression, Eqs. (14.142) and (14.143) may be rewritten as

$$\ddot{\Phi}_{bmn}(A1) + \ddot{\Phi}_{smn}(A2) + \dot{\Phi}_{bmn}(A3) + \Phi_{bm}(A4) - Q^*_{bmn} = 0 \tag{14.146}$$

$$\ddot{\Phi}_{bmn}(B1) + \ddot{\Phi}_{smn}(A2) + \Phi_{smn}(B3) - Q^*_{smn} = 0 \tag{14.147}$$

in which $A1, A2, A3, A4, B1, B2, B3$, and $B4$ indicate each coefficient of Eqs. (14.142) and (14.143) and they are defined as follows:

$$A1 = F_{bx}(m, 0; m, 0)F_{by}(n, 0; n, 0) - \frac{h_0^2}{12}[F_{bx}(m, 0; m, 2)F_{by}(n, 0; n, 0) + F_{bx}(m, 0; m, 0)F_{by}(n, 0; n, 2)] \tag{14.148}$$

$$A2 = F_{sx}(m, 0; m, 0)F_{sy}(n, 0; n, 0) \tag{14.149}$$

$$A3 = 2h_{mn}\omega_{mn}F_{bx}(m, 0; m, 0)F_{by}(n, 0; n, 0) \tag{14.150}$$

$$A4 = -\frac{\omega_{mn}^2 m_0 h_0^2}{12}[F_{bx}(m, 0; m, 2)F_{by}(n, 0; n, 0) + F_{bx}(m, 0; m, 0)F_{by}(n, 0; n, 2)] + \omega_{mn}^2[F_{bx}(m, 0; m, 0)F_{by}(n, 0; n, 0) + F_{sx}(m, 0; m, 0)F_{sy}(n, 0; n, 0)] \tag{14.151}$$

$$B1 = F_{bx}(m, 0; m, 0)F_{by}(n, 0; n, 0) \tag{14.152}$$

$$B3 = \omega_{mn}^2[F_{bx}(m, 0; m, 0)F_{by}(n, 0; n, 0) + F_{sx}(m, 0; m, 0)F_{sy}(n, 0; n, 0)] \tag{14.153}$$

Taking away the term including $\ddot{\Phi}_{smn}$ from the both equations, we have from Eqs. (14.146) and (14.147)

$$\Phi_{smn}(B3) = \ddot{\Phi}_{bmn}(A1 - B1) + \dot{\Phi}_{bmn}(A3) + \Phi_{bmn}(A4) - Q^*_{bmn} + Q^*_{smn} \tag{14.154}$$

The second derivative of the above equation with respect to t becomes

$$\ddot{\Phi}_{smn}(B3) = \ddot{\ddot{\Phi}}_{bmn}(A1 - B1) + \dddot{\Phi}_{bmn}(A3) + \ddot{\Phi}_{bmn}(A4) - \ddot{Q}^*_{bmn} + \ddot{Q}^*_{smn} \tag{14.155}$$

Substituting Eq. (14.155) into Eq. (14.146), we have an uncoupled equation concerning only Φ_{bmn}.

$$\ddot{\ddot{\Phi}}_{bmn}(A2/B3)(A1 - B1) + \dddot{\Phi}_{bmn}(A2/B3)(A3) + \ddot{\Phi}_{bmn}(A1 + A4 \cdot A2/B3) + \dot{\Phi}_{bmn}(A3) + \Phi_{bmn}(A4) - Q^*_{bmn} + (-\ddot{Q}^*_{bmn} + \ddot{Q}^*_{smn})(A2/B3) = 0 \tag{14.156}$$

The above equation is the fourth-order differential equation with constant coefficients. Φ_{bmn} may be given in closed-form solution or in the numerical calculation from Eq. (14.156). Then Φ_{smn} is obtained from Eq. (14.154). The dynamic displacement components w_b and w_s are determined from Eqs. (14.123) and (14.124), respectively. The displacements U and V may be obtained from Eqs. (14.12) and (14.13).

14.11 Advanced RPT

In the preceding section, we have proposed a methodology which reduces the two variables in simplified RPT to one variable. This section introduces a methodology proposed by Thai et al. [15]. They establish the relationship between the bending and shear parts, based on the equilibrium equations of three-dimensional elasticity theory, and reduce the number of unknowns in the refined plate theory from two to one.

The equilibrium equations of three-dimensional elasticity theory in the absence of body forces are given as

$$\sigma_{x,x} + \sigma_{xy,y} + \sigma_{xz,z} = \rho\ddot{U} \tag{14.157}$$

$$\sigma_{xy,x} + \sigma_{y,y} + \sigma_{yz,z} = \rho\ddot{V} \tag{14.158}$$

$$\sigma_{xz,x} + \sigma_{yz,y} + \sigma_{z,z} = \rho\ddot{W} \tag{14.159}$$

Substituting Eqs. (14.6) into Eqs. (14.157) to (14.159); and multiplying the first two equations by z; and integrating these equations with respect to z; and applying the following boundary conditions, (i.e., the transverse shear stresses σ_{xz} and σ_{yz} equal to zero at $z = \pm h_0/2$), then Eqs. (14.157) and (14.158) become

$$M_{x,x} + M_{xx,y} - Q_x = -I_z\ddot{w}_{b,x} \tag{14.160}$$

$$M_{xy,x} + M_{y,y} - Q_y = -I_z\ddot{w}_{b,y} \tag{14.161}$$

in which mass inertia I_z are defined as

$$I_z = \int_{-h_0/2}^{h_0/2} \rho z^2 \mathrm{d}z = \frac{\rho h_0^3}{12} \tag{14.162}$$

Substituting the relationships for the moments and shear forces given by Eqs. (14.36) to (14.38) into Eqs. (14.160) and (14.161), we have obtained the following relationship between the bending component w_b and shear component w_s:

$$w_s = \frac{I_z}{A_s}\ddot{w}_b - \frac{D_0}{A_s}\nabla^2 w_b \tag{14.163}$$

in which A_s is defined as

$$A_s = \frac{5Eh_0}{12(1+\nu)} \tag{14.164}$$

Meanwhile, integrating Eq. (14.159) with respect to z and considering the boundary condition that the normal stress through the thickness $\sigma_z = 0$ at $z = h_0/2$ and $\sigma_z = p$ at $z = -h_0/2$, because lateral loads p act on the upper surface of the plate, we have

$$Q_{x,x} + Q_{y,y} + p = I_0(\ddot{w}_b + \ddot{w}_s) \tag{14.165}$$

Substituting Eqs. (14.160) and (14.161) into Eq. (14.165), the equation of motion is obtained as

$$M_{,xx} + 2M_{xy,xy} + M_{y,yy} + p = m_0(\ddot{w}_b + \ddot{w}_s) - I_z\nabla^2\ddot{w}_b \tag{14.166}$$

The above equation is derived from the equilibrium equations of three-dimensional theory and is well known. Equation (14.166) can be represented in terms of bending displacement w_b by substituting Eqs. (14.36), (14.44), and (14.163) into Eq. (14.166):

$$-D_0\nabla^4 w_b + p = m_0\ddot{w}_b - \left(I_z + \frac{m_0 D_0}{A_s}\right)\nabla^2\ddot{w}_b + \frac{m_0 I_z}{A_s}\ddddot{w}_b \tag{14.167}$$

For static problem, the above equation equals to well-known equation:

$$\nabla^4 w_b = p \tag{14.168}$$

Equation (14.167) formulated in advanced RPT corresponds to Eqs. (14.51) and (14.52) in RPT and Eqs. (14.63) and (14.64) in simplified RPT.

The static and dynamic analyses based on Eq. (14.167) in advanced RPT can be developed by the use of the Galerkin method presented in the preceding chapters.

14.12 Conclusions

This chapter has presented recent topics of plate analysis. The explanation is restricted isotropic uniform rectangular plates in order to understand easily the advanced theory. The recent analytical methods are based on rationally considering the transverse shear deformation within the category of two-dimensional approach. Since the main theory is two-variable refined plate theory (RPT), the static and dynamic analytical methods based on RPT have been presented.

Further developments to various plates including the variable stiffness and variable mass, discussed in the preceding chapters, will be possible by means of the extended Dirac function in RPT.

References

1. Vasilév, V.V.: The theory thin plates. Izvestiya AN SSSR, Mekhanika Tverdogo Tela **27**(3), 26–47 (1992). [Mechanics of Solids, **27**(3), 22–42 (1992)]
2. Altenbach, J., Altenbach, H.: Trends in engineering plate theories. Eksploatacja I Niezawodność **4**, 21–30 (2001)
3. Ghugal, Y.M., Shimpi, R.P.: A review of refined shear deformation theories for isotropic and anisotropic laminated plates. J. Reinf. Plast. Compos. **21**(9), 775–813 (2002)
4. Vijayakumar, K.: Review of a few selected theories of plates in bending. Department of Aerospace Engineering, Indian Institute of Science, Bangalore 560012, India, **28**(10), 1–9 (2014)
5. Carrera, E., Brischetto, S.: A survey with numerical assessment of classical and refined theories for the analysis of sandwich plates. Appl. Mech. Rev. **62**(1), 1–17 (2009)
6. Reddy, J.N., Lin, C.F.: A higher-order shear deformation theory of laminated elastic shells. Int. J. Eng. Sci. **23**(3), 319–330 (1985)
7. Liew, K.M., Xiang, Y., Kitipornchai, S.: Research on thick plate vibration: a literature survey. J. Sound Vib. **180**(1), 163–176 (1995)
8. Shimpi, R.P.: Refined plate theory and its variants. AIAA J. **40**(1), 137–146 (2002)
9. Shimpi, R.P., Patel, H.G.: A two variable refined plate theory for orthotropic plate analysis. Int. J. Solids Struct. **43**(22–23), 6783–6799 (2006)
10. Shimpi, R.P., Patel, H.G.: New first-order shear deformation plate theories. J. Appl. Mech. **74**(3), 523–533 (2007)
11. Shimpi, R.P., Patel, H.G.: Free vibrations of plate using two variable refined plate theory. J. Sound Vib. **296**(4–5), 979–999 (2006)
12. Sayyad, I.I., Chikalthankar, S.B., Nandedkar, V.M.: Bending and free vibration analysis of isotropic plate using refined plate theory. Bonfring Int. J. Ind. Eng. Manag. Sci. **3**(2), 40–46 (2013)
13. Ismail, M., Hassen, A.A., Abdlouahed, T., Hichem, A.B., El, A.A.B.: A two variable refined plate theory for the bending analysis of functionally graded plates. Acta. Mech. Sin. **26**(6), 941–949 (2010)
14. Rouzegar, J., Sharifpoor, R.A.: Flexure of thick plates resting on elastic foundation using two-variable refined plate theory. Arch. Mech. Eng., LXII **2**, 181–203 (2015)
15. Thai, H.T., Nguyen, T.K., Vo, T.P., Ngo, T.: A new simple shear deformation plate theory. Compos. Struct. **171**(1), 277–285 (2017)
16. Ashour, A.S.: Vibration of elastically restrained cross-ply laminated plates with variable thickness. J. Sound Vib. **288**(1–2), 33–42 (2005)
17. Auricchio, F., Sacco, E.: Refined first-order shear deformation theory models for composite laminates. J. Appl. Mech. **70**(3), 381–390 (2003)
18. Civalek, Ö.: Nonlinear analysis of thin rectangular plates on Winkler-Pasternak elastic foundations by DSC–HDQ methods. Appl. Math. Model. **31**(3), 606–624 (2007)
19. Fares, M.E., Elmarghany, M.K., Atta, D.: An efficient and simple refined theory for bending and vibration of functionally graded plates. Compos. Struct. **91**(3), 296–305 (2009)
20. Ghugal, Y.M., Gajbhiye, P.D.: Bending analysis of thick isotropic plates by using 5th order shear deformation theory. J. Appl. Comput. Mech. **2**(2), 80–95 (2016)
21. Hanna, N.F., Leissa, A.W.: A higher order shear deformation theory for the vibration of thick plates. J. Sound Vib. **170**(4), 545–555 (1994)
22. Hasheminejad, S.M., Gheshlaghi, B.: Three-dimensional elastodynamic solution for an arbitrary thick FGM rectangular plate resting on a two parameter viscoelastic foundation. Compos. Struct. **94**(9), 2746–2755 (2012)
23. Mantari, J.L., Ore, M.: Free vibration of single and sandwich laminated composite plates by using a simplified FSDT. Compos. Struct. **132**(15), 952–959 (2015)
24. Mokhtar, B., Fodil, H., Mostapha, K.: Bending analysis of symmetrically laminated plates. Leonardo J. Sci. **16**, 105–116 (2010)

25. Reddy, J.N.: A refined nonlinear theory of plates with transverse shear deformation. Int. J. Solids Struct. **20**(9–10), 881–896 (1984)
26. Reddy, J.N.: On the generalization of displacement-based laminate theory. Appl. Mech. Rev. **42**(11S), S213–S222 (1989)
27. Reddy, J.N.: A general non-linear third-order theory of plates with moderate thickness. Int. J. Non-Linear Mech. **25**(6), 677–686 (1990)
28. Sayyad, S.A., Ghugal, M.Y.: Bending and free vibration analysis of thick isotropic plates by using exponential shear deformation theory. Appl. and Comput. Mech. **6**(1), 65–82 (2012)
29. Sayyad, S.A., Ghugal, M.Y.: Effect of stress concentration on laminated plates. J. Mech. **29**(2), 241–252 (2013)
30. Sayyad, S.A., Ghugal, M.Y.: On the free vibration analysis of laminated composite and sandwich plates: a review of recent literature with some numerical results. Compos. Struct. **129**(1), 177–201 (2015)
31. Shen, H.S.: Nonlinear analysis of simply supported Reissner-Mindlin plates subjected to lateral pressure and thermal loading and resting on two-parameter elastic foundations. Eng. Struct. **22**(11), 1481–1493 (2000)
32. Shen, H.S., Chen, Y., Yang, J.: Bending and vibration characteristics of a strengthened plate under various boundary conditions. Eng. Struct. **25**(9), 1157–1168 (2003)
33. Shi, G.: A new simple third-order shear deformation theory of plates. Int. J. Solids Struct. **44**(13), 4399–4417 (2007)
34. Shinde, B.M., Sayyad, A.S., Ghumare, S.M.: A refined shear deformation theory for bending analysis of isotropic and orthotropic plates under various loading conditions. J. Mater. Eng. Struct. **2**(1), 3–15 (2015)
35. Silva, A.R.D., Silveira, R.A.M., Gonçalves, P.B.: Numerical methods for analysis of plates on tensionless elastic foundations. Int. J. Solids Struct. **38**(10–13), 2083–2100 (2001)
36. Srinivas, S., Joga, C.V., Rao, A.K.: An exact analysis for vibration of simply-supported homogeneous and laminated thick rectangular plates. J. Sound Vib. **12**(2), 187–199 (1970)
37. Wang, C.M., Lim, G.T., Reddy, J.N., Lee, K.H.: Relationships between bending solutions of Reissner and Mindlin plate theories. Eng. Struct. **23**(7), 838–849 (2001)
38. Yazdani, S., Ribeiro, S.: Geometrically non-linear static analysis of unsymmetric composite plates with curvilinear fibres: p-version layerwise approach. Compos. Struct. **118**, 74–85 (2014)
39. Yuwaraj, M.G., Rameshchandra, P.S.: A review of refined shear deformation theories of isotropic and anisotropic laminated plates. J. Reinf. Plast. Compos. **21**(9), 775–813 (2002)
40. Zhang, B., Li, L.: Nonlinear vibration of rotating pre-deformed blade with thermal gradient. Nonlinear Dyn. **86**(1), 459–478 (2016)
41. Reissner, E.: The effect of transverse shear deformation on the bending of elastic plates. J. Appl. Mech. **12**, (Transactions ASME **67**), A 69–77 (1945)
42. Mindlin, R.D.: Influence of theory rotary inertia and shear on flexural motions of isotropic, elastic plates. J. Appl. Mech. **18**, (Transactions ASME **73**), 31–35 (1951)

Printed by Printforce, the Netherlands